CW00321881

Uncertainty in Remote Sensing and GIS

Uncertainty in Remote Sensing and GIS

Edited by

GILES M. FOODY and PETER M. ATKINSON

Department of Geography, University of Southampton, UK

Email (for orders and customer service enquiries): cs-books@wiley.co.uk
Visit our Home Page on www.wileyeurope.com or www.wiley.com

Other Wiley Editorial Offices

John Wiley & Sons Inc., 111 River Street,
Hoboken, NJ 07030, USA

Jossey-Bass, 989 Market Street,
San Francisco, CA 94103–1741, USA

Wiley-VCH Verlag GmbH, Boschstr. 12,
D-69469 Weinheim, Germany

John Wiley & Sons Australia Ltd, 33 Park Road, Milton,
Queensland 4064, Australia

John Wiley & Sons (Asia) Pte Ltd, 2 Clementi Loop #02–01,
Jin Xing Distripark, Singapore 129809

John Wiley & Sons Canada Ltd, 22 Worcester Road,
Etobicoke, Ontario, Canada M9W 1L1

Wiley also publishes its books in a variety of electronic formats. Some content that appears in print may not be available in electronic books.

British Library Cataloguing in Publication Data

A catalogue record for this book is available from the British Library

ISBN 0-470-84408-6

Typeset in 10/12 pt Times by Kolam Information Services Pvt. Ltd, Pondicherry, India
Printed and bound in Great Britain by Antony Rowe Ltd, Chippenham, Wiltshire
This book is printed on acid-free paper responsibly manufactured from sustainable forestry
in which at least two trees are planted for each one used for paper production

Contents

List of Contributors

P. M. Atkinson, Department of Geography, University of Southampton, Highfield, Southampton, SO17 1BJ, UK

D. S. Boyd, School of Earth Science and Geography, Kingston University, Penrhyn Road, Kingston-upon-Thames, KT1 2EE, UK

M. J. Collins, Department of Geomatics Engineering, University of Calgary, Calgary, Alberta, T2N 1N4, Canada

P. J. Curran, Department of Geography, University of Southampton, Highfield, Southampton, SO17 1BJ, UK

J. L. Dungan, NASA Ames Research Center, MS 242-4, Moffett Field, CA 94035-1000, USA

W. J. Duane, School of Geography, Portsmouth University, Portsmouth, PO1 3HE, UK

P. F. Ehlers, Department of Mathematics and Statistics, University of Calgary, Calgary, Alberta, T2N 1N4, Canada

G. M. Foody, Department of Geography, University of Southampton, Highfield, Southampton, SO17 1BJ, UK

R. M. Fuller, Unit for Landscape Modelling, University of Cambridge, Free School Lane, Cambridge, CB2 3RF, UK

J. R. Hall, Centre for Ecology and Hydrology, Monks Wood, Abbots Ripton, Huntingdon, PE28 2LS, UK

G. B. M. Heuvelink, Institute for Biodiversity and Ecosystem Dynamics, Universiteit van Amsterdam, Nieuwe Achtergracht 166, 1018 WV Amsterdam, The Netherlands

E. Heywood, Pollution and Ecotoxicology Section, Centre for Ecology and Hydrology, Monks Wood, Abbots Ripton, Huntingdon, PE28 2LS, UK

A. M. Jakomulska†, Remote Sensing of Environment Laboratory, Faculty of Geography and Regional Studies, Warsaw University, Krawkowskie Prezedmiescie 30, 00-927 Warsaw, Poland

E. A. Johnson, Department of Biological Sciences, University of Calgary, Calgary, Alberta, T2N 1N4, Canada

H. G. Lewis, Department of Aeronautical and Astronautical Engineering, University of Southampton, Highfield, Southampton, SO17 1BJ, UK

C. D. Lloyd, School of Geography, Queen's University, Belfast, Northern Ireland, BT7 1NN, UK

A. Lucieer, International Institute for Geo-Information Science and Earth Observation (ITC), PO Box 6, 7500 AA Enschede, The Netherlands

J. F. Manslow, Department of Electronics and Computer Science, University of Southampton, Highfield, Southampton, SO17 1BJ, UK

M. Molenaar, International Institute for Geo-Information Science and Earth Observation (ITC), PO Box 6, 7500 AA Enschede, The Netherlands

M. S. Nixon, Department of Electronics and Computer Science, University of Southampton, Highfield, Southampton, SO17 1BJ, UK

P. Nonin, Istar, 2600 route des Crètes, BP282 06905 Sophia-Antipolis Cedex, France

I. O. A. Odeh, Cotton Research Co-operation (CRC), Department of Agricultural Chemistry and Soil Science, Ross Street, Building AO3, University of Sydney, New South Wales, Australia

M. A. Oliver, Department of Soil Science, University of Reading, Whiteknights, Reading, RG6 6AB, UK

P. C. Phipps, School of Earth Science and Geography, Kingston University, Penrhyn Road, Kingston-upon-Thames, KT1 2EE, UK

J. P. Radomski, Interdisciplinary Center for Modeling, Warsaw University, Pawinskiego 5A, 02-106 Warsaw, Poland

† Deceased

M. Schmitt, Centre de Géostatistique, Ecole des Mines de Paris, 35 rue Saint-Honoré, 77305 Fontainebleau Cedex, France

J. Sénégas, Centre de Géostatistique, Ecole des Mines de Paris, 35 rue Saint-Honoré, 77305 Fontainebleau Cedex, France

G. M. Smith, Section for Earth Observation, Centre for Ecology and Hydrology, Monks Wood, Abbots Ripton, Huntingdon, PE28 2LS, UK

A. J. Tatem, TALA Research Group, Department of Zoology, University of Oxford, South Parks Road, Oxford, OX1 3PS, UK

R. A. Wadsworth, Centre for Ecology and Hydrology, Monks Wood, Abbots Ripton, Huntingdon, PE28, 2LS, UK

A. Wameling, Institute for Forest Biometrics and Applied Computer Science, University of Goettingen, Buesgenweg 4, 37077 Goettingen, Germany

B. Warr, Department of Soil Science, University of Reading, Whiteknights, Reading, RG6 6AB, UK

A. J. Warren, Department of Geomatics Engineering, University of Calgary, Calgary, Alberta, T2N 1N4, Canada

C. E. Woodcock, Department of Geography, Boston University, 675 Commonwealth Avenue, Boston, MA 02130, USA

Q. Zhan, International Institute for Geo-Information Science and Earth Observation (ITC), PO Box 6, 7500 AA Enschede, The Netherlands

Foreword

FORWARD WITH UNCERTAINTY

'It is certain because it is impossible' (Tertullianus, 210, Ch5).

This much quoted paradox describes one end of an uncertainty continuum. For instance, we are certain that we will not be able to measure cows jumping over the moon. However, if we knew from previous measurements that 1% of all cats were co-located with fiddles then we could say how uncertain any of our measurements of cat-fiddle pairings would be. These two measurement uncertainty examples, although apparently trivial, are also examples of uncertainty in understanding and so are worth exploring further. In the former we think we understand the processes involved (cows can't jump that high) while in the latter we do not understand the processes involved (why are some cats co-located with fiddles?). The world we think we understand we can try to manage or even control via such disciplines as engineering and planning. The world we do not understand spans from reasonably certain to very uncertain (May, 2001a) and falls within the realm of science (Krebs, 2002). To avoid confusion between uncertainty in measurement and uncertainty in understanding it is helpful to first define uncertainty in understanding in no uncertain (although rather gender-specific) terms.

> 'When a scientist doesn't know the answer to a problem he is ignorant. When he has a hunch as to what the result is, he is uncertain. And when he is pretty darn sure of what the result is going to be, he is still in some doubt. We have found it of paramount importance that in order to progress we must recognise our ignorance and leave room for doubt. Scientific knowledge is a body of statements of varying degrees of certainty – some most unsure, some nearly sure, but none *absolutely* certain' (Feynman, 1988, p. 245).

Regardless of our degree of understanding the knowledge upon which our understanding is built is derived increasingly from measurement. However, measurement alone is not enough:

> 'to serve the purposes of science the measurements must be susceptible to comparison. And comparability of measurement requires some common

understanding of their accuracy, some way of measuring and expressing the uncertainty in their values and the inferential statements derived from them' (Stigler, 1986, p. 1).

This measurement uncertainty can be described probabilistically by ascribing either the degree of uncertainty (e.g. error) or degree of certainty (e.g. accuracy) to measurements we make (Foody, 2001; Zhang and Goodchild, 2002). This is relatively simple to do for that part of the world that can be described by laws (e.g. gravity in cow illustration) or where that part has been measured already (e.g. cat and fiddle illustration). Paradoxically though, the poorly understood and the unmeasured parts of our world are where remote sensing and geographical information systems (GIS) are likely to be of most practical value. For example, remotely sensed data may provide 'measurements' of sea-surface temperature (SST), and GIS, via spatial data modelling, may provide socioeconomic 'measurements' for a region. In both cases we have no known data at the same location, time and size of support with which to determine the uncertainty of these remote sensing or GIS 'measurements'. In practice, we can predict uncertainty by comparing remote sensing and GIS measurement data to known measurement data from other locations. For example, without a law to help, information on the uncertainty of our remotely sensed SST measurements can be provided by the difference between remotely sensed SST 'measurements' and buoy SST measurements for a different time, place and size of support. Likewise, without a law to help, information on the uncertainty of our socioeconomic 'measurements' can be provided by the difference between these socioeconomic measurements and surrogate measurements (e.g. average income) made again, for a different time, place and size of support.

To turn again to issues of uncertainty in understanding: some of the most exciting areas in remote sensing are where our understanding is uncertain and other measurements (i.e. ground data or sea data) are not available. This is where remote sensing can be used as an inductive tool of exploration to 'find things' and to provide a base on which to develop a deterministic understanding of how the world works (Curran, 1987). Remote sensing is one of the few fields that make vastly (orders of magnitude) more measurements than are needed. These huge sets of measurements are, like library archives, meteorological records or space x-ray data repositories, just ripe for mining. However, for reasons outlined before, the uncertainty associated with these archives of remotely sensed 'measurements' of our environment is unknown. For example, the influential papers that have used global time series of AVHRR NDVI data to understand how our planet functions are based on measurements of vegetation characteristics with unknown levels of accuracy.

So far this Foreword has treated measurement uncertainty and understanding uncertainty as if they were separate. This is the stance taken by many disciplines. Economists, since the early part of the twentieth century, have given the name *risk* to measurement uncertainty (or chance) and *uncertainty* to understanding uncertainty (or the unexpected) (Knight, 1921; Keynes, 1936). In environmental impact studies *risk* is measurement uncertainty when you 'know the odds' and *uncertainty* is understanding uncertainty when you 'don't know the odds' and are ignorant (Wynne, 1992). By way of illustration, we know that the chance of the average westerner dying in any one year of heart disease is currently 1:385, compared to 1:519 for cancer and 1:1752 for stroke (HCRA, 2002) but we have no idea what the chances are of there being a catastrophic international event of some sort during our lifetime. Likewise, the measurement uncertainty associated with the release of dichlorodiphenyltrychloroethane (DDT) and chlorofluorocarbons (CFCs) into our environment was reasonably low but the understanding uncertainty was alarmingly high (Hoffmann-Riem and Wynne, 2002). However, in remote sensing, GIS and a wide range of related fields (e.g. those covered by the *International Journal of Uncertainty, Fuzzyness and Knowledge-Based Systems*) the term uncertainty, often seeks to incorporate measurement uncertainty and understanding uncertainty as it is their interplay that determines the efficacy of what we are trying to do (Goodchild and Case, 2001). It is not just error in x, y (location) but also understanding of z (variable) that is important. One of the better documented environmental examples of the interplay between the two forms of uncertainty occurred in the UK following the 1986 Chernobyl nuclear accident. Very accurate chemical immobilization measurements (made previously on lowland soil in the UK) were used to predict that the radioactive material would be immobilized within three weeks and as a result upland sheep sales were banned for that period. However, the underpinning understanding was uncertain. Organic upland soil did not respond like lowland soil and the soil immobilization of radioactive material and the reopening of a market in sheep took not weeks but years (Wynne, 1992). In this case, measurements with little and known levels of uncertainty were not that useful! However, in many of the cases we face in remote sensing and GIS both the measurement and understanding are uncertain. For example, the 2000 talks on mechanisms for implementing the Kyoto Protocol collapsed due to uncertainty in both our measurement and understanding of terrestrial carbon sinks (Read *et al.*, 2001; Bégni *et al.*, 2002). What was required was an accurate means of predicting carbon sink strength over land. These predictions needed to be accurate enough to enable them to be both traded in a world market and input to models of future carbon sink strength. The only practical methods are a judicious mix of remote sensing, carbon flux measurements, GIS and ecosystem simulation models. The outputs of such methods are therefore subject to both measurement uncertainty and understanding uncertainty, as we not only have difficulty in measuring accurately (e.g. standing biomass of forest) but also we do not understand many of the processes involved (e.g. interactions between vegetation, soil and climate). This mix of measurement uncertainty and understanding uncertainty is difficult to deal with as the former can be quantified readily while the latter is unquantifiable. As a result,

> 'the uncertainties associated with estimates (of carbon sink strength) derived from all of these methods is considerable. A further fundamental problem is that the magnitude of this uncertainty is unknown' (Read *et al.*, 2001, p. 12).

A way around the problem is to treat the phenomena we are interested in as a model and understanding uncertainty as a model imperfection. Therefore, understanding uncertainty can be quantified as the difference between measurement and model prediction once allowance has been made for measurement uncertainty. This is very difficult to implement in practice as understanding uncertainty has a considerably greater magnitude but (fortunately) a considerably lower frequency than measurement uncertainty. In climate change research an alternative solution has been to use 'expert opinion' to ascribe probabilities to the degree of understanding uncertainty (Schneider, 2001). However, this neat solution whilst popular with policy makers remains controversial (Grübler and Nakicenovik, 2001; Pittock *et al.*, 2001).

This Foreword has focused on the two types of uncertainty that are of most concern. In addition, these uncertainties are underpinned by a range of task-specific uncertainties. Uncertainty may arise because of incomplete information – what will the climate of China be in 2005? or because of disagreement over information sources – what was the DN to radiance calibration for a given sensor and date? Uncertainty may arise from linguistic imprecision – what exactly is meant by Case II water? or the inappropriate use of jargon – 'validating the retrieval of large scale parameters'. It may refer to variability – what exactly is the atmospheric ozone concentration over the poles? Uncertainty may be about a quality – slope of the variogram, or the type of model – shape of a variogram. Even if we were to have no measurement uncertainty we may still be uncertain because of simplifications or approximations used to make the analysis feasible – global NDVI data sets. If that were not enough we also add uncertainty by how we decide to analyse and then represent any analysis we do and as a result we are always uncertain, even about our uncertainty (Morgan and Henrion, 1992). Perhaps not surprisingly discussions of uncertainty in remote sensing and GIS have tended to focus on that which is tractable (i.e. has a relatively small magnitude but relatively large frequency). As a result much of what will follow in this book is a discussion of probability applied to random error, systematic error and thereby measurement uncertainty. This focus on measurement uncertainty, as has been noted earlier, is very necessary although not sufficient if measurements are to be the building blocks for understanding.

> 'Difficulties arise when the uncertainties are not caused by probabilistic predictions, but rather derive from a fundamental lack of understanding of a new phenomenon at or beyond the frontiers of present knowledge' (May, 2001b, p. 891).

or to put it more bluntly,

> 'we need to recognise and address the crucial distinction between uncertainty and ignorance' (Hoffmann-Riem and Wynne, 2002, p. 123).

In remote sensing and GIS we are striving to refine and better characterize our probabilistic predictions of such variables as land cover and biomass, ocean temperature and chlorophyll concentration and understand for example, how measurement uncertainties in inputs are propagated through spatial models (Heuvelink and Burrough, 2002). A clear focus on measurement uncertainty has, until recently, been neglected in the environmental (although not statistical/mathematical) literature. Such a focus will be an aid to ensuring the vitality/creditability of remote sensing and GIS and is the aim of what I anticipate will be a most influential book.

Individually, the chapters that follow serve to highlight impressive strides that are being made in the study of measurement uncertainty over a wide range of scales. Collectively, these chapters set down both an uncertainty research agenda and an intriguing *challenge for the future*. This challenge, and the topic of this Foreword, is that of not being lulled into believing that confidence of our measurement uncertainty (however important), is the same as confidence of our understanding uncertainty (Morgan and Henrion, 1992; Gupta, 2001). A sentiment that is captured most elegantly by the (very) recently deceased Douglas Adams.

> 'It sounded quite a sensible voice, but it just said, "Two to the power of one hundred thousand to one against and falling," and that was all.
>
> Ford skidded down a beam of light and spun around but could see nothing he could seriously believe in.
>
> "What was that voice?" shouted Arthur.
>
> "I don't know," yelled Ford, "I don't know. It sounded like a measurement of probability."
>
> "Probability? What do you mean?"
>
> "Probability. You know, like two to one, three to one, five to four against. It said two to the power of one hundred thousand to one against. That's pretty improbable, you know."
>
> A million-gallon vat of custard unended itself over them without warning.
>
> "But what does it mean?" cried Arthur.
>
> "What, the custard?"
>
> "No, the measurement of improbability?"
>
> "I don't know. I don't know at all." '
>
> (Adams, 1986, p. 241).

References

Adams, D., 1986, *The Hitch-Hiker's Guide to the Galaxy: A Trilogy in Four Parts* (London: William Heinman).

Bégni, G., Darras, S. and Belward, A., 2002, The Kyoto Protocol: Legal statements, associated phenomenon and potential impacts, in G. Bégni (ed.), *Observing our Environment from Space: New Solutions for a New Millennium* (Lisse: A. A. Balkema), pp. 9–21.

Curran, P. J., 1987, Remote sensing methodologies and geography, *International Journal of Remote Sensing*, **8**, 1255–75.

Feynman, R. P., 1988, *What do you Care What Other People Think?* (London: Harper Collins).

Foody G. M., 2001, GIS: the accuracy of spatial data revisited, *Progress in Physical Geography*, **25**, 389–98.

Goodchild, M. F. and Case, T. J., 2001, Introduction, in C. T. Hunsaker, M. F. Goodchild, M. A. Friedl and T. J. Case (eds), *Spatial Uncertainty in Ecology* (New York: Springer), pp. 3–10.

Grübler, A. and Nakicenovik, N., 2001, Identifying dangers in an uncertain climate, *Nature*, **412**, 15.

Gupta, S., 2001, Avoiding ambiguity: Scientists sometimes use mathematics to give the illusion of certainty, *Nature*, **412**, 589.

HCRA, 2002, Risk quiz, *The Harvard Center for Risk Analysis* (*www.hcra.harvard.edu*).

Heuvelink, G. B. M. and Burrough, P. A., 2002, Developments in statistical approaches to spatial uncertainty and its propagation, *International Journal of Geographical Information Science*, **16**, 111–13.

Hoffmann-Riem, H. and Wynne, B., 2002, In risk assessment, one has to admit ignorance, *Nature*, **416**, 123.

Keynes, J. M., 1936, *The General Theory of Employment, Interest and Money* (London: Macmillan).

Knight, F. H., 1921, *Risk, Uncertainty and Profit* (Cambridge, MA: Houghton Mifflin).

Krebs, J., 2002, Why natural may not equal healthy, *Nature*, **415**, 117.

May, R. M., 2001a, Counting on science: scientific uncertainties relating to the social issues of today, *Nature*, **414**, 17–18.

May, R. M., 2001b, Risking uncertainty: at the frontiers of science we don't always know what may happen, *Nature*, **411**, 891.

Morgan, M. G. and Henrion, M., 1992, *Uncertainty. A Guide to Dealing with Uncertainty in Quantative Risk and Policy Analysis* (Cambridge: Cambridge University Press).

Pittock, A. B., Jones, R. N. and Mitchell, C. D., 2001, Probabilities will help us plan for climate change, *Nature*, **413**, 249.

Read, D., Beerling, D., Cannell, M., Cox, P., Curran, P., Grace, J., Ineson, P., Jarvis, P., Malhi, Y., Powlson, D., Shepherd, J. and Woodward, I., 2001, *The Role of Land Carbon Sinks in Mitigating Global Climate Change* (London: The Royal Society).

Schneider, S. H., 2001, What is 'dangerous' climate change? *Nature*, **411**, 17–19.

Stigler, S. M., 1986, *The History of Statistics: The Measurement of Uncertainty before 1900* (Cambridge, MA: Harvard University Press).

Tertullianus, Q. S. F. (210), *De Carne Christi*. Unpublished.

Wynne, B., 1992, Science and social responsibility, in J. Ansell and F. Wharton (eds), *Risk Analysis Assessment and Management* (Chichester: Wiley), pp. 137–52.

Zhang, J. and Goodchild, M., 2002, *Uncertainty in Geographical Information* (London: Taylor and Francis).

Paul J. Curran
University of Southampton

Preface

Uncertainty is a topic of considerable importance but one which has only recently attracted major attention within the remote sensing and GIS communities. We, therefore, felt that a short conference to bring together researchers in remote sensing and GIS to discuss uncertainty would be useful. The conference was held at the University of Southampton in July 2001 and designed to fulfil two major objectives. First, it was to encourage a dialogue on uncertainty and provide a guide to the current status of uncertainty related research within remote sensing and GIS. Here, it is worth noting that, like others, we use the term GIS to mean both geographical information systems and geographical information science. In general the distinction between these two meanings is made where appropriate in the book but sometimes either interpretation is valid. Second, saddened by the decay in short, highly focused, research meetings in the UK, we wanted to run a meeting that was academically strong, fun and involved contributions from junior researchers through to the established leaders in the field that were drawn from a variety of backgrounds and nationalities to help plug a gap that appears to have arisen in our calendars. In total 25 papers were presented at the conference that was attended by 61 delegates from eight countries.

Our attempt to achieve the key objectives was aided greatly by sponsorship that allowed us to attract the involvement of a large number of postgraduate students as well as three keynote speakers, Jennifer Dungan (NASA, USA), Curtis Woodcock (Boston University, USA) and Gerard Heuvelink (Amsterdam University, The Netherlands). Perhaps more importantly, the sponsorship also ensured we could have an informal social evening at the Mill Arms public house. From the feedback we received after the conference this seems to have been a great success, with far more social interaction amongst delegates than normal. Indeed the evening went so well that nobody seemed to mind that the skittles competition was won by a team from Southampton! Most important of all, however, the conference contained some excellent material on uncertainty in remote sensing and GIS. All of the presenters were encouraged to write up their paper for possible publication as a means of extending and communicating the material discussed at the conference and this book contains some of those papers together with contributions to fill apparent gaps in coverage. Each article received was rigorously reviewed and edited. Typically, each chapter was reviewed by three people: a delegate with relevant expertise, an established researcher in the specific topic discussed as well as at least one of the editors. Those articles that appeared to both fit the general aims of the conference and pass the quality criteria set were accepted and follow. Several articles did not get past the review stage. In most cases, the articles that we were unable to include were judged to require further work that was not compatible with our strict timetable. We confidently hope and expect that many of these articles will appear in the open literature in the near future as the work matures.

Staging a conference and producing a book are large tasks and cannot be achieved without the inputs of a vast and varied team of helpers. Here, we wish to thank those who have helped us. This is not out of any formal requirement to do so but rather a genuine desire to thank the many people and organizations that provided invaluable assistance. Five groups deserving explicit thanks can be identified. First, we are extremely grateful to our various sponsors. We received kind and generous support from a number of sources, notably: the Remote Sensing and Photogrammetry Society's Edinburgh Award, Ordnance Survey, Taylor and Francis, Erdas (UK), PCI Geomatics, Research Systems, Wiley and the Quantitative Methods Research Group of the Royal Geographical Society (with the Institute of British Geographers). Our gratitude to these bodies is very real. Without them we simply could not have staged the meeting and undoubtedly the warm feedback we received from delegates after the conference is in part a reflection of the support we received to stage it. Second, many people helped us with the organization and smooth running of the conference. In particular we wish to thank Karen Anderson, Reno Choi, Adam Corres, Aidy Muslim, Nick Hamm, Tatiana Kuplich, Gary Llewellyn, Nick Odoni, Michael Riedmann, Valeria Salvatori, Isabel Sargent, Caiti Steele, Andy Tatem, Debbie Wilson and Matt Wilson (for those who attended the meeting, many of these were the folk in the yellow tee-shirts who provided invaluable assistance before, during and after the conference – perhaps that is why we forgave them for winning the skittles competition that they helped organize?). Third, the production of the book benefited enormously from the inputs of the referees. Refereeing book chapters is an important task and one that we believe should be recognized publicly. As we must, however, maintain their confidentiality we will not identify the referees but simply express here our genuine gratitude for their constructive input. Fourth, we are grateful to the authors themselves. Not only for providing the chapters to a tight schedule but also for their positive attitude, most noticeable when responding to sometimes sharp comment from the referees. Fifth, we wish to thank all others who have played a role in bringing this book to print. These include Lyn Roberts, Susan Barclay and Keily Larkins of Wiley for guidance, Matt Wilson for help in constructing the index and Denise Pompa in our Departmental Office who had the horrible job of converting a pile of annotated manuscripts from us into something the publishers could handle.

We hope that this book provides an interesting and useful guide to uncertainty in remote sensing and GIS and that all of those connected with it in any way are pleased with its final form.

Finally, it is with much sadness that we must report the death in August 2002 of Anna Jakomulska, lead author of Chapter 7 and a valued colleague and friend. Anna worked tirelessly in the fields of remote sensing and GIS, often going beyond even the highest expectations to support her research. Anna will be greatly missed. We dedicate this book to her memory.

Giles M. Foody and Peter M. Atkinson
University of Southampton

1

Uncertainty in Remote Sensing and GIS: Fundamentals

P. M. Atkinson and G. M. Foody

1 Introduction

Uncertainty has been an important topic in geographical information science for over a decade (e.g. Goodchild and Gopal, 1989; Veregin, 1996; Kiiveri, 1997; Heuvelink, 1998; Zhang and Goodchild, 2002). This focus has led researchers to recommend over many years that the *spatial* output of a geographical information system (GIS) should be (at least) twofold: (i) a map of the variable of interest and (ii) some assessment of uncertainty in that map. Uncertainty has also been the subject of much research in remote sensing (e.g. Congalton, 1991; Fisher, 1994; Canters, 1997; Foody, 2002) with the main focus of such work being the assessment (and increase) of prediction or classification accuracy. However, uncertainty in remote sensing seems to have been a less explicit focus of research effort than uncertainty in GIS, with perhaps fewer calls for the production of maps depicting uncertainty.

This book aims to (i) discuss the nature of uncertainty; its interpretation, assessment and implications, (ii) assemble a range of up-to-date research on uncertainty in remote sensing and geographical information science to show the current status of the field and (iii) indicate future directions for research. We seek, in this introductory chapter, to define uncertainty and related terms. Such early definition of terms should help to (i) provide a common platform upon which each individual chapter can build, thus, promoting coherence and (ii) avoid duplication of basic material, thus, promoting efficiency in this book. We have done this also in the hope that others will adopt and use the definitions provided here to avoid confusion in publications in the future. We start with uncertainty itself.

Uncertainty in Remote Sensing and GIS. Edited by G.M. Foody and P.M. Atkinson.
 ISBN: 0-470-84408-6

Uncertainty arises from many sources ranging from ignorance (e.g. in understanding), through measurement to prediction. The *Oxford English Dictionary* gives the following definition of the word uncertain: 'not known or not knowing certainly; not to be depended on; changeable'. Uncertainty (the noun) is, thus, a general concept and one that actually conveys little information. For example, if someone is uncertain, we know that they are not 100% sure, but we do not know more than that, for example, how uncertain they are or should be. If uncertainty is our general interest then clearly we shall need a vocabulary that provides greater information and meaning, and facilitates greater communication of that information. The vocabulary that we use in this chapter involves clearly defined terms such as error, accuracy, bias and precision. Before defining such terms, two distinctions need to be made.

First, and paradoxically, to be uncertain one has to know something (Hoffman-Riem and Wynne, 2002). It is impossible to ascribe an uncertainty to something of which you are completely ignorant. This interesting paradox is explored by Curran in the Preface to this book. Essentially, it is possible to distinguish that which is known (and to which an uncertainty can be ascribed) and that which is simply not known. In this book, we deal with uncertainty and, therefore, that which is known (but not perfectly).

Second, dealing solely with that which is known, it is possible to divide uncertainty into ambiguity and vagueness (Klir and Folger, 1988). Ambiguity is the uncertainty associated with crisp sets. For example, in remote sensing, land cover is often mapped using hard classification (in which each pixel in an image is allocated to one of several classes). Each hard allocation is made with some ambiguity. This ambiguity is most often (and most usefully) expressed as a probability. Vagueness relates to sets that are not crisp, but rather are fuzzy or rough. For example, in remote sensing, land cover often varies continuously from place to place (e.g. ecotones, transition zones). In these circumstances, the classes should be defined as fuzzy, not crisp. This fuzziness is an expression of vagueness. In this opening chapter, following the majority of chapters in this book, we concentrate on ambiguity.

2 Error

An error $e(\mathbf{x}_0)$ at location $\mathbf{x}_0$ may be defined as the difference between the true value $z(\mathbf{x}_0)$ and the prediction $\hat{z}(\mathbf{x}_0)$

$$e(\mathbf{x}_0) = \hat{z}(\mathbf{x}_0) - z(\mathbf{x}_0) \tag{1}$$

such that

$$\hat{z}(\mathbf{x}_0) = z(\mathbf{x}_0) + e(\mathbf{x}_0). \tag{2}$$

Thus, errors relate to individual measured or predicted values (equations 1 and 2) and are essentially data-based. There is no statistical model involved. Thus, if we measure the soil pH at a point in two-dimensional space and obtain a value $\hat{z}(\mathbf{x}_0)$ of 5.6, where the true value $z(\mathbf{x}_0)$ is 5.62, the error $e(\mathbf{x}_0)$ is −0.02.

Further, error has a very different meaning to uncertainty. In the example above, consider that both the predicted value (5.6) and the error (−0.02) are known perfectly. Then, an error exists, but there is no uncertainty. Uncertainty relates to what is 'not known or not known certainly'. Thus, uncertainty is associated with (statistical) inference and, in particular, (statistical) prediction. Such statistical prediction is most often of the unknown true value, but it can equally be of the unknown error. Attention is now turned to three properties that do relate information on uncertainty: accuracy, bias and precision.

3 Accuracy, Bias and Precision

Accuracy may be defined in terms of two further terms: bias and precision (Figure 1.1). Accuracy, bias and precision are frequently misused terms. For example, researchers often use precision to mean accuracy and *vice versa*. Therefore, it is important to define these terms clearly. Let us, first, deal with bias and precision, before discussing accuracy.

3.1 Bias

Unlike error, bias is model-based in that it depends on a statistical model fitted to an ensemble of data values. Bias is most often predicted by the mean error, perhaps the simplest measure of agreement between a set of known values $z(\mathbf{x}_i)$ and a set of predicted values $\hat{z}(\mathbf{x}_i)$

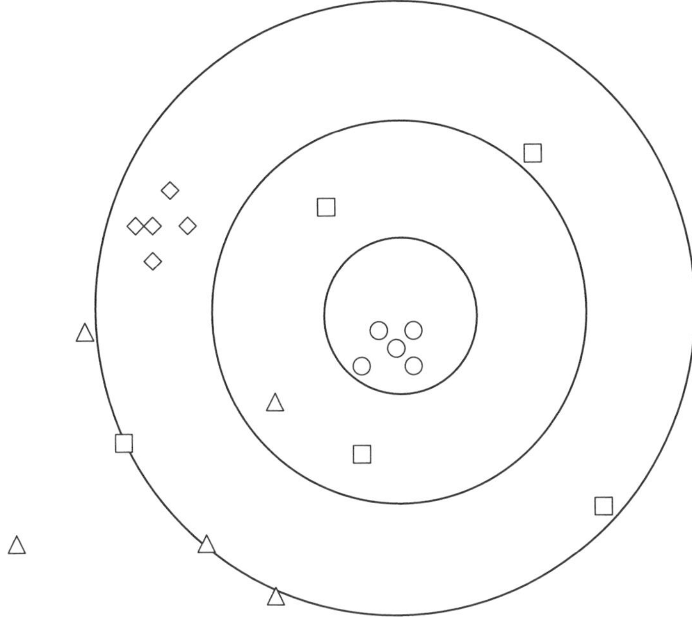

Figure 1.1 Bull's eye target, represented by the set of concentric rings, with four sets of predictions, only one of which is accurate: (triangles) biased and imprecise, therefore, inaccurate; (squares) unbiased but imprecise, therefore, inaccurate; (diamonds) biased but precise, therefore, inaccurate; (circles) unbiased and precise, therefore, accurate

$$ME = \frac{1}{n}\sum_{i=1}^{n}(\hat{z}(\mathbf{x}_i) - z(\mathbf{x}_i)). \tag{3}$$

Thus, bias is an *expectation* of over- or under-prediction based on some statistical model. The larger the (systematic) errors, the greater the bias.

Take a single measured value with a given error (equation 2). Suppose that the error is +0.1 units. True, the measurement has over-predicted the true value. However, this does not imply bias: nor can bias be inferred from a single over-prediction. In simple terms, bias is what we might expect for a long run of measurements, or in the long term. Therefore, bias is a statistical expectation and to predict it, generally, we need an ensemble of values. If for a large sample the true value was consistently over-predicted we could infer that the measurement process was biased. Such inference is based on a statistical model.

3.2 Precision

Like bias, precision is model-based in that it depends on a statistical model fitted to an ensemble of values. Precision is most often predicted using some measure of the spread of errors around the mean error, for example, the standard deviation of the error, often termed the prediction error (where the prediction error is the square root of the prediction variance).

The standard deviation of the error can be predicted directly from the errors themselves if they are known

$$s_e = \sqrt{\frac{\sum_{i=1}^{n}(\bar{e}_i - \hat{e}_i)^2}{n-1}}. \tag{4}$$

Thus, (im)precision is an expectation of the spread of errors. The smaller the (random) errors, the greater the precision. Such statistical inference is, again, model-based.

3.3 Accuracy

Accuracy is the sum of unbias (i.e. the opposite of bias) and precision:

$$\text{Accuracy} = \text{Unbias} + \text{Precision} \tag{5}$$

This simple equation defining accuracy in terms of unbias and precision is fundamentally important in research into uncertainty.

Where an independent data set is used to assess uncertainty, accuracy may be predicted directly. In particular, the root mean square error (RMSE) which is sensitive to both systematic and random errors, can be used to predict accuracy

$$RMSE = \sqrt{\frac{\sum_{i=1}^{n}(z_i - \hat{z}_i)^2}{n}}. \tag{6}$$

Accuracy, like bias and precision, depends on a statistical model. It is an expectation of the overall error.

It is important to distinguish between error and accuracy. Essentially, error relates to a single value and is data-based. Accuracy relates to the average (and, thereby, statistical expectation) of an ensemble of values: it is model-based. An actual observed error depends on no such underlying model. This distinction between model and data is fundamental to statistical analysis.

4 Model-based Prediction and Estimation

In this book, and more generally in the literature relating to remote sensing and geographical information science, it is acknowledged that many authors use both estimation and prediction in relation to variables. However, in this chapter, a distinction is made between the terms prediction and estimation. Prediction is used in relation to variables and estimation is used in relation to parameters. This is the scheme adopted by most statisticians and we shall adopt it here in the hope of promoting consistency in the use of terms.

4.1 Variables and parameters

Variables are simply measured or predicted properties such as leaf area index (LAI) and the normalized difference vegetation index (NDVI). Parameters are either constants (in a classical statistical framework) or vary (in a Bayesian framework), but in both cases define models. For example, two parameters in a simple regression model are the slope, β, and intercept, α. In the classical framework, once a regression model is fitted to data, a and b, which estimate α and β, are fixed and are referred to as model coefficients (i.e. estimated parameters). However, it is possible to treat parameters as varying. For example, within a Bayesian framework, parameters vary and their statistical distributions (i.e. their cumulative distribution functions, cdfs) are estimated. Thus, in the Bayesian framework, the distinction between estimation (of parameters) and prediction (of variables) disappears.

It is very difficult to achieve standardization on the use of the terms prediction and estimation. For example, authors often use estimate in place of predict. Such interchange of terms has been tolerated in this book.

4.2 Prediction variance

Much effort has been directed at designing techniques for predicting efficiently, that is, with the greatest possible accuracy (unbiased prediction with minimum prediction variance). An equally large effort has been directed at predicting the prediction variance and other characteristics of the error cdf. Let us concentrate on the prediction of continuous variables. In particular, consider the prediction of the mean of some variable, to illustrate the basic concept of prediction variance.

Given a sample $\hat{z}(\mathbf{x}_i)$ for all $i = 1, 2, 3, \ldots, n$ of some variable z, the unknown mean μ may be predicted using

$$m = \frac{\sum_{i=1}^{n} \hat{z}(\mathbf{x}_i)}{n} \quad (7)$$

The above model (equation 7) has been fitted to the available sample data, and the model used to predict the unknown mean.

From central limit theorem, for large sample size n the conditional cdf (i.e. the distribution of the error) should be approximately Gaussian (or normal) (see Figure 1.2) meaning that it is possible to predict the prediction error (and, therefore, imprecision) using the standard error, SE

$$SE = \frac{s}{\sqrt{n}} \quad (8)$$

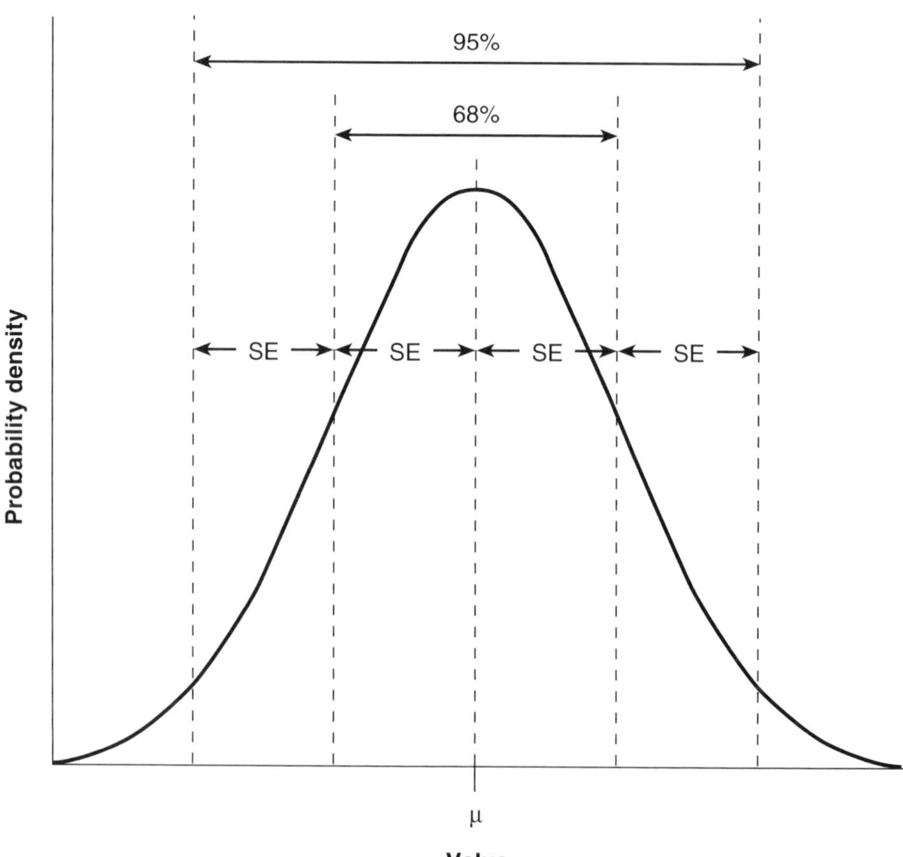

Figure 1.2 *Schematic representation of probability density function for prediction of the mean μ, with given standard error (SE)*

where,

$$s = \sqrt{\frac{\sum_{i=1}^{n} (\bar{z}_i - \hat{z}_i)^2}{n - 1}} \tag{9}$$

where, s is the standard deviation of the variable z and n is the number of data used to predict, in this case, the mean value (Figure 1.2). The prediction variance is simply the square of the standard error. As stated above, the standard error, and the prediction variance, are measures of precision. More strictly, in statistical terms, precision may be defined as the inverse of the prediction variance, although the term precision is more often used (and is used in this chapter) less formally.

For a Gaussian model (i.e. a normal distribution), the mean and variance are the only two parameters. Therefore, given:

1. unbiasedness (i.e. the predicted mean tends towards the true mean as the sample size increases, such that the error can be assumed to be random) and
2. that the conditional cdf (i.e. the error distribution) is known to be Gaussian (e.g. via central limit theorem above)

then

3. the prediction variance (equivalent to the square of the standard error) is sufficient to predict the full distribution of the error.

For most statistical analyses, it is this *error distribution* (i.e. the full conditional cdf) that is the focus of attention.

It is useful to note that generally most statistical models can be used to infer precision, but not bias (and, therefore, not accuracy). For the example above involving predicting the mean, the SE conveys information only on precision. Nothing can be said about bias (or, therefore, accuracy) because we do not have any information by which to judge the predicted mean relative to the true mean (we can only look at the spread of values relative to the predicted mean). Statistical models are often constructed for unbiased prediction, meaning that the prediction error (or its square, the prediction variance), which measures imprecision (not bias), is sufficient to assess accuracy.

The example above involving predicting the regional mean can be extended to other standard statistical predictors. For example, linear regression (prediction of continuous variables) and maximum likelihood classification, MLC (prediction of categorical variables) are now common in remote sensing and geographical information science (Thomas *et al.*, 1987; Tso and Mather, 2001). In both cases (i.e. regression or classification), the objective is prediction (of a continuous or categorical variable), but the model can be used to predict also the prediction error. The model must, first, be fitted to 'training' data. In regression, for example, the regression model is fitted to the scatterplot of the predicted variable y on the explanatory variable z (or more generally, $\mathbf{z}$) in the sense that the sum of the squared differences between the line and the data (in the y-axis direction) is minimized (the least-squares regression estimator). For MLC, the multivariate Gaussian distribution is fitted to clusters of pixels representing each class in spectral feature space. Once the model has been fitted it may be used to predict unknown values given z (simple regression) or unknown classes given the multispectral values $\mathbf{z}$ (classification). However, the (Gaussian) model may be used (in both cases) to predict the prediction error. As with the SE, the prediction error is based on the variance in the original data that the models are fitted to.

There is not space here to elaborate further on standard statistical predictors. However, several of the chapters in this book deal with the geostatistical analysis of spatial data (e.g. Dungan, Chapter 3; Warr *et al.*, Chapter 14; Lloyd, Chapter 15). Therefore, a brief introduction to geostatistical prediction is provided here for the reader who may be unfamiliar with the field.

4.3 Spatial prediction

In geostatistics, the variogram is a central tool. Generally, the experimental variogram is predicted from sample data and a continuous mathematical model is fitted to it (Figure 1.3). This variogram model can then be used in the geostatistical technique for spatial prediction known as kriging (i.e. best linear unbiased prediction, BLUP). The variogram can, alternatively, be used in conditional simulation (Journel, 1996) or for a range of related objectives (e.g. optimal sampling design, van Groenigen *et al.*, 1997; van Groenigen and Stein, 1998; van Groenigen, 1999).

Kriging is often used within a GIS framework to predict unknown values spatially from sample data; that is, to interpolate. Kriging has several important properties that distinguish it from other interpolators such as inverse distance weighting. In particular, since kriging is a least-squares regression-type predictor (Goovaerts, 1997; Deutsch and Journel, 1998; Chilès and Delfiner, 1999), it is able to predict with minimum prediction variance (referred to as kriging variance). Kriging predicts this kriging variance as a byproduct of the least-squares fitting (i.e. the regression). The kriging variance for a given predicted value, therefore, plays the same role as the square of the standard error for the predicted mean of a sample. However, the kriging variance is predicted for all locations (usually constituting a map). Thus, kriging is able to provide both (i) a map of the predicted values and (ii) a map of prediction variance: exactly what is required by users of GIS. A caveat is that the prediction variance is only minimum for the selected linear model (i.e. it is model-based).

A problem with the kriging variance is that it depends only on the variogram and the spatial configuration of the sample in relation to the prediction: it is not dependent on real local data values. However, for situations where local spatial variation is similar from place to place, the kriging variance provides a valuable tool for optimizing sampling design (Burgess *et al.*, 1981). For example, for a fixed variogram, the sample configuration can be optimized for a given criterion without actual sample or

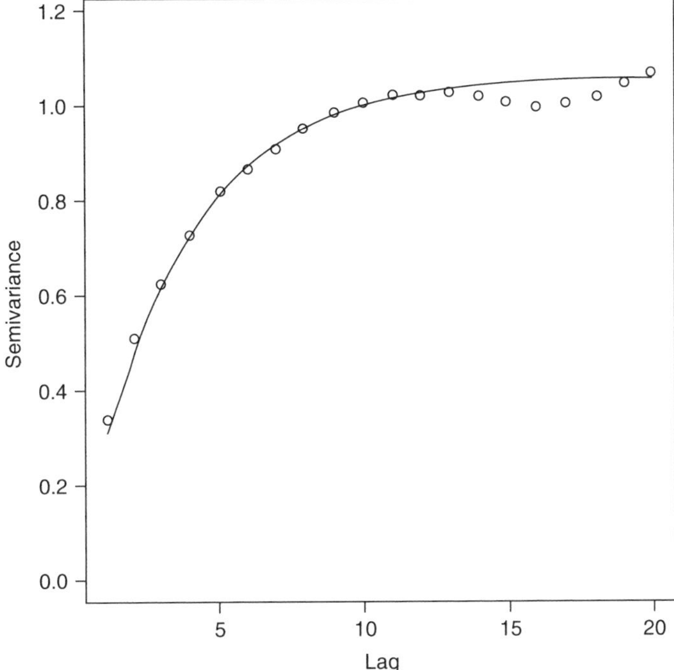

Figure 1.3 Exponential variogram model (line) fitted to an experimental variogram (symbols). The experimental variogram was obtained from an unconditional simulation of a random function model parameterized with an exponential covariance. The estimated exponential model parameters are as follows: non-linear parameter, r_1, is 3.5 pixels; sill variance, c_1, is 1.06 units. The model was fitted without a nugget variance, c_0

survey data (van Groenigen, 1999). Wameling provides an example of the use of the kriging variance in Chapter 13 of this book.

4.4 A note on confidence intervals

Knowledge of the standard error allows the researcher to make statements about the limits within which the true value is expected to lie. In particular, the true value is expected to lie within ±1 SE with a 68% confidence, within ±2 SE with a 95% confidence and within ±3 SE with (approximately) a 99% confidence.

There has been a move within statistics in recent years away from the use of confidence intervals. This is partly because confidence intervals are chosen arbitrarily. The problem arises mostly in relation to statistical significance testing. For example, in the case of the simple *t*-test for a significant difference between the means of two data sets the result of the test (is there a significant difference?) depends entirely on the (essentially arbitrary) choice of confidence level. A confidence of 68% may lead to a significant difference whereas one of 95% may not. Statistical significance testing is increasingly being replaced by probability statements.

5 Accuracy Assessment

The methods discussed so far for predicting the prediction error (e.g. SE of the mean) have been model-based. Thus, they depend on model choice (the particular model chosen and fitted). Approaches for assessing the accuracy of predictions based primarily on data and, in particular, an 'independent' testing data set (i.e. data not used to fit the model and estimate its parameters) are now considered.

5.1 Retrieval, verification and validation

Before examining accuracy assessment it is important to highlight the misuse of some terms. First, it is common, particularly in the literature relating to physical modelling, to read of parameters being 'retrieved'. This term is misleading (implying that a parameter is known *a priori* and that it has somehow been lost and must be 'fetched' or 'brought back'). Parameters are estimated. Second, it is common in the context of accuracy assessment to read the words 'verify' and 'validate'. To verify means to show to be true (i.e. without error). This is rarely the case with modelling in statistics and even in physical modelling (strictly, verification is scientifically impossible). No one model is the 'correct' model. Rather, all models compete in terms of various criteria. One model may be the preferred model in terms of some criterion, but nothing more. For example, if we define the RMSE as a criterion on which models are to be judged, then it should be possible to select a model that has the smallest RMSE. However, this does not imply that the selected model (e.g. simple regression) is in any sense 'true'.

To validate has a similar, although somewhat less strict meaning (it is possible for something to be valid without being 'true'). Its use is widespread (e.g. in relation to remote sensing, researchers often refer to 'validation data' and physical models are often 'validated' against real data). This use of the word validation is often inappropriate. For example, we do not show a classification to be valid. Rather, we assess its accuracy. In these circumstances, the correct term is 'accuracy assessment', and that is the subject to which we now return.

Once a model has been fitted and used to predict unknown values, it is important that an accuracy assessment is executed, and there are various ways that this can be achieved. The following sections describe some of the alternatives.

5.2 Cross-validation

A popular method of assessing accuracy that requires no independent data set is known as cross-validation. This involves fitting a model to all available data and then predicting each value in turn using all *other* data (that is, omitting the value to be predicted). The advantage of this approach is that all available data are used in model fitting (training). The main disadvantage is that the assessment is biased in favour of the model. That is, cross-validation may predict too high an accuracy because there is an inherent circularity in using the same data to fit and test the model.

5.3 Independent data for accuracy assessment

An alternative, and far preferable (if it can be afforded), approach is to use an independent data set (i.e. one not used in model fitting) to assess accuracy. In general terms, this data set should have some important properties. For example, the data should be representative of either the data to which the model was fitted (to provide a fair test of the model) or the data that actually need to be predicted (to assess the prediction). For continuous variables, given an independent testing data set, the full error cdf can be fitted and the parameters of that distribution estimated. Also, summary statistics such as the mean error (bias) and RMSE (accuracy) can be estimated readily. Further, the 'known' values may be plotted against the predicted values and the linear model used to estimate the correlation coefficient and other summaries of the bivariate distribution. The correlation coefficient, in this context, measures the association (i.e. precision)

$$r = \frac{c_{y \cdot \hat{y}}}{s_y \cdot s_{\hat{y}}}, \tag{10}$$

where $c_{y \cdot \hat{y}}$ is the covariance between y and z defined as

$$c_{y \cdot \hat{y}} = \frac{\sum_{i=1}^{n} (\bar{y} - y_i) \cdot (\bar{\hat{y}} - \hat{y}_i)}{n - 1} \tag{11}$$

and s_y and $s_{\hat{y}}$ are the standard deviations of y and $\hat{y}$.

For categorical data, the accuracy assessment is based usually on a confusion matrix (or contingency table) from which a variety of summary statistics may be derived. These include the overall accuracy, the users' and producers' accuracies and the kappa coefficient, which attempts to compensate for chance agreement (Congalton, 1991; Janssen and van der Wel, 1994).

5.4 Spatially distributed error

An accuracy assessment based on an independent data set may provide a spatially distributed set of errors (i.e. an error for every location that is predicted and for which independent testing data are available). These errors can be analysed statistically without the locational information. However, increasingly researchers are interested in the spatial information associated with error. For example, spatially, one might be interested in the local pattern of over- or under-prediction: the map of errors might highlight areas of over-prediction (positive errors) and under-prediction (negative errors). It has been found that in many cases such spatially distributed errors (or differences) are spatially autocorrelated (e.g. Fisher, 1998). That is, errors that are proximate in space are more likely to take similar values than errors that are mutually distant. In many cases, the investigator will be interested in spatial autocorrelation (e.g. the autocorrelation function could be estimated). Spatially autocorrelated error, which implies local bias, has important implications for spatial analysis.

5.5 Repeated measurement

Often, the objective is not to measure the accuracy of prediction, but rather to assess the precision associated with a given measurement process. A valid question is 'how might one predict the error $e(\mathbf{x}_0)$ given only $\hat{z}(\mathbf{x}_0)$?'. One possible solution is to measure repeatedly at the same location $\mathbf{x}_0$, with the mean of the set of values predicting the true value, and the distribution of values conveying information on the magnitude of the likely error. For a single location $\mathbf{x}_0$ a sample of $i = 1, 2, \ldots, n = 100$ values $\hat{z}(\mathbf{x}_{0i})$ could be obtained. For example, suppose the objective was to quantify the precision in measurements of red reflectance from a heathland canopy made using a spectroradiometer (Figure 1.4). Repeated measurement at the same location would provide information on the spread of values around the prediction of the mean and, thereby, the variance associated with measurement.

The remainder of this chapter is devoted to providing a gentle introduction to several important issues that are investigated in some depth within this book.

6 Further Issues I: Spatial Resolution

Spatial resolution is a fundamental property of the sampling framework for both remote sensing and geographical information science. It provides a limit to the scale

Figure 1.4 Repeated measurement of a heathland canopy (z) at a single location $\mathbf{x}_0$ using a Spectron spectroradiometer

(or spatial frequency, in Fourier terms) of spatial variation that is detectable. It is the interplay between the scale(s) of spatial variation in the property of interest and the spatial resolution that determines the spatial nature of the variable observed. Importantly, all data (whether a remotely sensed image or data in a GIS) are a function of two things: reality and the sampling framework (Atkinson and Tate, 2000). One of the most important parameters of the sampling framework (amongst the sampling scheme, the sample size, sampling density) is the support. The support is the size, geometry and orientation of the space on which each observation or prediction is made. The support is a first-order property since it relates to individual observations. The spatial resolution, on the other hand, is second order because it relates to an ensemble of observations.

6.1 Scaling-up and scaling-down

The above link between spatial resolution and observed spatial variation makes the comparison of data obtained with different spatial resolutions a major problem for geographical information science. Before sensible comparison can be made it is important to ensure that the data are represented with the same sampling framework (e.g. two co-located raster grids with equal pixel size). In remote sensing, such comparison is often necessary to relate ground data (made typically with a small support) to image pixels that are relatively large (e.g. support of 30 m by 30 m for a Landsat Thematic Mapper (TM)). The question is 'how should the ground data be "scaled-up" to match the support of the image pixels?'. In a GIS, where several variables may need to be compared (e.g. overlay for habitat suitability analysis) the problem may be compounded by a lack of information on the level of generalization in each variable (or data layer).

Much of the literature related to scaling-up and scaling-down (e.g. Foody and Curran, 1994; Quattrochi and Goodchild, 1997) actually relates to spatial resolution. The chapter by Dungan (Chapter 3) describes the problems associated with scaling variables. In particular, where (i) non-linear transforms of variables are to be made along with (ii) a change of spatial resolution, the order in which the two operations (i.e. non-linear transform and spatial averaging) are undertaken will affect the final prediction. The two options lead to different results. Interestingly, there is not much guidance on what is the most appropriate order for the two operations (for some advice, see Bierkens *et al.*, 2000).

6.2 Spatial resolution and classification accuracy

A reported, but surprising, phenomenon is that classification accuracy is often inversely related to spatial resolution (Townshend, 1981). Thus, a decrease in pixel size can result in a decrease in classification accuracy. However, there are two important points that mitigate against this result. The first is that the relation holds only for (i) a given level of *spatial* generalization and (ii) a given level of generalization in the possible hierarchical classification scheme (see Figure 1.5). Thus, the

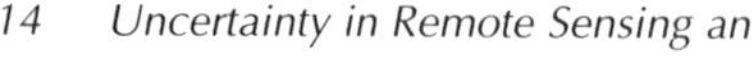

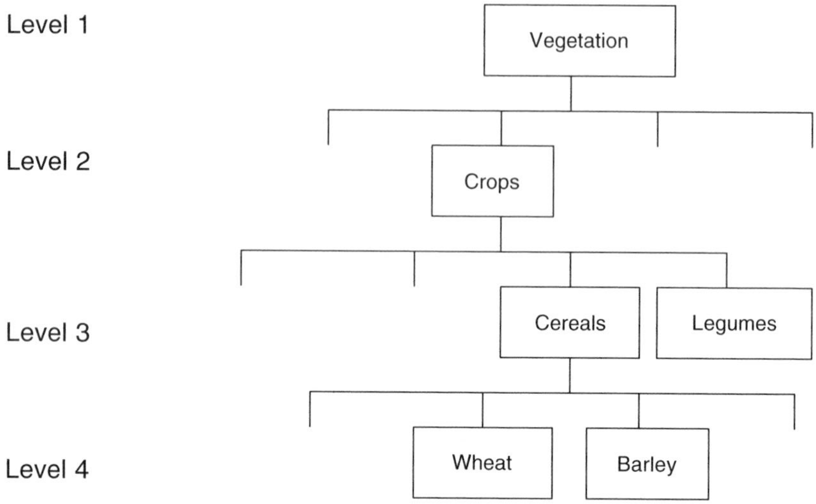

Level 1 Vegetation

Level 2 Crops

Level 3 Cereals Legumes

Level 4 Wheat Barley

Figure 1.5 *Different levels of generalization in a classification hierarchy. Given a classification scheme at level 4, the separation of a field of barley into barley and bare soil may result in misclassification either (i) because the investigator wishes the whole field to be classified as barley or (ii) because the classification scheme does not permit it (e.g. there is no bare soil class)*

classification of patches within an agricultural field of barley into bare soil (which are in fact bare soil) will be regarded as misclassification if the investigator wishes the whole field to be classified as barley. This is a direct result of the level of spatial generalization. Further, if bare soil is not one of the land cover classes to be predicted then misclassification may again occur (as a result of, arguably, too general a classification scheme; see Figure 1.5).

The second, and more fundamental, point is that accuracy is only at most half of the picture. Of greater importance is information. As spatial resolution increases (pixels become smaller), so the amount of spatial variation, and thereby, the amount of information revealed increases (Woodcock and Strahler, 1987; Atkinson and Curran, 1997; Curran and Atkinson, 1999). Thus, to compare the accuracy of classification for different spatial resolutions without reference to their information content amounts to an unfair comparison. As an example, consider the classification of land cover (into classes such as asphalt, grassland and woodland) based on Landsat TM imagery (pixels of 30 m by 30 m) and IKONOS multispectral imagery (pixels of 4 m by 4 m). The classification accuracy might be greater for the Landsat TM imagery, but only on a per-30 m pixel basis. What if the information required was a map of the locations of small buildings (around 10 m by 10 m) within the region? The Landsat TM imagery would completely fail to provide this level of information. What is the point of predicting more accurately if the spatial information provided is inadequate?

The issues presented above are addressed by several of the chapters of this book. For example, these are the sort of issues that providers of remotely sensed information have to face. The chapter by Smith and Fuller (Chapter 9) describes the problems surrounding defining the spatial resolution of the Land Cover Map of

Great Britain 2000 and, in particular, the decision to map using parcel-based units rather than pixel units. The chapter by Manslow and Nixon (Chapter 4) goes beyond issues of spatial resolution by considering the effects of the point-spread function of the sensor on land cover classification. Further, Jakomulska and Radomski (Chapter 7) describe the issues involved in choosing an appropriate window size for texture-based classification of land cover using remote sensing.

7 Further Issues II: Soft Classification

It has been assumed so far that remote sensing classification is 'hard' in the sense that each pixel is allocated to a single class (e.g. woodland). However, in most circumstances soft classification, in which pixels are allocated to many classes in proportion to the area of the pixel that each class represents, provides a more informative, and potentially more accurate, alternative. For example, where the spatial resolution is coarse relative to the spatial frequency of land cover, many pixels will represent more than one land cover class (referred to as mixed pixels). It makes little sense to allocate a mixed pixel to a single class. Even where the spatial resolution is fine relative to the spatial frequency of land cover, soft classification may be more informative than hard classification at the boundaries between land cover objects (where mixed pixels inevitably occur). Thus, since the mid-1980s (Adams *et al.*, 1985), soft classification has seen increasingly widespread use (Bezdek *et al.*, 1984; Foody *et al.*, 1992; Gillespie, 1992; Foody and Cox, 1994; Atkinson *et al.*, 1997).

Despite the clear advantages of soft classification of land cover (land cover proportions are predicted), usually no information is provided about where within each pixel the land cover actually exists. That is, land cover is not mapped within each pixel. Two chapters within this book make super-resolution land cover mapping their objective (Tatem *et al.*, Chapter 6 and Zhan *et al.*, Chapter 5). These, and related techniques, allow the prediction of land cover (and potentially other variables) at a finer spatial resolution than that of the original data.

8 Further Issues III: Conditional Simulation

Warr *et al.* (Chapter 14) use conditional simulation to solve the problem of predicting soil horizon lateral extent. In this section, conditional simulation is introduced as a valuable tool for spatial uncertainty assessment.

A problem with kriging is that the predicted variable is likely to be smoothed in relation to the original sample data. That is, some of the variance in the original variable may have been lost through kriging. Most importantly, the variogram of the predictions is likely to be different to that of the original variable meaning that the kriged predictions (map) could not exist in reality ('not a possible reality') (Journel, 1996). A solution is provided by conditional simulation which seeks to predict unknown values while maintaining, in the predicted variable, the original variogram (see Dungan, 1998, for a comparison of kriging and conditional simulation in remote sensing). A simple technique for achieving a conditional simulation is known as sequential Gaussian simulation (sGs). Given the multi-Gaussian case (i.e. that a multi-Gaussian model is sufficiently appropriate for the data) the kriging variance is sufficient (note that kriging ensures unbiasedness) to characterize fully the conditional cdf (ccdf) of the prediction. All nodes to be simulated are visited in (usually random) order. The ccdf of any location to be simulated is conditional on all data and previously simulated nodes such that any draw from the generated set of ccdfs will reproduce the original variogram.

Interestingly, while the variogram (i.e. the spatial character) is reproduced, the prediction variance of any simulation is double that for kriging. Thus, conditional simulation represents the decision to put greater emphasis (and less uncertainty) on the reproduction of spatial character and less emphasis (and greater uncertainty) on prediction.

9 Further Issues IV: Non-stationarity

Lloyd (Chapter 15) provides an example of a non-stationary modelling approach in geostatistics. Many of the models that are applied in remote sensing and geographical information science are stationary in a spatial sense. Stationarity means that a single model is fitted to all data and is applied equally over the whole geographic space of interest. For example, to predict LAI from NDVI a regression model might be fitted to coincident data (i.e. where both variables are known) and applied to predict LAI for pixels where only NDVI is known. The regression model and its coefficients are constant across space meaning that the model is stationary. The same is true of geostatistics where a variogram is estimated from all available data and the single fitted model (McBratney and Webster, 1986) used to predict unobserved locations.

An interesting and potentially efficient alternative is to allow the parameters of the model to vary with space (non-stationary approach). Such non-stationary modelling has the potential for greater prediction precision because the model is more tuned to local circumstances, although clearly a greater number of data (and parameters) is required to allow reliable local fitting. In some cases, for example, geographically weighted regression (Fotheringham *et al.*, 2000), the spatial distribution of the estimated parameters is as much the interest as the predicted variable.

10 Conclusions

Uncertainty is a complex and multi-faceted issue that is at the core of (spatial) statistics, is of widely recognized importance in geographical information science and is of increasing importance in remote sensing. It is fundamentally important to agree terms and their definitions to facilitate the communication of desired meanings. While it is impossibly difficult to hold constant any language, particularly one in rapidly changing fields such as remote sensing and geographical information science, we believe that it is an important and valid objective that the meanings of terms should be shared by all at any one point in time. The view taken in this chapter is that the remote sensing and geographical information science communities should, where relevant, adopt the terminology used within statistics (e.g. Cressie, 1991), and otherwise should adopt terms that convey clearly the author's meaning.

In this chapter, we have covered a wide range of terms, their definitions, and their wider meanings. The authors of chapters in this book have been urged to use these terms, and it is hoped that in the future others will follow suit.

Acknowledgements

Professor Paul Curran and Mr Nicholas Hamm are thanked for commenting constructively on an earlier version of this chapter.

References

Adams, J. B., Smith, M. O. and Johnson, P. E., 1985, Spectral mixture modelling: a new analysis of rock and soil types at the Viking Lander 1 site, *Journal of Geophysical Research*, **91**, 8098–112.

Atkinson, P. M. and Curran, P. J., 1997, Choosing an appropriate spatial resolution for remote sensing investigations, *Photogrammetric Engineering and Remote Sensing*, **63**, 1345–51.

Atkinson, P. M., Cutler, M. E. J. and Lewis, H., 1997, Mapping sub-pixel proportional land cover with AVHRR imagery, *International Journal of Remote Sensing*, **18**, 917–35.

Atkinson, P. M. and Tate, N. J., 2000, Spatial scale problems and geostatistical solutions: a review, *Professional Geographer*, **52**, 607–23.

Bezdek, J. C., Ehrlich, R. and Full, W., 1984, FCM: the fuzzy *c*-means clustering algorithm, *Computers and Geosciences*, **10**, 191–203.

Bierkens, M. F. P., Finke, P. A. and de Willigen, P., 2000, *Upscaling and Downscaling Methods for Environmental Research* (Dordrecht: Kluwer Academic).

Burgess, T. M., Webster, R. and McBratney, A. B., 1981, Optimal interpolation and isarithmic mapping of soil properties. IV. Sampling strategy, *Journal of Soil Science*, **32**, 643–59.

Canters, F., 1997, Evaluating the uncertainty of area estimates derived from fuzzy land cover classification, *Photogrammetric Engineering and Remote Sensing*, **55**, 1613–18.

Chilès, J.-P. and Delfiner, P., 1999, *Geostatistics: Modelling Spatial Uncertainty* (New York: Wiley).

Congalton, R. G., 1991, A review of assessing the accuracy of classifications of remotely sensed data, *Remote Sensing of Environment*, **37**, 35–46.

Cressie, N. A. C., 1991, *Statistics for Spatial Data* (New York: Wiley).

Curran, P. J. and Atkinson, P. M., 1999, Issues of scale and optimal pixel size, in A. Stein, F. van der Meer and B. Gorte, (eds), *Spatial Statistics for Remote Sensing* (Dordrecht: Kluwer), pp. 115–33.

Deutsch, C. V. and Journel, A. G., 1998, *GSLIB: Geostatistical Software and User's Guide*, Second Edition (Oxford: Oxford University Press).

Dungan, J., 1998, Spatial prediction of vegetation quantities using ground and image data, *International Journal of Remote Sensing*, **19**, 267–85.

Fisher, P. F., 1994, Visualization of the reliability in classified remotely sensed images, *Photogrammetric Engineering and Remote Sensing*, **60**, 905–10.

Fisher, P. F., 1998, Improved modelling of elevation error with geostatistics, *Geoinformatica*, **2**, 215–33.

Foody, G. M., 2002, Status of land cover classification accuracy assessment, *Remote Sensing of Environment*, **80**, 185–201.

Foody, G. M., Campbell, N. A., Trodd, N. M. and Wood, T. F., 1992, Derivation and applications of probabilistic measures of class membership from the maximum likelihood classification, *Photogrammetric Engineering and Remote* Sensing, **58**, 1335–41.

Foody, G. M. and Cox, D. P., 1994, Sub-pixel land cover composition estimation using a linear mixture model and fuzzy membership functions, *International Journal of Remote Sensing*, **15**, 619–31.

Foody, G. M. and Curran, P. J., 1994, Scale and environmental remote sensing, in G. M. Foody and P. J. Curran, (eds), *Environmental Remote Sensing from Regional to Global Scales* (Chichester: Wiley), pp. 223–32.

Fotheringham, A. S., Brunsdon, C. and Charlton, M., 2000, *Quantitative Geography. Perspectives on Spatial Data Analysis* (London: Sage).

Gillespie, A. R., 1992, Spectral mixture analysis of multi-spectral thermal infrared images, *Remote Sensing of Environment*, **42**, 137–45.

Goodchild, M. and Gopal, S. (eds.), 1989, *Accuracy of Spatial Databases* (London: Taylor and Francis).

Goovaerts, P., 1997, *Geostatistics for Natural Resources Evaluation* (New York: Oxford University Press).

Heuvelink, G. B. M., 1998, *Error Propagation in Environmental Modelling with GIS* (London: Taylor and Francis).

Hoffman-Riem, H. and Wynne, B., 2002, In risk assessment, one has to admit ignorance, *Nature*, **416**, 123.

Janssen, L. L. F. and Vanderwel, F. J. M., 1994, Accuracy assessment of satellite-derived land-cover data -a review, *Photogrammetric Engineering and Remote Sensing*, **60**, 419–26.

Journel, A. G., 1996, Modelling uncertainty and spatial dependence: stochastic imaging, *International Journal of Geographical Information Systems*, **10**, 517–22.

Kiiveri, H. T., 1997, Assessing, representing and transmitting positional uncertainty in maps, *International Journal of Geographical Information Science*, **11**, 33–52.

Klir, G. J. and Folger, T. A., 1988, *Fuzzy Sets, Uncertainty and Information* (London: Prentice Hall International).

McBratney, A. B. and Webster, R., 1986, Choosing functions for semi-variograms of soil properties and fitting them to sampling estimates, *Journal of Soil Science*, **37**, 617–39.

Quattrochi, D. A., and Goodchild, M. F. (eds), 1997, *Scale in Remote Sensing and GIS* (New York: CRC Press).

Thomas, I., Benning, V. and Ching, N. P., 1987, *Classification of Remotely Sensed Images* (Bristol: Adam Hilger).

Townshend, J. R. G., 1981, The spatial resolving power of Earth resources satellites, *Progress in Physical Geography*, **5**, 32–55.

Tso, B. and Mather, P. M., 2001, *Classification Methods for Remotely Sensed Data* (London: Taylor and Francis).

Van Groenigen, J.-W., 1999, *Constrained Optimisation Of Spatial Sampling. A Geostatistical Approach*, ITC Publication Series No. 65 (Enschede: International Institute for Aerospace Survey and Earth Sciences).

Van Groenigen, J. W. and Stein, A., 1998, Constrained optimization of spatial sampling using continuous simulated annealing, *Journal of Environmental Quality*, **27**, 1078–86.

Van Groenigen, J. W., Stein, A. and Zuurbier, R., 1997, Optimization of environmental sampling using interactive GIS, *Soil Technology*, **10**, 83–97.

Veregin, H., 1996, Error propagation through the buffer operation for probability surfaces, *Photogrammetric Engineering and Remote Sensing*, **62**, 419–28.

Woodcock, C. E., and Strahler, A. H., 1987, The factor of scale in remote sensing, *Remote Sensing of Environment*, **21**, 311–32.

Zhang, J. and Goodchild, M. F., 2002, *Uncertainty in Geographical Information* (London: Taylor and Francis).

2

Uncertainty in Remote Sensing

Curtis E. Woodcock

1 Introduction

The goal of remote sensing is to infer information about objects from measurements made from a remote location, frequently from space. The inference process is always less than perfect and thus there is an element of uncertainty regarding the results produced using remote sensing. When viewed from this perspective, the problem of uncertainty is central to remote sensing. However, the topic of uncertainty in remote sensing gets a relatively modest amount of attention. The reasons for this can be debated, but it is somewhat natural for people involved with remote sensing to focus on what can be done well with remote sensing rather than what cannot be done well. However, as a matter of the natural growth and maturity of the field of remote sensing, it is essential that the topic of uncertainty begins to demand more attention. The recent conference devoted to questions of uncertainty in remote sensing and GIS that formed the foundation for this book was a good start toward raising awareness of the importance of uncertainty in remote sensing. This chapter is devoted to outlining some of the issues related to uncertainty in remote sensing. As such, this chapter is more about raising questions than attempting to answer them. The hope is that by making the issues more explicit, future work in remote sensing will devote more attention to uncertainty.

To start this discussion, consider as an example the following question, which is typical of the kind frequently asked by people in remote sensing:

Is it possible to map agricultural lands using remote sensing?

Of course, it is *possible* to make a map of agricultural lands using remote sensing. However, that is not really what is being asked. What is being asked is whether or

Uncertainty in Remote Sensing and GIS. Edited by G.M. Foody and P.M. Atkinson.
 ISBN: 0–470–84408–6

not remote sensing can be used to provide someone with the information about agricultural lands they desire with suitable precision and at a level of accuracy which makes the map useful for their purposes. As such, this kind of question is inherently about uncertainty! As a result, answering such a question requires considerable clarification, careful qualification and the ability to make judgements about uncertainty. One reason it is difficult to answer such a question is that there are many dimensions to the remote sensing problem that contribute to the resulting uncertainty in a derived map. First, what kind of information about the agricultural area is desired and at what spatial scale? Is it enough to identify which areas are cultivated, or is more detailed information required about which crops are present in the individual fields, or the stage of crop development, or the moisture status of the crop, or any of many other characteristics of agricultural environments of interest. For thematic maps, as the categorical detail in the map (or the precision of the map) increases, the accuracy of the map typically decreases (or the uncertainty increases). While this general relationship is well known, the rate of decrease associated with increasing categorical detail is typically not well known.

A second factor concerns the nature and quality of the inputs used in the mapping process. In this context, both the remote sensing imagery to be used as well as other kinds of data, such as field data or laboratory spectra are important and might influence the ability to derive the desired information. As the quality of the imagery and other inputs increases, so will the accuracy of the resulting map.

A third consideration concerns the area to be mapped. In general, as the size of the area to be mapped increases, the accuracy of a map will decrease. And finally, answering the question posed requires a solid understanding of the requirements for the accuracy of the resulting map. Frequently, map users are, to some degree, willing to trade off the precision (detail) and accuracy of maps. Also, are the accuracy requirements the same for all categories in the map? For thematic maps, the accuracy of a few categories is often far more important than others.

While this list of factors that need consideration prior to answering our question is certainly not exhaustive, it does help illustrate that there are many factors involved and these factors are not independent. Thus, to answer a question about the feasibility of using remote sensing for various applications involves a complex calculus of many factors. One could say it is an exercise in evaluating uncertainty.

2 Characterizing Uncertainty

Perhaps the first challenge related to uncertainty in remote sensing is to adequately characterize uncertainty in maps derived from remote sensing. For thematic maps, the most conventional approach has been to provide estimates of overall map accuracy, where the percentage of sites mapped accurately is the most common metric used. However, there are a number of other metrics that have been developed and applied which attempt to improve on overall map accuracy such as the kappa statistic. Additionally, map accuracies are frequently reported for the individual classes as both user's and producer's accuracies, which is a significant improvement relative to overall map accuracies in many instances (Congalton, 1991).

In addition to knowing, on average, how likely errors are for the whole map, map users frequently would like to know where these errors are most likely to occur. Several recent studies have begun to explore the possibility of providing spatially explicit data on uncertainty, or mapping confidence (see for examples Foody *et al.*, 1992; Carpenter *et al.*, 1999; McIver and Friedl, 2001). In such approaches each pixel is assigned both a class and then some measure of the confidence that the pixel actually belongs in the class. Many classification algorithms can be configured to provide such indications. This attention to providing spatially explicit uncertainty data is a very positive step and it is clear there is still much to be done and learned in this domain.

The methods used to report accuracy of maps representing continuously measured variables are inherently somewhat different from those for thematic maps. The most commonly used metric is the root mean square error (RMSE), although it is common for R^2 values to be reported in the literature. It is worth noting that R^2 values measure the strength of the relationship between an underlying remote sensing variable and the desired map variable and thus do not directly characterize accuracy. Thus, for the purposes of characterizing uncertainty in maps, the RMSE may be preferable. The RMSE is analogous to the overall map accuracy reported for thematic maps as it is a single value intended to characterize the accuracy for the entire map. Development of methods for spatially explicit characterization of uncertainty in maps of continuously measured variables would be highly desirable.

3 Understanding Uncertainty

A significant challenge is to better understand the nature and sources of uncertainty in maps derived from remote sensing. For example, a thorough understanding of the causes of uncertainty would allow for an informed answer to our hypothetical question at the beginning of this chapter about the use of remote sensing to map agricultural lands. For a variety of reasons, remote sensing has been plagued by the tendency to study problems in 'one place and one time' rather than from a comprehensive perspective. Thus when interested in a specific application of remote sensing, one may find many relevant articles in the literature, but each is typically done in a single place using a single kind of imagery and a single set of methods. As a result it may be difficult to determine the best kind of imagery or methods to use in a new place for a similar application as a comprehensive understanding of the causes of uncertainty is unavailable.

One model for the remote sensing process is the image chain approach, as proposed by Schott (1997) and illustrated in Figure 2.1. One fundamental idea underlying the image chain approach is that the entire remote sensing process as applied in any situation is only as strong as the weakest link. In this context, the links in the chain represent various steps in the remote sensing process from image capture to image processing to image display or representation. The image chain approach is useful because it helps identify the many steps in the remote sensing process (or links in the chain) and illustrate that these steps are interrelated. Once the remote sensing process is viewed in this manner, it becomes easier to understand that limitations at

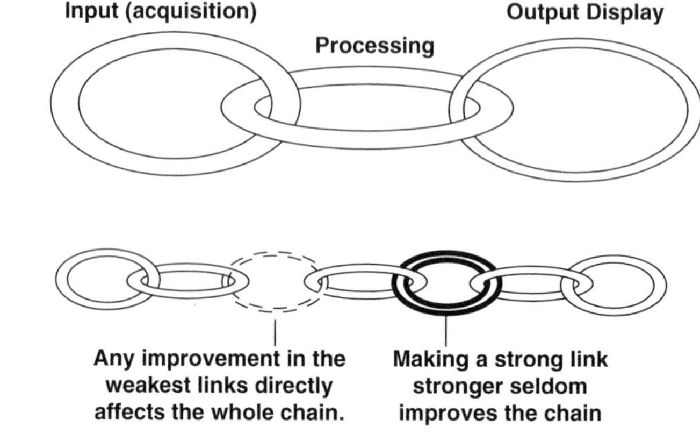

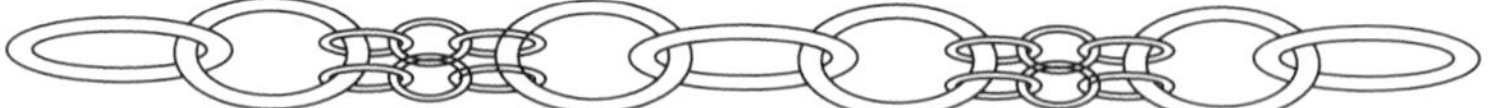

Figure 2.1 Simplified image chain (reproduced from Schott, J. R. 1997)

each step limit the entire process. Improvements in the weakest links stand to improve the entire process the most. In the context of our discussion of uncertainty, a thorough understanding of the links in the chain should help us evaluate the various sources of uncertainty in a particular application.

It is interesting to consider the nature of the efforts in the remote sensing research community concerning uncertainty from the perspective of the image chain approach. Overwhelmingly, the primary step that has been evaluated with respect to its contribution to uncertainty in the final results (or map) has been the image classification step, and most particularly the effect of various classification algorithms on uncertainty. While this step is extremely important and certainly merits much attention, other steps in the remote sensing process may also contribute greatly to map uncertainty. For example, Song *et al.* (2001) have evaluated the effect of atmospheric correction on classification and change detection in cases where these mapping processes are calibrated using images from a different place and/or time to the images used to make the final maps. In such a situation, there are several steps in the remote sensing process (or links in the chain) that could limit the entire process, including image calibration and radiometric processing. As a community, we need to make sure that we are identifying the steps in the remote sensing process that contribute most to uncertainty (or the weakest links in the chain) and focusing our attention accordingly.

4 Future Challenges

What are some of the remaining challenges with respect to uncertainty in remote sensing?

(i) *Reduce uncertainty in remote sensing products (i.e. make better maps!).*

In one respect, this challenge is so obvious that it does not require mentioning. This goal is central to the entire field of remote sensing and researchers are constantly trying to find better ways to make better maps. However, it is worth noting here, from the perspective of the discussion of sources of uncertainty and the image chain approach, that one tendency of researchers is to focus their attention on the steps of the remote sensing process over which they have the most control, such as the selection of an image classification algorithm. However, to most effectively contribute to the larger goal of producing better maps from remote sensing we must find the weakest links in the image chain and devote our efforts to improving them. This charge may include more research on ways to improve sensing systems and the preprocessing of sensor data.

(ii) *Improve the nature and quality of uncertainty information provided with maps derived from remote sensing.*

The value, use and acceptance of maps derived from remote sensing will increase greatly if they have accompanying quantitative information on uncertainty.

Similarly, if spatially explicit data on uncertainty can be provided it will greatly benefit efforts to model uncertainty in analyses undertaken within a geographical information system.

(iii) *Develop ways to characterize the sources of uncertainty in maps derived from remote sensing.*

Again drawing from the image chain analogy, it would be very beneficial to know the sources of uncertainty in remote sensing maps. One can imagine trying to decompose the uncertainty in a map into the errors due to sensor limitations, preprocessing for calibration, radiometry or atmospheric effects, and mapping algorithms. Knowledge of this kind could greatly aid efforts to answer questions about the expected uncertainties for new maps of similar areas based on the proposed sensor and processing methods.

5 Concluding Comments

The question of uncertainty is central to the field of remote sensing. Essential to the process of increasing the quality of maps made using remote sensing is an understanding of the nature and sources of errors in maps. The image chain approach provides a valuable model for the remote sensing process that directs us to devote our efforts to strengthening the weakest link in the image chain. From the perspective of map users, accurate characterization of uncertainty greatly enhances the value of remote sensing products. In particular, spatially explicit data on map uncertainty promises to increase the utility of remote sensing products used in geographic information systems.

References

Carpenter, G. A., Gopal, S., Macomber, S., Martens, S., Woodcock, C. E., and Franklin, J., 1999, A neural network method for efficient vegetation mapping, *Remote Sensing of Environment*, **70**, 326–38.

Congalton, R. G., 1991, A review of assessing the accuracy of classifications of remotely sensed data, *Remote Sensing of Environment*, **37**, 35–46.

Foody, G. M., Campbell, N. A., Trodd, N. M., and Wood, T. F., 1992, Derivation and applications of probabilistic measures of class membership from the maximum likelihood classifications, *Photogrammetric Engineering and Remote Sensing*, **58**, 1335–41.

McIver, D. K., and Friedl, M. A., 2001, Estimating pixel-scale land cover classification confidence using nonparametric machine learning methods, *IEEE Transactions on Geoscience and Remote Sensing*, **39**, 1959–68.

Schott, J. R., 1997, *Remote Sensing: The Image Chain Approach* (New York: Oxford University Press).

Song, C., Woodcock, C. E., Seto, K., Pax Lenney, M., and Macomber, S. A., 2001, Classification and change detection using Landsat TM data: when and how to correct atmospheric effects?, *Remote Sensing of Environment*, **75**, 230–44.

3

Toward a Comprehensive View of Uncertainty in Remote Sensing Analysis

Jennifer L. Dungan

1 Introduction

The topic of uncertainty has been receiving increasing attention in the geographical sciences (Kundzewicz, 1995; Mowrer and Congalton, 2000; Hunsaker *et al.*, 2001; Odeh and McBratney, 2001). One might expect that an outgrowth of an increasingly mature science is a thorough quantification of uncertainty and its eventual reduction. However, if the scientific objective is to increase our understanding of the biogeophysical world through remote sensing, does it necessarily make sense to strive for reduced uncertainty? After all, an argument can be made that, while a reduction in uncertainty may imply progress, an increase in uncertainty may also represent advancing understanding. When new observational evidence is acquired and is incompatible with the results of the currently accepted model, uncertainty has just increased. Uncertainty increases despite the informative nature of the new evidence. Nonetheless, more generally it is logical to expect that reduced uncertainty gauges success in applying models to data.

Descriptions of uncertainty are routine in statistical practice. Despite the fact that the analysis of remotely sensed data often relies on statistical methods, these descriptions have been often neglected in the presentation of remote sensing results. Perhaps part of the reason for this neglect is that describing uncertainty for geographical data is particularly difficult (Gahegan and Ehlers, 2000). Putting a prediction interval on

Uncertainty in Remote Sensing and GIS. Edited by G.M. Foody and P.M. Atkinson.
 ISBN: 0-470-84408-6

the value at every spatial unit represented in a map generated from remotely sensed data has at least three challenges:

1. Spatial units are not generally independent, so statistical models that assume errors to be independent may be inadequate.
2. There are not enough reference data, measured using ground-based methods or other independent means, available to adequately test these intervals. Reference data for variables of interest to remote sensing analyses are often expensive and time-consuming to acquire and are not usually plentiful in a statistical sense.
3. It is a challenge to represent or visualize such spatially varying intervals or even distributions. In a non-spatial context, such intervals can be represented as stem-and-whisker plots – distributions can be represented as histograms. Representing uncertainty for a 2D map requires at least four or five dimensions and, therefore, becomes a problem in visualization.

In addition, uncertainty may change when one is talking about a single pixel or multiple pixels. That is, a confidence statement about the limited area represented by a single pixel may be different to a confidence statement about a larger area of which that pixel forms only a part.

The purpose of this chapter is to advance a classification of the sources of uncertainty in the analysis of remote sensing data by incorporating statistical and geostatistical concepts.

2 What is Uncertainty?

The word uncertainty has resisted a narrow definition, possibly because unlike related words such as bias, precision, error and accuracy, the generic meaning of uncertainty deals with the subjective. That is, while two individuals may arrive at the same answer to a question, one individual may be more certain than the other about that answer. This subjectivity would suggest that uncertainty would escape rigorous scientific examination. However, given that measures of uncertainty are important for gauging progress, the most appropriate way to proceed is to use well-documented probability models, Bayesian or frequentist, that are agreed upon with some degree of consensus and that can be shown over time to be useful.

The definition of uncertainty to be used in this chapter is the simple 'quantitative statement about the probability of error'. A reasonable extrapolation of this definition is that accurately measured, estimated or predicted values will have small uncertainty; inaccurate measurements, estimates or predictions should be associated with large uncertainty. This explanation is not very helpful until the term accuracy is discussed further, because it turns out that there are two conflicting interpretations in the current literature of how accuracy relates to error.

Something that is accurate should lack error (Taylor and Kuyatt, 1994). Accuracy has traditionally been defined (Johnston, 1978) to mean more specifically that

1. on average, the error is insignificantly small – values are unbiased, and
2. the spread of those errors is also small – values are precise.

Cochran (1953) clearly stated the orthodox view in statistics about these two components of accuracy in parameter estimation theory. In his discussion of estimating the mean parameter, he states, 'Because of the difficulty of ensuring that no unsuspected bias enters into estimates, we shall usually speak of the precision of an estimate rather than its accuracy. Accuracy refers to the size of the deviations from the true mean (μ), whereas precision refers to the size of deviations from the mean (m) obtained by repeated application of the sampling procedure' (Cochran, 1953, p. 15). Olea (1990, p. 2) also reflects this view, stating that, 'Accuracy presumes precision, but the converse is not true'. Both Cochran and Olea mention mean square error (MSE) as a typical quantifier of accuracy:

$$\begin{aligned} MSE(\hat{\mu}) &= E(\hat{\mu} - \mu) \\ &= E[(\hat{\mu} - m) + (m - \mu)]^2 \\ &= \mathrm{var}(\hat{\mu}) + (m - \mu)^2 \\ &= precision + (bias)^2 \end{aligned} \tag{1}$$

where $\hat{\mu}$ is the estimate of the true value of the parameter μ. $\hat{\mu}$ is distributed around the potentially biased m, where m is calculated from a sample of the population. By this reasoning, inaccurate values may, therefore, occur if they are biased though precise, unbiased and imprecise or, most obviously, biased and imprecise.

A contrasting definition has been used by others (e.g. Maling, 1989; Goovaerts, 1997; Mowrer and Congalton, 2000) who equate accuracy directly with unbiasedness. This implies that a set of imprecise values, as long as it is unbiased, is declared accurate. This alternative interpretation of accuracy is rejected here because a definition of uncertainty that ignores imprecision or random error would not be terribly useful. Therefore, the probability of error incorporated into uncertainty bounds should represent both the random and systematic components of error.

3 Sources of Uncertainty

To explicate the sources of uncertainty in remote sensing analyses, consider that remotely sensed data are observations on a variable or, more often because of the multi-spectral nature of most sensors, multiple variables; the objective of analysis is to transform these variables to obtain new variables of interest. In equation form, this can be expressed as:

$$z_\nu(\mathbf{u}) = f(\mathbf{y}_\nu(\mathbf{x}), \boldsymbol{\theta}) \tag{2}$$

where z is a variable with values on some nominal, ordinal, interval or ratio scale; $\mathbf{y}$ is a vector of input variables; ν is the spatial support of $\mathbf{y}$ and z; $\mathbf{x}$ and $\mathbf{u}$ are the spatial locations of $\mathbf{y}$ and z, respectively; f is a model and $\boldsymbol{\theta}$ is the vector of the parameters of this model. The variables $\mathbf{y}_\nu(\mathbf{x})$ and $z_\nu(\mathbf{u})$ are both examples of the geographical datum defined by Gahegan (1996); they each have a value and a spatial extent. The geostatistical term 'support' is used here to refer to this spatial extent. Strictly, these variables also have a temporal extent and a time for which they are relevant. For simplicity, temporal aspects are neglected herein.

An important distinction (conventional in statistics) is made in equation (2) between variables, which here are a function of location, and parameters, which are not a function of location but a part of the model. The job of analysis is, therefore, to predict values of the variable z with support ν at every location $\mathbf{u}$ using a model and its parameters.

An example of the application of equation (2) is the transformation of digital numbers ($\mathbf{y} = \mathbf{DN}$) to surface reflectance ($z = \rho$). In this case, f would be a non-linear function and parameters $\boldsymbol{\theta}$ might include atmospheric optical depth, total irradiance and the backscatter ratio if the region was flat and these were considered spatially constant. Another example would be the identification of a nominal land cover class at each pixel based on **DN**. The model in this case is a classification algorithm, whose parameters include the means and covariances of spectral classes. z may also be defined as a 'fuzzy' class (Foody, 1999), usually represented as a continuous variable ranging between 0 and 100% and predicted using any one of the fuzzy classification algorithms (Atkinson *et al.*, 1997). Friedl *et al.* (2001) describe another family of examples, called 'inversion techniques' – the transformation of spectral variables into continuous variables related to vegetation structure or density. The models for these analyses have included empirically derived regression models where the shape of the regression is chosen from the data to be analysed, semi-empirical models where the shape of the regression model is chosen based on physical theory, or canopy reflectance models.

Putting analyses into the framework of equation (2) implies a different perspective on uncertainty to that of Friedl *et al.* (2001, p. 269), who stated, 'It is important to note that the errors introduced to maps produced by [...] remote sensing inversion techniques are of an entirely different nature and origin than thematic errors present in classification maps'. In equation (2), classification is placed into the same context as inversion techniques, where the probability of error is influenced by the spectral variables used in the process, the classification algorithm and the parameters selected in that algorithm.

The perspective adopted is similar to that of the 'image chain approach' described by Schott (1997). A complex analysis would be broken down by the image chain approach into a sequence of applications of simple models f, where the z variable from each step in the sequence would become a y variable in the next step. Expressing remote sensing analysis using this framework makes it clear that uncertainty can arise about input variables, their locations and spatial supports, the model and parameters of the model (Table 3.1). Each of these sources of uncertainty is discussed below.

Table 3.1 *Sources of uncertainty in remote sensing analysis. Refer to equation (2) for definition of symbols*

Symbol(s)	Source
$\mathbf{x}, \mathbf{u}$	positional
ν	support
θ	parametric
f	structural
z, y	variable

3.1 Uncertainty about parameters

Parametric uncertainty is perhaps the most familiar type because of the emphasis on parameter estimation in basic statistics courses. Methods to estimate parameters usually incorporate unbiasedness criteria and maximum precision (the least squares criterion). Characterizing parametric uncertainty is typically accomplished using training data for the input variables. Statements about parametric uncertainty assume that the model associated with the parameters is known with certainty.

In the above example about the conversion of **DN** to reflectance, a decision may be made to treat optical depth as a parameter. Uncertainty about the value of this optical depth parameter might then be informed by the variation among dark targets in relevant imagery. On the other hand, if optical depth is treated as a variable and so is a y rather than a θ, it is expected to vary spatially. In this case, the variation among dark targets would pertain to both the actual values of optical depth and its uncertainty.

It is an easier problem to estimate parameters than to predict variables; in general, there is less uncertainty about them than about variables. This is because parameters, by definition, are constant throughout the image domain to which the model is applied. Therefore, all of the data available for training or calibrating the model can be used to obtain the expected value and its distribution for the parameter. The relative certainty of parameters can be seen in the familiar contrast (Draper and Smith, 1998) between confidence intervals for a mean and prediction intervals for unmeasured values. Confidence intervals are in general narrower than prediction intervals.

3.2 Uncertainty about models

Less often acknowledged, but always present, is uncertainty about the form or structure of the model used to transform the input variables to the desired output variables. Draper (1995) calls this structural uncertainty. Structural uncertainty comes from the fact that several models are potentially useful to describe the phenomenon being studied. For instance, a nearest mean and a maximum likelihood algorithm are each based on reasonable models for a classification problem. If a set of models are considered, each of which gives different results, uncertainty is attached to the fact that the most accurate model is unknown. The type of error likely to occur is bias error. Because it is time-consuming and labour intensive to accomplish many analyses, usually it has not been profitable for multiple models to be tried. If they are tried, a common occurrence is that one preferred analysis is chosen based on the results on a very small sample and other analyses are discarded.

Datcu *et al.* (1998) deal explicitly with the ideas of model fitting, where parametric uncertainty is relevant, and model selection. They suggest that the choice of a model may be one of the biggest sources of uncertainty. More generally, the precise path of analysis contains many decision points. Alternatives at any one of these decision points may generate uncertainty. For instance, results of thematic classifications may vary greatly when independent analysts construct them (Hansen and Reed, 2000). An influential study in geostatistics (Englund, 1990) showed that the spread of outcomes from alternative analysis paths can result in large uncertainty, with bias as the major concern.

3.3 Uncertainty about support

Spatial support is a concept from geostatistics that refers to the area over which a variable is measured or predicted (Journel and Huijbregts, 1978; Atkinson and Curran, 1997; Dungan, 2001). For remotely sensed data, the support can be considered that area covered by the effective resolution element (ERE), a function of the instantaneous field of view, flight variables such as altitude, velocity and attitude and atmospheric effects (Forshaw *et al.*, 1983). The variables for which data are available, the **y**, have a spatial support ν defined by this ERE. It may or may not be the aim to predict the z variable on the same spatial support.

The support of remotely sensed data is never known precisely. There are no real boundaries of a pixel and the value represented at a pixel is not a strict spatial average of that unknown area (Fisher, 1997). The point spread function used to model the sensor response describes the y value as an unequally weighted integral of its spatial extent. Yet more often than not, approximations to the support are made using the grid cell size reported in image metadata. Efforts have, therefore, been made to increase the precision with which this value is known (e.g. Schowengerdt *et al.*, 1985).

3.4 Uncertainty about position

Uncertainty about **x** and **u**, the locations of the data values, is routinely addressed in image analysis. The 'raw' images delivered for analysis by a data provider incorporate positional errors because of the process of converting a sensor's responses to a raster data structure. This delivered raster is usually registered or georectified prior to the application of any form of equation (2). Registration can be accomplished using different approaches; those which use only ground control points (GCPs) that can also be identified in the image or those which model the scene and its observation geometry. Regardless of the approach used, uncertainty metrics can be usefully derived from a set of numerous 'test' points, or what Schowengerdt (1997) calls ground points (GPs other than the GCPs used to train or calibrate the registration process). Customarily, only a root-mean-square error on GPs is reported when it would be even more useful to report the distribution of errors – unsuspected bias might sometimes be uncovered (Lewis and Hutchinson, 2000).

Numerous studies have examined the relationships between values of remotely sensed spectral variables (say, $\mathbf{y}(\mathbf{x})$) at locations within images and ground-based observations of variables (say, $z(\mathbf{u})$) such as temperature, vegetation amount and pigment concentration. A way to think of this type of study is to regard it as a search for a model f that fits the observed data. Because precise co-location of the remotely sensed and ground-measured variables is important in this type of study, a common practice in these studies is to use an average of values from a 2×2, 3×3 or 4×4

pixel window area surrounding the estimated position of the ground-based observation to represent the spectral variable. The average is taken in an attempt to reduce the positional uncertainty of **u**. This practice essentially increases the support of the spectral variables **y**. The aggregation effect will act to reduce variance of the **y** and the statistics used as criteria for model-fitting will differ from what they would be if **y** and z were on the same support.

3.5 Uncertainty about variables

Uncertainty about the input variables **y** exists as a function of the measurement process (i.e. measurement error) or whatever transformation came prior to the current one in the image chain. In addition, parametric, structural, support and positional uncertainty lead to uncertainty about variables predicted by the model. This can be understood as an error or uncertainty propagation problem. Techniques exist to handle some aspects of the propagation problem (Heuvelink and Burrough, 1993), but a fuller treatment of all five aspects has yet to be formulated (Gahegan and Ehlers, 2000).

4 The Spatial Support Factor

Why make support explicit in equation (2)? This way of characterizing analyses implies that a variable, for example the normalized difference vegetation index (NDVI), defined on a specific support, say 900 m^2, is fundamentally different than the NDVI defined at another support, say 6400 m^2. Framing variables this way makes it evident that uncertainty is influenced by the size of ν. Because of the aggregation effect, the values from a set of large pixels covering a region will have smaller variance than the values from a set of small pixels covering the same region. And since the sample variance is used as an estimate of population variance in formulas for confidence and prediction intervals, variable uncertainty should be smaller (or precision should be greater) for large rather than small pixels, all other factors remaining equal. Of course, all other factors rarely remain equal!

Another reason to make the support explicit is to think about the uncertainty that may arise when the same model f is used in analyses with different variable supports ν. This situation has often been met in environmental modelling with geographic data derived from remote sensing. Take the case of a z variable with support ν that is to be predicted using the model f. Further, specify that this model is non-linear, such as a ratio transform, exponential or higher order polynomial. If the model is not changed, but ν is increased or decreased, this will affect the values of z that are predicted and add uncertainty about model results, most likely in the form of bias. Several investigators using deterministic models of ecosystem processes that require remotely sensed inputs have studied this phenomenon. Over a range of input spatial unit sizes that differed by up to over 100 fold (from 1 km to 1°), Pierce and Running (1995) reported that estimates of net primary production (NPP) increased up to 30% using the BIOME-BGC model and remotely sensed leaf area index values. Lammers *et al.* (1997), using a related model (RHESSys), also reported significantly larger values of photosynthesis and evapotranspiration when moving from smaller grid cells to larger (comparing 100 m to 1 km) grid cells. Over an input pixel size range of 25 m to 1 km, Turner *et al.* (2000) found a decrease in NPP of 12% and an increase in total biomass estimates of 30% using a 'measure and multiply' method. Looking at a much smaller range of pixel sizes (30 m to 80 m, less than three-fold), Kyriakidis and Dungan (2001) found that aggregated input data reduced NPP estimates slightly using the CASA (Potter *et al.* 1993) model. While all of these studies identified bias in model results, they did not find the same direction or magnitude of the effect. Heuvelink and Pebesma (1999) provide an excellent summary of a similar situation in the application of pedometric transforms.

5 Multiple Pixel Uncertainty

To this point, only uncertainty at an individual location **u** for the variable z has been discussed. Many applications require predictions about multiple-pixel regions, and issues of uncertainty become more complicated in such circumstances. It stands to reason that uncertainty about attribute values at a single location is a different issue to that about an attribute value at a number of locations. It is because of this difference that geostatisticians have developed the concepts of local versus spatial uncertainty (Goovaerts, 1997). Local uncertainty considers one location at a time whereas spatial uncertainty considers contiguous locations jointly. Kyriakidis (2001) describes a method for estimating spatial uncertainty about the variable z at a set of J locations $\mathbf{u}_j$ using L simulations of the conditional probability

$$\Pr\{z(u_j) \leq z_c, j = 1, \ldots, J|(n)\} \cong \frac{1}{L}\sum_{l=1}^{L}\prod_{j=1}^{J} i^{(l)}(u_j;\, z_c) \tag{3}$$

where z_c is some threshold value, (n) are all the observed (reference) data, $i^{(l)}(u_j;\, z_c)$ is an indicator variable (see van der Meer, 1996), $\prod$ denotes the product and L is the number of simulations. This model has yet to be employed extensively in remote sensing.

Both parametric and model uncertainty can be considered global, at least (and sometimes at the most!) for the region to be analysed. There is, therefore, no need to make the multiple-pixel distinction for these two sources of uncertainty. Positional uncertainty, on the other hand, has a multiple-pixel aspect. For example, characterizing the uncertainty in the position of a linear feature (Jazouli *et al.*, 1994) or an areal feature is more problematic than characterizing uncertainty about the locations of single pixels.

An aspect of uncertainty in remote sensing analysis that is not captured by viewing problems as forms of equation (2) is when the ultimate objective is to estimate the area covered by z equal to a value or range of values. Several investigators have turned their attention to this problem (e.g. Crapper, 1980; Dymond, 1992; Mayaux and Lambin, 1997; Hlavka and Dungan, 2002). As support changes, bias in area estimation is likely to change. The trend depends on the real spatial distribution of the feature whose area is being estimated and how the support of the remotely sensed data compares with this distribution. For example, Benjamin and Gaydos (1990) found that bias in estimating the area of roads decreased with supports ranging from approximately 1 to 3 m^2 and then increased with supports ranging from approximately 3 to 6 m^2. Area estimation is obviously a multiple-pixel problem.

6 Conclusion

This chapter has presented a perspective that allows many sources of uncertainty to be identified and distinguished. It has highlighted two sources of uncertainty that have received little attention in the remote sensing literature: model (or structural) uncertainty and uncertainty about spatial support. The framework of equation (2) shows how uncertainty from all sources are inextricably linked. Quantifying parametric, structural, support and positional uncertainty are all intermediate goals toward understanding uncertainty about variables. It is the values of variables that represent the ultimate purpose of many remote sensing analyses.

Remotely sensed data are a special kind of geographical data, and as such share many properties. Sinton's (1978) categorization of uncertainty for geographical data included value, spatial location, temporal location, consistency and completeness. This chapter has discussed value and spatial location, but neglected temporal location. As for the fourth category, consistency, the premise of this chapter is that the outcome of a remote sensing analysis can never be truly consistent in the sense that uncertainty will vary spatially. This 'inconsistency' is exacerbated by the data scarcity problem, since the number of reference data available relative to the size of the whole field being predicted is usually miniscule. However, analyses done with the same spatial supports on both sides of the equal sign in equation (2) should be more consistent than those with unequal supports. Finally, remotely sensed data sets are usually considered complete, at least for the region represented by the images at hand.

Further research, perhaps using Bayesian models (Datcu *et al.*, 1998) that make support an explicit factor, should enhance methods of quantifying uncertainty. Doing so will increase the ability to identify the largest sources of uncertainty and help focus research efforts on these 'weakest links' in the image analysis chain.

References

Atkinson, P. M. and Curran, P. J., 1997, Choosing an appropriate spatial resolution for remote sensing investigations, *Photogrammetric Engineering and Remote Sensing*, **63**, 1345–51.

Atkinson, P. M., Cutler, M. E. J. and Lewis, H., 1997, Mapping sub-pixel proportional land cover with AVHRR imagery, *International Journal of Remote Sensing*, **18**, 917–35.

Benjamin, S. and Gaydos, L., 1990, Spatial resolution requirements for automated cartographic road extraction, *Photogrammetric Engineering and Remote Sensing*, **56**, 93–100.

Cochran, W. G., 1953, *Sampling Techniques* (New York: Wiley).

Crapper, P., 1980, Errors incurred in estimating an area of uniform land cover using Landsat, *Photogrammetric Engineering and Remote Sensing*, **46**, 1295–301.

Datcu, M., Seidel, K. and Walessa, M., 1998, Spatial information retrieval from remote-sensing images – Part 1: Information theoretical perspective, *IEEE Transactions on Geoscience and Remote Sensing*, **36**, 1431–45.

Draper, D., 1995, Assessment and propagation of model uncertainty, *Journal of the Royal Statistical Society Series B*, **57**, 45–98.

Draper, N. and Smith, H., 1998, *Applied Regression Analysis*, 3rd edition (New York: Wiley).

Dungan, J., 2001, Scaling up and scaling down: The relevance of the support effect on remote sensing of vegetation. In N. Tate and P. Atkinson (eds), *Modelling Scale in Geographical Information Science* (Chichester: Wiley), pp. 221–35.

Dymond, J. R., 1992, How accurately do image classifiers estimate area? *International Journal of Remote Sensing*, **13**, 1735–42.

Englund, E., 1990, A variance of geostatisticians, *Mathematical Geology*, **22**, 417–55.

Fisher, P. F., 1997, The pixel – a snare and a delusion, *International Journal of Remote Sensing*, **18**, 679–85.

Foody, G. M., 1999, The continuum of classification fuzziness in thematic mapping, *Photogrammetric Engineering and Remote Sensing*, **65**, 443–51.

Forshaw, M. R. B., Haskell, A., Miller, P. F., Stanley, D. J. and Townshend, J. R. G., 1983, Spatial resolution of remotely sensed imagery: A review paper, *International Journal of Remote Sensing*, **4**, 371–83.

Friedl, M., McGwire, K. C. and McIver, D. K., 2001, An overview of uncertainty in optical remotely sensed data for ecological applications, in C. T. Hunsaker, M. F. Goodchild, M. A. Friedl and T. Case (eds), *Spatial Uncertainty in Ecology* (Berlin: Springer-Verlag), pp. 258–83.

Gahegan, M., 1996, Specifying the transformations within and between geographic data models, *Transactions in GIS*, **1**, 137–52.

Gahegan, M. and Ehlers, M., 2000, A framework for the modelling of uncertainty between remote sensing and geographic information systems, *ISPRS Journal of Photogrammetry and Remote Sensing*, **55**, 176–88.

Goovaerts, P., 1997, *Geostatistics for Natural Resources Evaluation* (New York: Oxford University Press).

Hansen, M. C. and Reed, B., 2000, A comparison of the IGBP DISCover and University of Maryland 1 km global land cover products, *International Journal of Remote Sensing*, **21**, 1365–73.

Heuvelink, G. and Burrough, P., 1993, Error propagation in cartographic modelling using boolean logic and continuous classification, *International Journal of Geographical Information Systems*, **7**, 231–46.

Heuvelink, G. and Pebesma, E., 1999, Spatial aggregation and soil process modelling, *Geoderma*, **89**, 47–65.

Hlavka, C. and Dungan, J., 2002, Areal estimates of fragmented land cover – effects of pixel size and model-based corrections, *International Journal of Remote Sensing*, **23**, 711–21.

Hunsaker, C. T., Goodchild, M. F., Friedl, M. A. and Case, T. (eds), 2001, *Spatial Uncertainty in Ecology* (Berlin: Springer-Verlag).

Jazouli, R., Verbyla, D. and Murphy, D., 1994, Evaluation of SPOT panchromatic digital imagery for updating road locations in a harvested forest area, *Photogrammetric Engineering and Remote Sensing*, **60**, 1449–52.

Johnston, R. J., 1978, *Multivariate Statistical Analysis in Geography: A Primer on the General Linear Model* (London: Longman).

Journel, A. G. and Huijbregts, C. J., 1978, *Mining Geostatistics* (London: Academic Press).

Kundzewicz, Z. W., 1995, Hydrological uncertainty in perspective, in Z. W. Kundzewicz (ed.), *New Uncertainty Concepts in Hydrology and Water Resources* (Cambridge: Cambridge University Press), pp. 3–10.

Kyriakidis, P. C., 2001, Geostatistical models of uncertainty for ecological data, in C. T. Hunsaker, M. F. Goodchild, M. A. Friedl and T. Case (eds), *Spatial Uncertainty in Ecology* (Berlin: Springer-Verlag), pp. 175–213.

Kyriakidis, P. C. and Dungan, J. L., 2001, Stochastic simulation for assessing thematic classification accuracy and the impact of inaccurate spatial data on ecological model predictions, *Environmental and Ecological Statistics*, **8**, 311–30.

Lammers, R. B., Band, L. E. and Tague, C. L., 1997, Scaling behaviour of watershed processes, in P. R. van Gardingen, G. M. Foody and P. J. Curran (eds), *Scaling-up* (Cambridge: Cambridge University Press), pp. 295–317.

Lewis, A. and Huchinson, M. F., 2000, From data accuracy to data quality: Using spatial statistics to predict the implications of spatial error in point data. In H. T. Mowrer and R. G. Congalton (eds), *Quantifying Spatial Uncertainty in Natural Resources: Theory and Applications for GIS and Remote Sensing* (Chelsea, MI: Ann Arbor), pp. 17–35.

Maling, D. H., 1989, *Measurement from Maps: Principles and Methods of Cartometry* (New York: Pergamon Press).

Mayaux, P. and Lambin, E., 1997, Tropical forest area measured from global land-cover classifications: Inverse calibration models based on spatial textures, *Remote Sensing of Environment*, **59**, 29–43.

Mowrer, H. T. and Congalton, R. G., 2000, *Quantifying Spatial Uncertainty in Natural Resources: Theory and Applications for GIS and Remote Sensing* (Chelsea, MI: Ann Arbor).

Odeh, I. O. A. and McBratney, A. B., 2001, Estimating uncertainty in soil models (Pedometrics '99), *Geoderma*, **103**, 1.

Olea, R. A., 1990, *Geostatistical Glossary and Multilingual Dictionary* (New York: Oxford University Press).

Pierce, L. L. and Running, S. W., 1995, The effects of aggregating sub-grid land surface variation on large-scale estimates of net primary production, *Landscape Ecology*, **10**, 239–53.

Potter, C. S., Randerson, J. T., Field, C. B., Matson, P. A., Vitousek, P. M., Mooney, H. A. and Klooster, S. A., 1993, Terrestrial ecosystem production – a process model based on global satellite and surface data, *Global Biogeochemical Cycles*, **7**, 811–41.

Schott, J. R., 1997, *Remote Sensing: The Image Chain Approach* (New York: Oxford University Press).

Schowengerdt, R. A., 1997, *Remote Sensing, Models and Methods for Image Processing*, 2nd edition (San Diego: Academic Press).

Schowengerdt, R. A., Archwamety, C. and Wrigley, R. C., 1985, Landsat Thematic Mapper image-derived MTF, *Photogrammetric Engineering and Remote Sensing*, **51**, 1395–406.

Sinton, D., 1978, The inherent structure of information as a constraint to analysis: Mapped thematic data as a case study, in G. Dutton (ed.) *Harvard Papers on Geographic Information Systems*, volume 6, (Reading, MA: Addison-Wesley), pp. 1–17.

Taylor, B. N. and Kuyatt, C. E., 1994, *Guidelines for Evaluating and Expressing the Uncertainty of NIST Measurement Results*, National Institute of Standards and Technology (NIST) Technical Note 1297 (Washington, DC: US Government Printing Office).

Turner, D. P., Cohen, W. B. and Kennedy, R. E., 2000, Alternative spatial resolutions and estimation of carbon flux over a managed forest landscape in western Oregon, *Landscape Ecology*, **15**, 441–52.

van der Meer, F., 1996, Classification of remotely-sensed imagery using an indicator kriging approach: Application to the problem of calcite-dolomite mineral mapping, *International Journal of Remote Sensing*, **17**, 1233–49.

4

On the Ambiguity Induced by a Remote Sensor's PSF

J. F. Manslow and M. S. Nixon

1 Introduction

Remote sensors, whether carried by aircraft or satellites, are widely used to provide synoptic information about large areas of the Earth's surface. As an information source, they are particularly attractive because they can acquire data for a large ground area almost instantaneously, regardless of geographic accessibility. The information they acquire has a wide range of applications within industry, government and academia, including guiding prospectors to important sources of natural resources, measuring crop acreage to allow quotas to be efficiently enforced, and monitoring rates of deforestation.

Remote sensing is a complex process in which, generally, electromagnetic radiation emitted by the Sun is reflected by objects on the ground, travels through the atmosphere, and, in the case of satellite-based sensors, through space, before arriving at the sensor. In the sensor, some physical change is caused, which is typically measured electronically and stored digitally before being transmitted back to Earth. There are thus many stages between the radiation leaving the Sun, and information about the sensor measurement reaching its final application within which uncertainty about the actual reflective properties of the ground cover can be introduced.

The overwhelming complexity of the remote sensing process has so far frustrated all attempts to produce comprehensive physical models of the uncertainty present in remotely sensed data. The first part of the research presented in this chapter attempts to increase the understanding of this uncertainty by examining the effect of one source of ambiguity – the transfer function of the sensor itself – within one domain –

Uncertainty in Remote Sensing and GIS. Edited by G.M. Foody and P.M. Atkinson.
 ISBN: 0-470-84408-6

estimating the areas of different land cover classes. From this starting point, the uncertainty introduced by the sensor can be characterized in some detail, and the results generalized to other applications.

Despite the advances that will be discussed, a full analytical characterization of the uncertainty in remotely sensed data (and quantities derived from it) is still not possible, and perhaps never will be. The second section of this chapter, therefore, focuses on a statistical technique that is capable of learning, by example, the exact form of the ambiguity present in quantities derived from remotely sensed data, thus avoiding the need to develop complex analytical models altogether. To demonstrate how the technique should be applied in practice, and what information it can be expected to produce, it was applied to a real-world, area estimation, problem.

2 A Theoretical Analysis of the Ambiguity Induced by the PSF

One of the great challenges in developing models of the uncertainty present in remotely sensed data, and, perhaps more importantly, information derived from them, is to make statements that are sufficiently general. For example, it is sometimes possible to gain insights into the sources of uncertainty in specific remote sensing activities that have occurred in the past. Unless these insights can be generalized to future activities, however, they are of little use.

One way to guarantee that research is of the broadest relevance (and which is taken in this chapter) is to examine some aspect of the remote sensing process that is common to a large proportion of applications. Perhaps the most obvious (but least studied) of these are to do with the characteristics of the sensor itself. Results pertaining to a particular type of sensor are likely to apply to every situation in which the sensor is used, and as the research in this chapter will show, the sensor is an important source of uncertainty, which places a fundamental limit on the amount of information that remotely sensed data can contain.

Traditionally, the output of a remote sensor is represented as a series of raster images, each containing information on the reflectance of the target area in one of the spectral bands in which the sensor operates (Fisher, 1997). These images consist of regular grids of rectangular pixels, and each pixel is usually considered to correspond to some well-defined area of the ground beneath the sensor, as shown for a single row of pixels in Figure 4.1. The reflectance information given by each pixel is assumed to be the average of the reflectance of the land cover within the corresponding area of the ground, and unaffected by the reflectance of land cover outside it, as shown for a single pixel in one row of an image in Figure 4.1.

In practice, the characteristics of remote sensors are far from ideal (Cracknell, 1998). One of their deficiencies is that the reflectance information contained within a single pixel is affected by land cover outside the area of ground normally associated with it. This can be seen by comparing Figures 4.1 and 4.2. In the former, the sensitivity of the ideal sensor is, for a single pixel, zero outside the area of ground that is associated with it, whereas the sensitivity of the real sensor is non-zero into the area covered by neighbouring pixels. This inter-pixel interference effect is discussed in greater detail in Townshend *et al.* (2000), where a method of compensating for it is described.

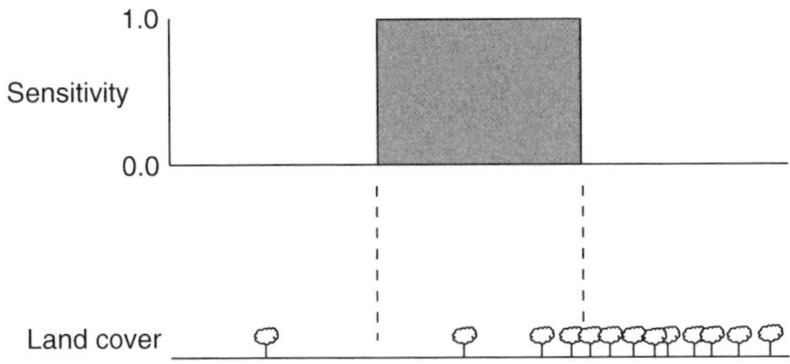

Figure 4.1 An idealized sensor is uniformly sensitive to the reflectance of all land cover within the pixel area (which lies between the dashed lines). Outside this area, the sensor has zero sensitivity and is unaffected by the reflectance of the land cover there

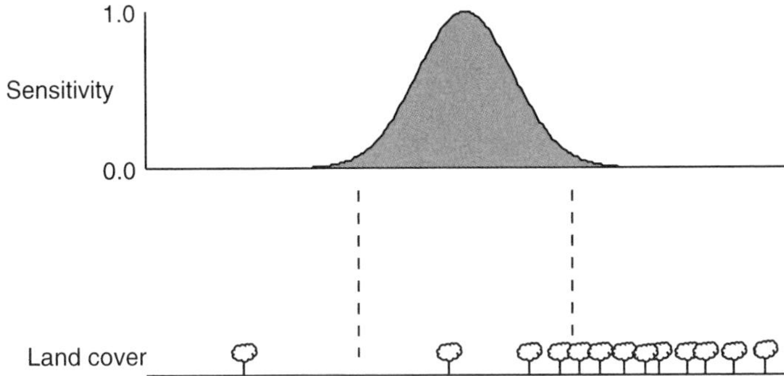

Figure 4.2 In practice, the sensitivity of real sensors varies within the pixel area (which lies between the dashed lines). Outside the pixel, the sensor has non-zero sensitivity and hence the reflectance measured for a pixel is affected by land cover that lies outside it

Another undesirable characteristic of real sensors is shown in Figure 4.2, where the sensor is not uniformly responsive to the reflectance of land cover within the area of ground associated with each pixel (Justice *et al.*, 1989). Most sensors exhibit sensitivity variation similar to that shown in Figure 4.2 – being most sensitive to land cover in the centre of a pixel's ground area, and least sensitive towards its edge. The reflectance information contained in a pixel thus tends to be most similar to the reflectance of land cover located towards the centre of the pixel's ground area, and least similar to cover towards its edge. This variation in sensitivity is described by a function, called the point spread function (PSF) of the sensor, and its practical effects – which have, until fairly recently, been largely ignored – are considered in this chapter.

Since a pixel has a two-dimensional 'footprint' on the ground, a real PSF has a shape similar to that shown in Figure 4.3. The horizontal axes correspond to the distance on the ground of a point from the centre of the area that a pixel covers on the ground, and for each point, the height of the function gives the sensitivity of the sensor relative to its maximum sensitivity. The PSF shown is Gaussian – that is, if r is the distance of a point (x,y) on the ground from the point of maximum sensitivity, the sensor's sensitivity $\Psi(\cdot)$ is given by:

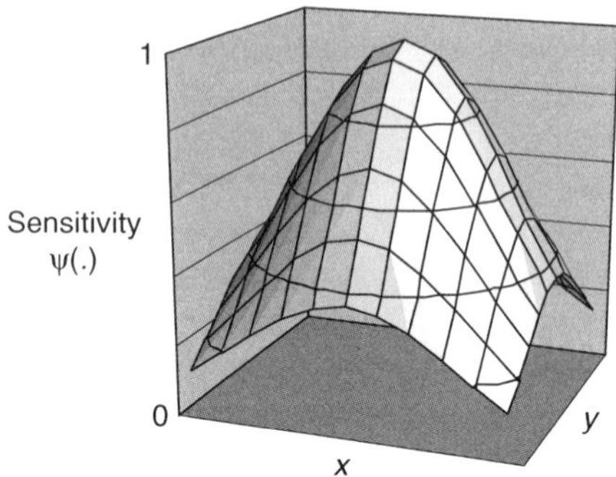

Figure 4.3 A Gaussian model of a PSF with parameter $\alpha = 2$. The dark grey square at the figure's base represents the ground area that the pixel would normally be considered to cover. The 'height' of the surface above a point on this square indicates the sensitivity of the remote sensor to the reflectance of ground cover at that point compared to that at the pixel's centre. For example, the sensitivity of the sensor to land cover located at the nearest corner of the pixel is less than one-fifth that at its centre

$$\Psi(r) = \exp(-\alpha r^2) \tag{1}$$

where α controls how quickly the sensitivity falls off as the distance from the point of maximum sensitivity increases. Although this PSF model is simplistic (for example, real PSFs do not usually possess exact circular symmetry), it is easy to analyse mathematically.

In order to make the present discussion of the effect of a sensor's PSF more concrete, it will be considered in the context of the problem of estimating the areas of land cover classes from remotely sensed imagery. This does not limit the generality of the results that are derived, however, since they relate fundamentally to the ambiguity that a sensor's PSF creates regarding all processes that occur at ground level. Estimates of the areal extent of land cover classes have been extracted from remotely sensed images for a long time. This has generally been achieved by classifying each pixel in an image into one of a predefined set of classes. Since the area covered by the image is usually known, counting the number of pixels belonging to each class yields an estimate of their areas.

This area estimation by pixel classification process produces area estimates with rather large errors – typically too large to be useful in most applications. One of the more promising developments that aimed to reduce the magnitudes of these errors recognized that the area of ground covered by a particular pixel may not be homogeneous – that is, there may be several different cover types within it. This critical realization led to the development of sub-pixel proportion estimation algorithms that attempt to estimate the proportion of the sub-pixel area covered by each of the classes of interest (Horwitz *et al.*, 1971; Foody, 1995), and can be considered to be the current state of the art.

In Manslow and Nixon (2000) it was shown that sensitivity variation within the PSF of a sensor introduces ambiguity into these proportion estimates. To illustrate this, Figure 4.4 shows an abstract plan-view representation of the area of ground

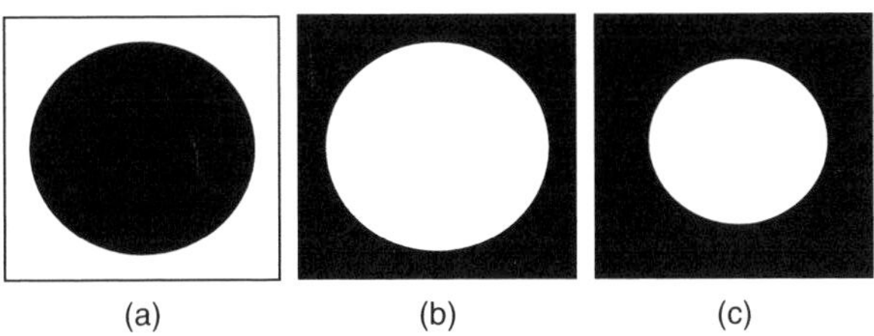

Figure 4.4 *Representations of the sub-pixel arrangement of land cover for three pixels. Each of the three squares represents a plan view of the ground area associated with one of three pixels in a remotely sensed image. The pixel areas are divided between two classes, represented in black and white. Pixel (a) contains the two classes in equal proportions, and the black class is arranged in the centre, and the white class towards its perimeter. Pixel (b) is identical to pixel (a) except that the locations of the classes are swapped. Pixel (c) is identical to pixel (b), except that the area of the black class has been increased so that the remote sensor observes the same reflectance as for pixel (a)*

covered by three pixels. In the first pixel, Figure 4.4(a), the sub-pixel area is divided equally between two classes, one located centrally (shown in black), and the other located towards the pixel's perimeter (shown in white). Since most PSFs display greatest sensitivity to the reflectance of land cover that is located centrally (as shown in Figure 4.3), the measured reflectance of the pixel will be most similar to the centrally located class – the black class.

The second pixel, shown in Figure 4.4(b), is identical to the first, except that the classes have been swapped over – the black class now lies towards the pixel's perimeter, and the white class, at its centre. Once again, because the PSF gives the sensor greatest sensitivity to centrally located land cover, the reflectance that is measured will be most similar to that of the centrally located class – in this case, the white class. Although the proportions with which the black and white classes are present in the pixels in Figures 4.4(a) and 4.4(b) are identical, a different reflectance is measured in each case.

Now consider taking the pixel in Figure 4.4(b), and gradually increasing the proportion of the black class, as shown in Figure 4.4(c). As more and more of the black class is added, the reflectance of the pixel would increasingly come to resemble that of the black class until, eventually, the pixel had the same reflectance as that in Figure 4.4(a). Thus, although a remote sensor would measure identical reflectances for the pixels in Figures 4.4(a) and 4.4(c), their sub-pixel compositions are distinct not simply in terms of the spatial arrangement of the sub-pixel cover, but also in terms of the proportions in which the classes are present. This indicates that a pixel's reflectance contains only ambiguous information about its composition.

The above effect was analysed in more detail by Manslow and Nixon (2000) using the Gaussian PSF model that was introduced earlier. Making use of the simplifying assumptions that a pixel has a circular footprint, area π, and that there are only two sub-pixel classes, which exhibit no spectral variation, it was possible to derive upper bounds on the ambiguity induced by the PSF. In particular, assuming a pixel has been observed, which contains two sub-pixel classes, which cover areas a_1 and a_2, pixels with identical reflectances may also exist where the first class covers an area anywhere from

$$-\frac{\pi}{\alpha}\ln|e^{-\alpha} - e^{-\alpha r_{min}^2} + 1| \tag{2}$$

to

$$\pi\left(1 + \frac{1}{\alpha}\ln|e^{-\alpha} - e^{-\alpha r_{max}^2} + 1|\right) \tag{3}$$

where

$$r_{min} = \sqrt{1 - a_1/\pi} \tag{4}$$

and

$$r_{max} = \sqrt{a_1/\pi} \tag{5}$$

These limits can be interpreted as upper bounds on the ambiguity induced by the Gaussian PSF because they define the range over which sub-pixel proportions can vary without any change in a pixel's measured reflectance. The bounds are plotted in Figure 4.5 and have a number of interesting properties:

1. The amount of ambiguity can be expressed as a function of the degree of sub-pixel mixing, making it easy to assess the impact of the PSF on sub-pixel area estimates.
2. The amount of ambiguity is not dependent on the level of spectral separation of the classes. The effect of the PSF is the same regardless of whether the classes are spectrally very similar, or highly distinct.
3. The amount of ambiguity induced by the PSF increases with the amount of sub-pixel mixing. That is, greatest ambiguity is induced when the sub-pixel area is partitioned equally between the two classes.

Figure 4.6 provides a simple illustration of how to interpret the bounds for a pixel that contains two sub-pixel classes present in equal proportions. To find the bounds on the ambiguity induced by the PSF, a line is drawn up from the x-axis from the point $x = 0.5$ (indicating that one of the classes covers 50% of the sub-pixel area).

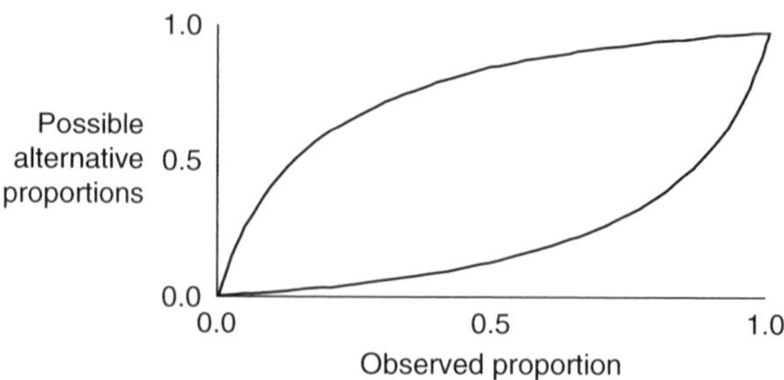

Figure 4.5 The bounds on the ambiguity induced by a Gaussian PSF with $\alpha = 2$ calculated analytically, as in this figure, are almost identical to those predicted by the computational technique. This suggests that the grid-based approximations used in the latter have minimal impact on the accuracy of the predicted bounds. Note how the bounds are widest when one of the sub-pixel classes covers a proportion of around 0.5 of the sub-pixel area

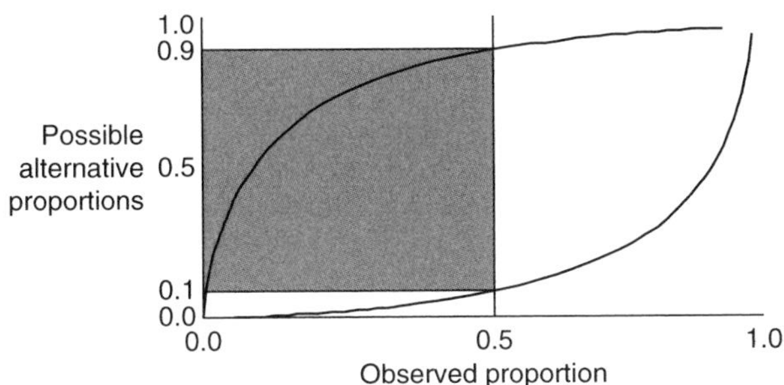

Figure 4.6 *Ambiguity induced by the Gaussian PSF with $\alpha = 2$ calculated by the computational technique. When a proportion of 0.5 of sub-pixel area is known to be covered by a class, other pixels could also exist, that cannot be distinguished by a remote sensor, where the same class covers anywhere between a proportion of 0.1 and 0.9 of the sub-pixel area*

The y-coordinate of the point where the line intersects the lower curve – in this case about 0.1 – gives the minimum proportion that the class could cover in other pixels with the same spectral properties as the one currently being considered. Similarly, the y-coordinate of the point where the line intersects the upper curve – in this case, around 0.9 – gives the maximum proportion that the class could cover. Thus, for the Gaussian model of the PSF that was described earlier, if a pixel is known to contain two classes in equal proportions, other pixels may also exist, which are spectrally identical, but where either of the classes cover anywhere between 10 to 90% of the sub-pixel area. Since the pixels are spectrally identical, they cannot be distinguished by a remote sensor, regardless of how spectrally distinct the individual classes are. Figure 4.7 shows another example where the same process has been applied to estimate the bounds when a pixel is known to be divided between two classes in the proportions 0.25 to 0.75. In this case, other, spectrally identical pixels exist where the first class covers anywhere between 5 and 80% of the sub-pixel area.

Before pursuing the implications of the above discussion further, it is important to consider how, and indeed whether, they can be extended to more realistic PSF models. Although the analytical examination of a Gaussian model of a PSF (presented

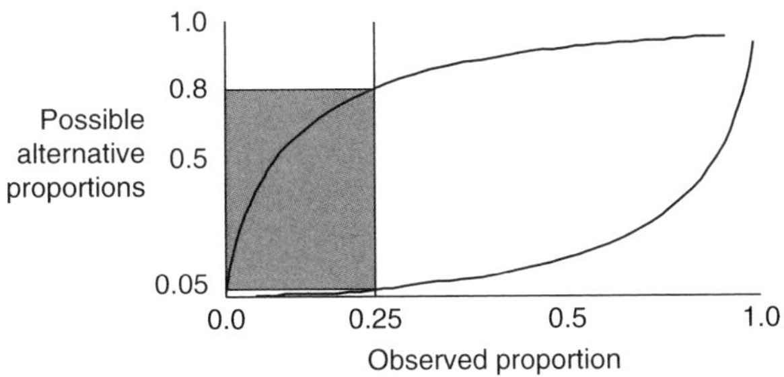

Figure 4.7 *Ambiguity induced by the Gaussian PSF with $\alpha = 2$ calculated by the computational technique. When a proportion of 0.25 of sub-pixel area is known to be covered by a class, other pixels could also exist, that cannot be distinguished by a remote sensor, where the same class covers anywhere between a proportion of 0.05 and 0.8 of the sub-pixel area*

in detail in Manslow and Nixon, 2000) produces important insights, it cannot easily be extended to other PSF models because of the intractability of the analysis that is required. The following section describes an efficient computational technique for deriving bounds on PSF-induced ambiguity that can be applied to almost any PSF model to arbitrary accuracy, provided that sufficient computing power is available.

3 A Computational Analysis of the Ambiguity Induced by the PSF

This section describes an efficient computational technique for estimating bounds on the ambiguity induced by a sensor's PSF. The technique can, in principle, estimate the bounds to arbitrary accuracy, and is limited in practice only by the computer power available to it. It works by alternately arranging one class in the area where the sensor is most responsive, moving it to the area where the sensor is least responsive, and measuring the increase or decrease in the area of the class that is required to leave the measured reflectance unchanged. As with the analytical technique discussed above, this assumes that the sub-pixel area is covered by only two classes, and that each exhibits no spectral variation. The validity and significance of these assumptions is discussed later.

To simulate different sub-pixel arrangements of land cover, the sub-pixel area is represented as a fine grid, as shown in Figure 4.8(a), and each grid cell assigned to one of the two sub-pixel classes, as illustrated in the example in Figure 4.8(b). This allows any arrangement of two non-intergrading classes to be represented arbitrarily closely, provided that a sufficiently fine grid is used. The sensitivity profile of the sensor is represented on the same grid by calculating the value of the PSF at the centre of each cell, as shown in Figure 4.8(c). Again, this produces an arbitrarily close approximation to any smooth PSF, provided that a sufficiently fine grid is used.

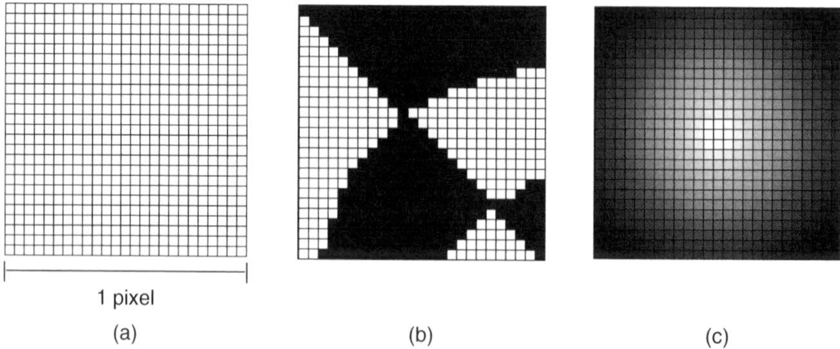

Figure 4.8 Components of the technique. (a) The sub-pixel area is shown divided into a regular grid of 25 × 25 cells. (b) Different arrangements of sub-pixel cover are approximated by assigning each cell to one of the sub-pixel classes. The arrangement shown could correspond to several fields divided between two crop types, for example. (c) The PSF of the sensor can also be approximated on the grid by assigning each cell a value equal to the PSF at the cell's centre. These values are represented as shades of grey in the figure, with white representing greatest sensitivity

Using the grid-based representation of the sub-pixel area, the lower bound on the PSF-induced ambiguity can be determined in the following way, assuming that the black class covers a proportion μ of the sub-pixel area:

1. Distribute the black class in the area where the PSF is least sensitive by assigning grid cells within the pixel to the black class, starting with the cells for which the PSF is least sensitive, and finishing once the class covers a proportion μ of the sub-pixel area.
2. Compute the apparent reflectance of the pixel by finding the sum of the PSF values for all cells that are covered by the black class.
3. Set all grid cells back to white.
4. Distribute the black class in the area where the PSF is most sensitive by assigning cells to it, starting with those for which the PSF is most sensitive, and finishing when the apparent reflectance of the pixel – computed in the same way as in step 2 – is the same as that computed in step 2.
5. The lower bound on the PSF-induced ambiguity for a pixel where a class covers a proportion μ of the sub-pixel area is given by the proportion of the pixel's area covered by the black class after step 4.

The upper bound may be computed by the reverse of the above procedure by, in step 1, distributing the black class where the PSF is most sensitive and moving it to where the PSF is least sensitive in step 4.

By repeating steps 1 to 5 for a range of values of μ between zero and one, the bounds can be plotted in the same way as the analytically derived ones were in the previous section. Figure 4.9 shows the bounds derived for the Gaussian PSF model using a grid size of 50 by 50 cells. The similarity of these bounds to those derived using the analytical method of Manslow and Nixon (2000) indicates that the grid-based representation of the arrangement of sub-pixel cover has minimal impact on the accuracy of the estimated bounds, at the selected spatial resolution, for the Gaussian PSF.

Figure 4.10 shows the bounds derived for a cosine PSF model, which is shown in Figure 4.11. Despite the differences between the Gaussian and cosine models, the shape of the bounds on the ambiguity they induce are almost identical. This is

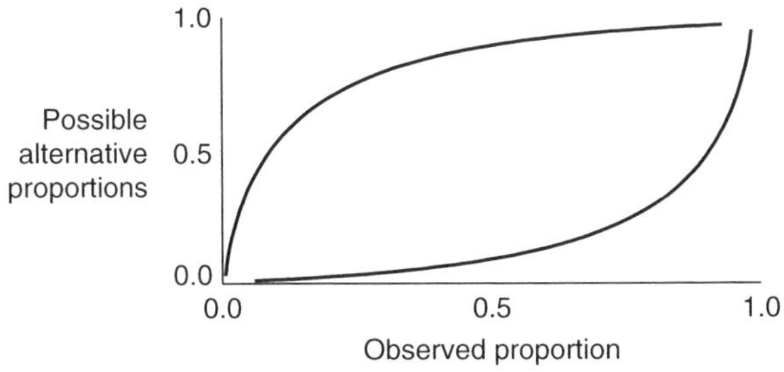

Figure 4.9 *Bounds on the ambiguity induced by the Gaussian PSF with parameter $\alpha = 2$ calculated by the computational technique*

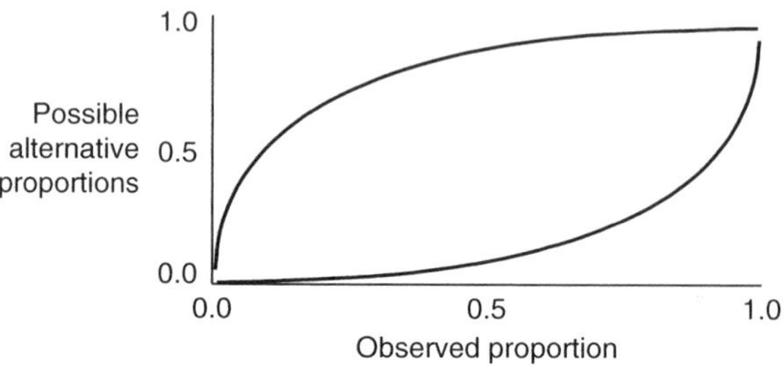

Figure 4.10 Bounds on the ambiguity induced by the cosine PSF predicted by the computational technique. It is interesting to note the close resemblance to those predicted for the Gaussian PSF

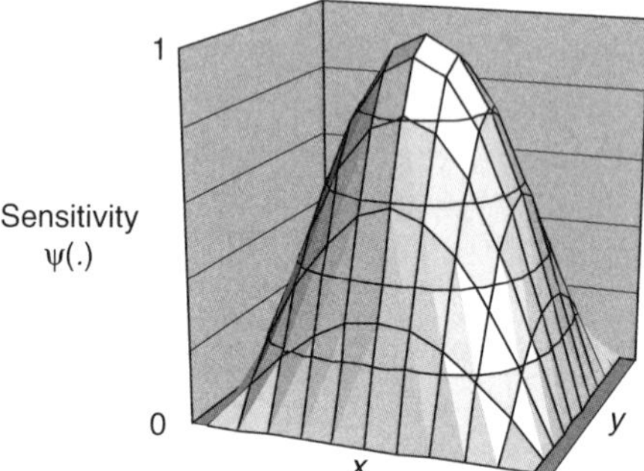

Figure 4.11 A cosine model of a PSF

because the main difference between the models occurs close to the pixel perimeter, where the sensitivity of the cosine model drops to zero. This causes the ambiguity induced by the cosine PSF to be larger for almost pure pixels – those that are covered almost completely by a single class – than for the Gaussian PSF.

The interpretation of PSF ambiguity bounds must be carried out with reference to the distribution of the proportion parameter μ that is likely to be observed in a particular application. As an example, consider estimating the area of two cereal crops from SPOT HRV data. In this case, because individual fields are much larger than the ground area covered by a pixel, most pixels will consist of a single crop type, and only the small number of pixels that straddle field boundaries will contain a mixture of classes (see Kitamoto and Takagi, 1999, for a discussion of the relationship between the distribution of μ and the spatial characteristics of different cover types). This means that μ will be close to zero or one for most pixels, which is where the bounds are closest together, indicating that the PSF will have minimal effect.

When a pixel contains an almost equal mixture of two classes, however, the proportion μ of each is close to 0.5, and hence the pixel lies towards the centre of the bound diagram where the upper and lower bounds are furthest apart, the PSF induces most ambiguity. In this case, the exact form of the ambiguity is a complex function of the spatial and spectral characteristics of the particular sub-pixel classes, which is difficult to derive analytically, and, in any case, is specific to each particular application. Rather than pursue this analysis in detail, a technique for modelling the ambiguity in information extracted from remotely sensed data, which can be applied in any application, is described in the next section.

The algorithm for estimating bounds on PSF-induced ambiguity that has been described in this section has made a number of assumptions and approximations. These have been necessary to reduce the problem to the point where analysis is actually possible, and to make the results of the analysis comprehensible. One source of ambiguity that has been ignored throughout this analysis is the overlap between the PSFs of neighbouring pixels, that was briefly described in an earlier section. Townshend *et al.* (2000) consider this issue in detail, and describe a technique that can be used, in part, to compensate for it.

In the analyses discussed above, it has been assumed that sub-pixel classes exhibit no spectral variation. In practice, however, real land cover classes do show substantial variation in reflectance, and this considerably complicates the analysis of a PSF's effect. In addition, the differences in reflectance variation between and within classes make it difficult to maintain the generality once these are included in the analysis, because the results that would be derived would be specific to particular combinations of classes. Spectral variation was ignored in the analyses presented in this chapter to maintain this generality, and to provide a clearer insight into the basic effect of the PSF.

Additional information on sources of ambiguity in remotely sensed data can be found in Manslow *et al.* (2000), which lists the conditions that must be satisfied for it to be possible, in principle, to extract error-free area information from remotely sensed data. The conditions must apply even in the case of the ideal PSFs, which were described in an earlier section, and require classes to exhibit no spectral variation, and for there to be at least as many spectral bands as classes in the area being sensed. Clearly, such conditions cannot be met in practice, indicating that additional sources of uncertainty cause the total ambiguity in area information derived from remote sensors to be greater than that originating within the PSF alone.

4 Extracting Conditional Probability Distributions from Remotely Sensed Data using the Mixture Density Network

The previous sections of this chapter have focused on developing analytical tools that can provide very precise descriptions of the effect of a sensor's PSF on the information the sensor acquires. Although it has been possible to gain new insights into the limitations imposed by the PSF, the PSF itself is only one of many sources of ambiguity, and indeed, one that is still not fully understood. The remote sensing process as a whole is so complex, that there will probably never be a general theoretical framework that allows the ambiguity in remotely sensed data to be quantified. This chapter, therefore, describes an empirical–statistical technique that is able to estimate the ambiguity directly.

To clearly illustrate the techniques discussed in this section, they were used to extract estimates of the proportions of the sub-pixel areas of pixels in a remotely sensed image that were covered by a class of cereal crops. The area that was used for this study was an agricultural region situated to the west of Leicester in the UK.

A remotely sensed image of this area was acquired by the Landsat TM, and the area was subject to a ground survey by a team from the University of Leicester as part of the European Union funded FLIERS (Fuzzy Land Information in Environmental Remote Sensing) project to determine the actual distribution of land cover classes on the ground.

From the ground survey, it was possible to provide an accurate indication of the true sub-pixel composition of every pixel in the remotely sensed image. Figure 4.12(a) shows band 4 of the image, and Figure 4.12(b) shows the proportions of each pixel that were found to be covered by cereal crops during the ground survey. In all, the data consisted of roughly 22 000 pixels, the compositions of which were known. Foody (1995) described a technique, which is still state of the art, for using a type of empirical–statistical model called a neural network to estimate the area of a class – such as cereal crops – from remotely sensed images using a data set such as the one that has just been described. This would be done by using the pixels of known composition to teach the neural network the relationship between a pixel's reflectance and its sub-pixel composition. Once the neural network has learned this relationship, it can be applied to pixels of unknown composition – the network is given the pixels' reflectances, and responds with estimates of their composition. The procedure for doing this, which was outlined in Foody (1995), was followed in this work, and produced the sub-pixel proportion estimates shown in Figure 4.13. Although there is a fairly reasonable correspondence between the actual proportions (shown in Figure 4.14) and the estimates, there are important differences.

The differences between the actual and estimated class proportions have many causes, not least of which is the ambiguity introduced by the sensor's PSF. One of the weaknesses of the neural network approach to estimating sub-pixel composition is that it can only associate a single proportion estimate with every pixel – even though remotely sensed data contains too little information to state unambiguously what

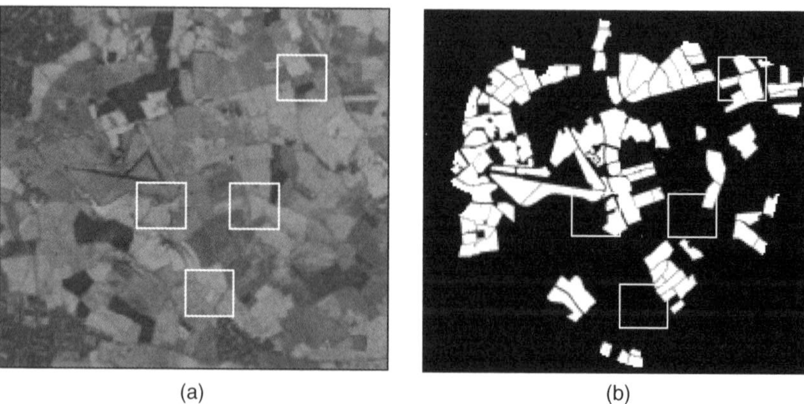

Figure 4.12 *The area from which data were collected for the experiments described in this chapter. (a) shows the remotely sensed area in band 4. The regions within the white squares are magnified to provide greater detail in later figures. (b) The proportions of sub-pixel area covered by cereal crops represented on a grey scale. White indicates that a pixel was purely cereal, and black indicates that no cereal was present*

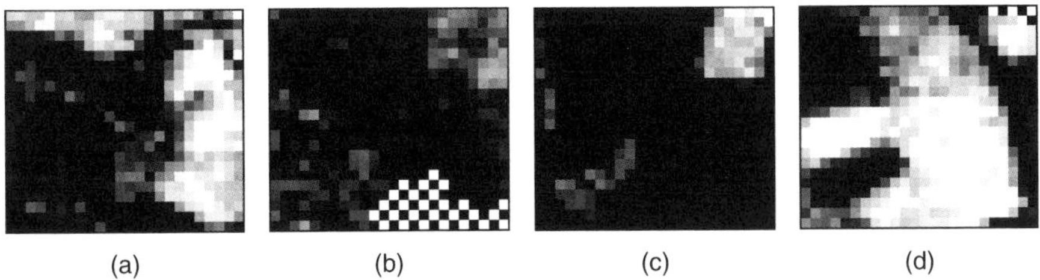

(a) (b) (c) (d)

Figure 4.13 *The proportions of sub-pixel areas that were predicted to be covered by cereal crops using a neural network for the four regions highlighted in Figure 4.12. White corresponds to a pixel consisting purely of cereal, whereas black corresponds to a pixel containing no cereal at all. Here, and in later figures, the checkered areas were outside the study area and were not used*

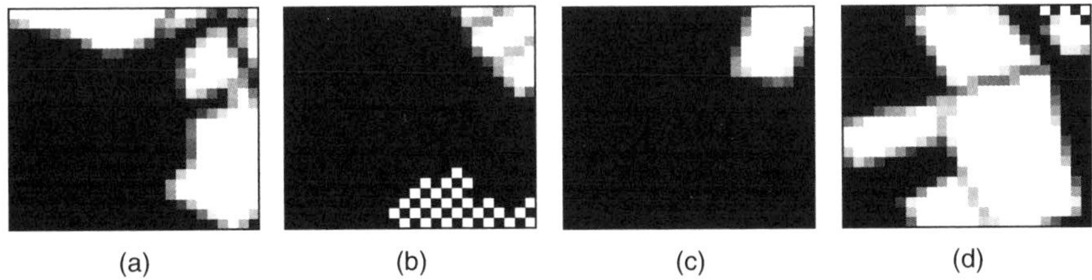

(a) (b) (c) (d)

Figure 4.14 *The real proportions of sub-pixel area that were covered by cereal crops, as determined by a ground survey*

proportion of a pixel's area is covered by any class. This limitation is not just a failure to express the ambiguity in the estimate, but also to express all the information contained in the remotely sensed data. For example, the neural network may state that 63% of the sub-pixel area is covered by cereal crops, but does not indicate that, in reality, cereal crops may cover anywhere between 12 and 84% of the area.

The above suggests that a more flexible representation is needed that is able to express fully not only the ambiguity, but also all the information in the remotely sensed data. Figure 4.15 shows a representation that can do this – the probability distribution over the sub-pixel proportion estimate. The principle behind this representation is that it not only represents the range of proportions that are consistent with the reflectance of the pixel, but also how consistent particular values are. For example, the distribution in Figure 4.15 indicates that, although the sub-pixel area could be covered by anything between about 0 and 50% by cereal crops, it is most likely to be around 0 to 30%. This type of representation is information rich, and can summarize all the ambiguity in the remote sensing process, regardless of its source.

In order to extract proportion distributions, a technique will be used that was first introduced by Bishop (1994) and is further described in Bishop (1995). The mixture density network (MDN) is similar to a neural network both in structure and in operation, and can use exactly the same data without the need for any special preparation. Like the neural network, the MDN must be taught the relationship between pixel reflectances and their sub-pixel composition by being shown a large number of pixels of known composition. Details of how this is done can be found in Bishop (1994) and Manslow (2001). Once the MDN has learned the relationship

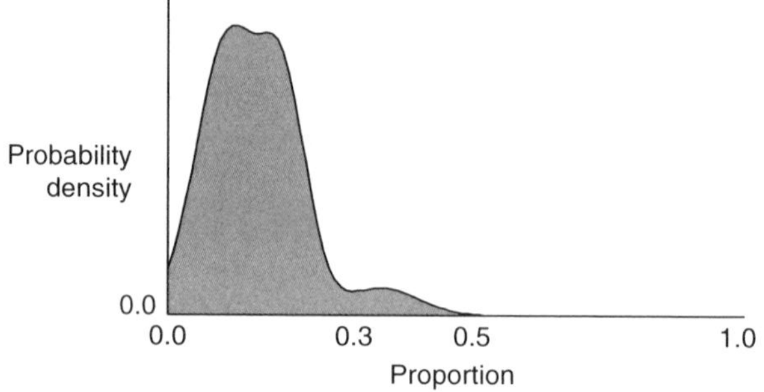

Figure 4.15 *An example of how a probability distribution can be used to represent information extracted from remotely sensed data. In this example, the proportion of the sub-pixel area covered by cereal crops is estimated. The distribution explicitly represents the ambiguity in the estimate – the actual proportion is most likely to lie between about 0.00 and 0.30, but could be anywhere between 0.00 and 0.50*

between pixels' reflectances and the proportions with which sub-pixel classes are present, it can be used to extract a proportion distribution for a pixel of unknown composition.

To do this, the pixel's reflectance is applied to the MDN's spectral inputs, and a range of proportions (such as 0.00, 0.01, ..., 0.99 and 1.00) applied to its proportion input. For each value of the proportion input, the MDN responds with the posterior probability (density) that a pixel with the specified reflectance contains the specified proportion of the class of interest – in this case, cereal crops. Plotting these probabilities, as shown in Figure 4.15, provides a convenient way of representing the distribution. The following section presents some results that were obtained by using a MDN that was taught using a data set acquired for the FLIERS project, to estimate the distribution of the proportion of sub-pixel area that was covered by cereal crops.

5 Results Illustrating the Utility of the Conditional Distribution Representation

The previous section introduced a new way of representing information derived from remotely sensed data. Rather than explicitly extracting the quantity of interest – in this case, the proportion of a sub-pixel's area covered by cereal crops – the new approach extracts a probability distribution that describes how consistent different values of the quantity of interest are with the reflectance that was observed. This new way of representing information extracted from remotely sensed data offers a wide range of benefits, which are discussed in this section, and fall under three headings, visualization, combination, and propagation.

5.1 Visualization

One of the most important benefits of representing estimated quantities as probability distributions is that the information about them that is contained in the remotely sensed data can be expressed fully. For example, the information in the distribution in Figure 4.15 would, in a more conventional representation, be approximated by the distribution's mean, or at most, its mean and variance. Clearly, this results in almost all of the detailed information in the distribution, and hence in the remotely sensed data from which it was derived, being lost.

One problem with the representation is that although it provides detailed information about the different possibilities as to the sub-pixel composition of individual pixels, in itself, it is inappropriate for the representation of synoptic information for large numbers of pixels at once. This problem can be overcome, however, by summarizing the distributions associated with individual pixels by their means and variances, both of which can be efficiently calculated using the formulae provided in Bishop (1995), and can be used to form images, as shown in Figures 4.16 and 4.17.

The distribution means provide useful synoptic information because it is the estimate of the quantity of interest that minimizes mean squared error measured over the probability distribution and, in this sense, is the single value that summarizes it most accurately. The variance is useful because it provides information on how well the mean summarizes the distribution. A variance close to zero, for example, means that the entire probability distribution is centred close to a single point – its mean – which thus describes the distribution very well, as shown in Figure 4.18. A distribution with large variance, however, contains substantial probability mass away from

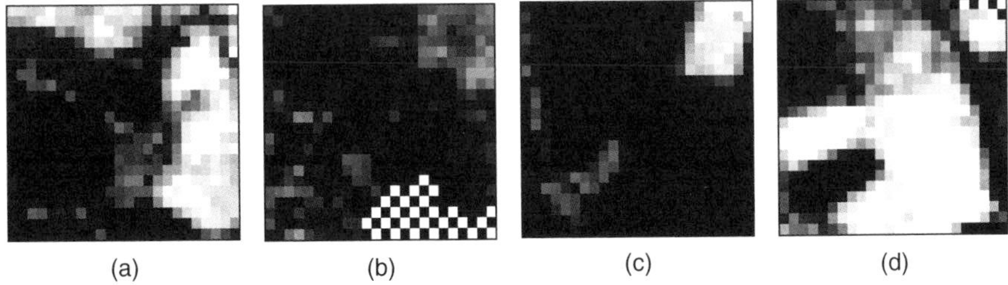

Figure 4.16 *The means of the distributions estimated by the MDN can be represented as images, and provide a convenient way of summarizing the information that they contain*

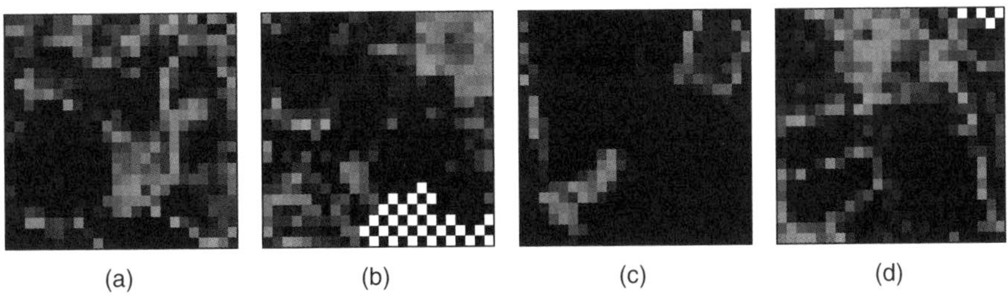

Figure 4.17 *The variances of the distributions estimated by the MDN can also be represented as images and provide a useful indication of how well the distributions are summarized by their means. Lighter pixels are associated with distributions that have large variances, and hence provide little information about sub-pixel composition. It is interesting to note that these occur mainly for mixed pixels along field boundaries where the PSF is expected to have a significant impact*

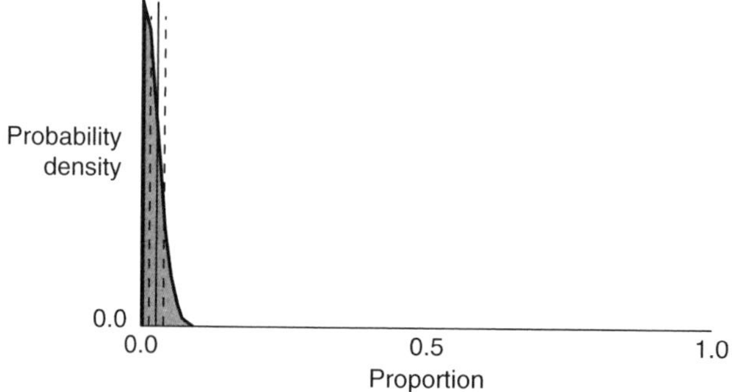

Figure 4.18 In a distribution with small variance, nearly all the probability mass occurs close to the distribution's mean. This indicates that, in most pixels, the class that the distribution was estimated for will occupy a proportion of the sub-pixel area very close to the distribution's mean. In the above example, the distribution's mean is represented by the solid vertical line, and its variance by the dashed lines

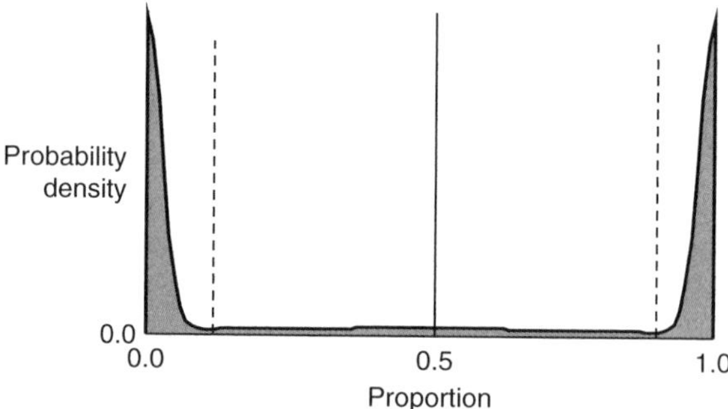

Figure 4.19 In a distribution with large variance, much of the probability mass occurs far from the distribution's mean, making it a poor guide as to what proportions can occur. In the above example, the distribution's mean is represented by the solid vertical line, and its variance by the dashed lines. It should be noted that, although the distribution's mean is 0.5, proportions close to 0.5 are unlikely to occur, and those close to 0.0 and 1.0 can be expected to be far more common

its mean, indicating that the mean itself inadequately summarizes the range of values of the quantity being estimated, as shown in Figure 4.19. When the probability distribution associated with a pixel has large variance, it is likely that a direct visualization of the entire distribution is likely to yield extra useful information. This concept is illustrated in Figure 4.20, where the distribution for a pixel with a large variance has been represented explicitly.

To reinforce the value of the synoptic information provided by images of the distribution means and variances, it is worth comparing Figures 4.16 and 4.17 with the proportions estimated by an MLP neural network (produced in accordance with the methods described in Foody (1995, 1997), shown in Figure 4.13, and the squared errors of these estimates when compared with the actual proportions, established by a ground survey, as shown in Figure 4.21. Not only are the distribution means very similar to the MLP's proportion estimates, but the distribution variances are large in areas where each of these estimates corresponds poorly with the actual proportions.

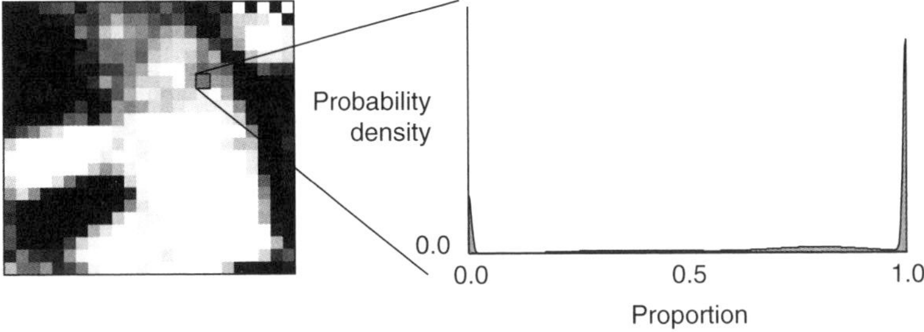

Figure 4.20 If estimates of the proportions of the sub-pixel areas covered by different classes are represented as probability distributions, images of the distribution means (such as that shown to the left) provide a useful summary of the information in the distributions. For any pixel in the image, however, the full probability distribution can be extracted (as shown on the right) to reveal the full range of alternative possibilities

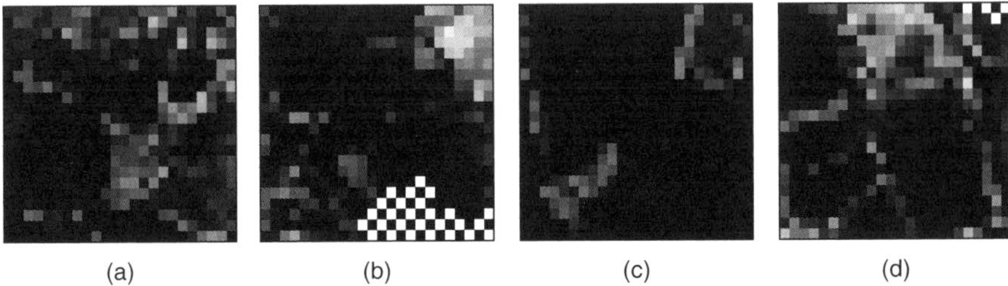

Figure 4.21 The magnitude of the squared differences between the distribution means and the sub-pixel proportions revealed by a ground survey. The similarity with Figure 4.17 suggests that the distribution variances provide a good indication of the likely accuracy of the distribution means as proportion estimates. Whiter pixels indicate greater difference

This suggests that the distribution variances provide a useful practical indication of the likely accuracy of the distribution means as estimates of the quantity of interest. As a final point, it is interesting to note that the distribution variances shown in Figure 4.17 tend to be large along field boundaries. These are the areas where pixels occur whose area is divided most evenly between classes, and where, from the earlier analyses, the sensor's PSF would be expected to introduce most ambiguity.

5.2 Combination

When trying to estimate a quantity of interest, such as the proportion of the area of a pixel that is covered by cereal crops, it is common that different estimates can be obtained from a variety of sources. For example, Figure 4.22 shows a system that contains two proportion estimators. The first of these, an MDN, uses pixel reflectances to estimate unique probability distributions over proportions for each one. The second segments the image according to the textures within it, and then, for each texture, estimates a unique probability distribution over proportions.

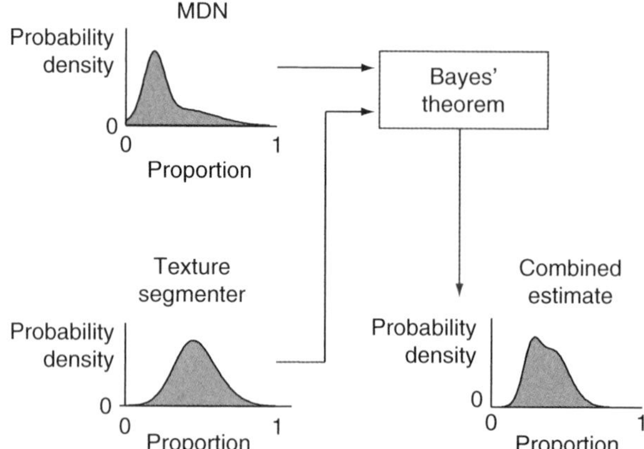

Figure 4.22 *Representing information from different sources as probability distributions allows it to be combined optimally. This example shows how proportion information from a texture segmentation algorithm that has been modified to estimate the proportions of classes in a pixel based on the texture that surrounds it, can be combined with information extracted using an MDN. Bayes' theorem provides a mathematical framework for combining sources of information without making assumptions or approximations. The result of a Bayesian combination is information that is always more specific than that contained in either of the original sources*

Each pixel in the image thus has two estimates associated with it, one originating from the MDN and the other from the texture segmenter, both of which are represented by probability distributions. Fortunately, probability theory states that these information sources can be combined optimally using either Bayes' theorem, or by marginalization (Bishop, 1995). If only the single 'best' proportion estimate was made available by either or both of the information sources, the estimates could not have been combined without making assumptions (typically, that the unknown distributions are Gaussian). Representing any estimated quantity by a probability distribution is thus extremely important as far as combining the estimate with information from other sources is concerned.

5.3 Propagation

Quantities that are extracted from remotely sensed data are frequently used as inputs to other systems and processes. For example, measurements of changes in the area of important cover types are often made by taking the difference between area estimates obtained from successive remotely sensed images. Since such estimates are ambiguous, however, it is reasonable to assume that the estimated change is also ambiguous. More generally, the output of any system or process that uses information from a remote sensor will be characterized by some ambiguity. Provided that all inputs to the system and process are represented by probability distributions, however, the ambiguity in the system's or process's output can be derived.

As an example, consider estimating the percentage change in the area of forest cover using successive remotely sensed images of the same region. To simplify the example, only a single pixel in each image is analysed and it is assumed that the two pixels sample exactly the same area of ground. If the same reflectance was associated with each pixel, state-of-the-art area estimation algorithms, such as neural networks, would estimate exactly the same area for the forest class in each case. This would inevitably lead to the conclusion that there had been no change in sub-pixel cover between the two observations.

Now consider analysing the pixels using an MDN. Like the neural network, the MDN makes the same estimate for the area of the forest class for each pixel, but represents its estimate as the probability distribution of Figure 4.23. This indicates that the sub-pixel area is likely to be exclusively forest, but could also contain no forest whatsoever. Using the techniques outlined in DeGroot (1989), it is possible to propagate the area estimate distributions produced by the MDN through the percentage change calculation so that the ambiguity in the percentage change estimate can also be represented as a distribution, as shown in Figure 4.24. As would be expected for an area that has not changed in reflectance, there is a large peak around 0% change, indicating that it is highly likely that no significant change in forest cover

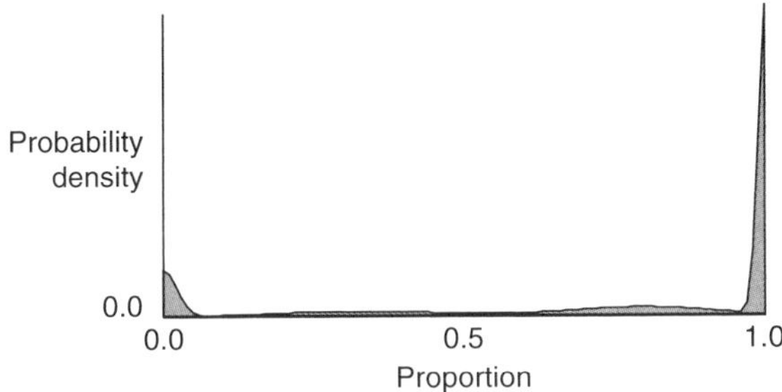

Figure 4.23 *Estimates for the proportion of sub-pixel area covered by forest obtained on the first and second observation of the target area. Because the remote sensor measured the same reflectance on each occasion, the MDN made the same estimates*

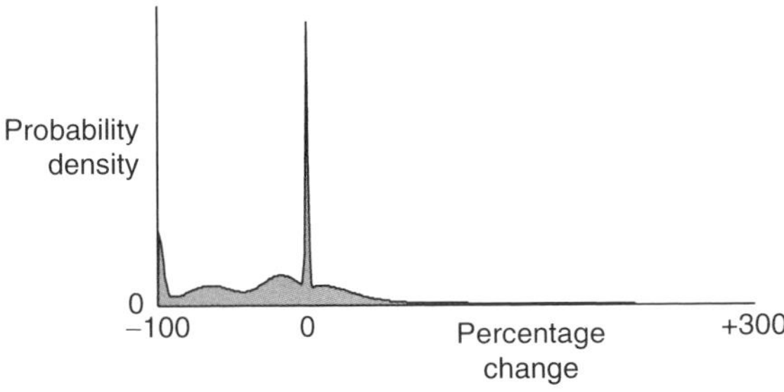

Figure 4.24 *The distribution over the percentage change in the sub-pixel forest area that could have occurred between the observations. As expected, the large peak at 0% indicates that it is most likely that no change in sub-pixel cover took place. The smaller peak at −100% indicates that the pixel could have changed from being completely covered by forest to containing no forest at all, while the non-zero probability up to (and beyond) +300% shows that the proportion of the sub-pixel area covered by forest could have more than trebled*

has occurred. It is interesting to note, however, that there is also a peak around −100%, suggesting that the sub-pixel area could have changed from being covered completely by forest to containing no forest at all. In fact, all percentage changes from −100% up have non-zero probabilities associated with them, and hence cannot be excluded on the basis of the remote observations.

It is important to emphasize that the uncertainty in the percentage change estimates in Figure 4.24 was also present in those obtained using the neural network based area estimates. In that case, however, since the neural network provided no representation of the ambiguity in its estimates, its effect could not be tracked through the percentage change calculation, and hence the ambiguity it induced in the percentage change estimate could not be represented. Provided that techniques that model explicitly the ambiguity in their estimates by probability distributions are used to extract information from remotely sensed data, the effect of the ambiguity implicit within such data can always be tracked and represented, regardless of where and how it is later applied.

6 Conclusions

This chapter has presented new research into the effect of a remote sensor's PSF on the information that it acquires. A sensor's PSF describes the variation in sensitivity of the sensor to the reflectance of land cover within and around individual pixels. It was shown that this sensitivity variation is an important source of a particular type of uncertainty, known as ambiguity. Specifically, sensitivity variation can lead to the remote sensor observing the same reflectance for radically different mixtures of sub-pixel classes. Such differences cannot, therefore, be distinguished by a remote sensor, and hence a remote sensor can only provide highly ambiguous information about processes that occur at ground level. Although these arguments were presented in terms of estimating the areas of different cover types, they apply equally well to all information extracted from data acquired by remote sensors.

Section 4 described an advanced neural-statistical technique called a mixture density network, that offers important benefits when extracting information from remotely sensed data. Because the technique uses a probability distribution to represent the information it extracts, it can represent all the information contained in remotely sensed data. This richness of the representation is important for visualization, making all the information contained in remotely sensed data available to analysts. Perhaps more importantly, without the representation, systems that use estimates (either in combination with information from other sources, or in isolation) cannot, on average, behave optimally.

The research presented in this chapter has implications for virtually all projects that use remotely sensed data. The techniques that have been presented provide a starting point for estimating the limits to the information that can be derived from remotely sensed data for any purpose (Wilkinson, 1996), and deriving such limits must form an important part of any research programme. For if it can be shown that the limit on the information that can be extracted from remotely sensed data has been reached, further research on more powerful analytical techniques cannot be justified.

At the same time, a wide range of new research opportunities emerge as a result of the possibility of propagating uncertainty, which, as was shown in section 5 can produce startling new results even in the simplest systems.

References

Bishop, C. M., 1994, Mixture Density Networks, *Technical Report* NCRG/94/004. Neural Computing Research Group, Aston University.

Bishop, C. M., 1995, *Neural Networks for Pattern Recognition* (Oxford: Oxford University Press).

Cracknell, A. P., 1998, Synergy in remote sensing – what's in a pixel? *International Journal of Remote Sensing*, **19**, 2025–47.

DeGroot, M. H., 1989, *Probability and Statistics* (Reading, MA: Addison-Wesley).

Fisher, P., 1997, The pixel: a snare and a delusion, *International Journal of Remote Sensing*, **18**, 679–85.

Foody, G. M., 1995, Fully fuzzy supervised image classification, *Proceedings of the 21st Annual Conference of the Remote Sensing Society* (Nottingham: Remote Sensing Society), pp. 1187–94.

Foody, G. M., 1997, Land cover mapping from remotely sensed data with a neural network: accommodating fuzziness, in I. Kanellopoulos, G. G. Wilkinson, F. Roli and J. Austin (eds), *Neural-computation in Remote Sensing Data Analysis* (Berlin: Springer-Verlag), pp. 28–37.

Horwitz, H. M., Nalepka, R. F., Hyde, P. S. and Morgenstern, J. P., 1971, Estimating the proportions of objects within a single resolution element of a multispectral scanner, *Proceedings 7th International Symposium on Remote Sensing of Environment* (Michigan: Ann Arbor), pp. 1307–20.

Justice, C. O., Markham, B. L., Townshend, J. R. G. and Kennard, R. L., 1989, Spatial degradation of satellite data, *International Journal of Remote Sensing*, **10**, 1539–61.

Kitamoto, A. and Takagi, M., 1999, Image classification using probabilistic models that reflect and internal structure of mixels, *Pattern Analysis and Applications*, **2**, 31–43.

Manslow, J. F., 2001, *On the Extraction and Representation of Land Cover Information Derived from Remotely Sensed Imagery*, Unpublished Ph D Thesis (Southampton: University of Southampton).

Manslow, J. F. and Nixon, M., 2000. On the representation of fuzzy land cover classifications, *Proceedings of the 26th Annual Conference of the Remote Sensing Society* (Nottingham: Remote Sensing Society).

Manslow, J. F., Brown, J. and Nixon, M., 2000, On the probabilistic interpretation of area based fuzzy land cover mixing proportions, in S. Lek and J. F. Guégan (eds), *Artificial Neuronal Networks: Application to Ecology and Evolution* (Berlin: Springer-Verlag), pp. 81–95.

Townshend, J. R. G., Huang, C., Kalluri, S. N. V., DeFries, R. S. and Liang, S., 2000, Beware of per-pixel characterisation of land cover, *International Journal of Remote Sensing*, **21**, 839–43.

Wilkinson, G. G., 1996, Classification algorithms – where next? in E. Binaghi, P. A. Brivio and A. Rampini (eds), *Soft Computing in Remote Sensing Data Analysis* (Singapore: World Scientific), pp. 93–9.

5

Pixel Unmixing at the Sub-pixel Scale Based on Land Cover Class Probabilities: Application to Urban Areas

Qingming Zhan, Martien Molenaar and Arko Lucieer

1 Introduction

Urban features (e.g. roads, buildings) often have sharp boundaries. Because of the coarse spatial resolution of most remotely sensed images relative to such features, many pixels (particularly those containing boundaries) will contain a mixture of the spectral responses from the different features. A number of techniques have been proposed for pixel unmixing (i.e. predicting sub-pixel land cover proportions; see, for example, Atkinson, 1997; Foody *et al.*, 1997; Schowengerdt, 1997; Foody, 1998; Steinwendner, 1999; Tatem *et al.*, 2001a, 2001b). These techniques are based on the assumption that the spectral value of each pixel is the composite spectral signature of the land cover types present within the pixel. For example, mixture modelling, neural networks and the fuzzy *c*-means classifier have been compared for predicting the proportions of different classes that a pixel may represent (Foody, 1996; Atkinson, 1997; Atkinson *et al.*, 1997; Bastin, 1997). While these 'soft' classifiers offer benefits over traditional 'hard' classifiers, no information is provided on the location of the land cover proportions within each pixel.

A per-field approach using detailed vector data can be used to increase the spatial resolution of the remote sensing classification (Aplin and Atkinson, 2001) and increase classification accuracy (Aplin *et al.*, 1999a, 1999b). In most cases, however, accurate

Uncertainty in Remote Sensing and GIS. Edited by G.M. Foody and P.M. Atkinson.
 ISBN: 0-470-84408-6

vector data sets are rarely available (Tatem *et al.*, 2001). Therefore, it is important to consider techniques for sub-pixel mapping that do not depend on vector data.

A method was applied previously to map the sub-pixel location of the land cover proportions predicted from a mixture model based only on the spatial dependence (the likelihood that observations close together are more similar than those further apart) inherent in the data (Atkinson, 1997). The approach was based on maximizing the spatial dependence in the resulting image at the sub-pixel scale. However, the resulting spatial arrangement conflicted with our understanding of what constitutes 'maximum spatial order', as the author indicated (Atkinson, 1997). In an attempt to circumvent this problem, the proposed approach was designed to take account of contributions to the sub-pixel classification from both the centre of the 'current' pixel and from its neighbouring pixels. These 'contributions' are based on the assumption that the land cover is spatially dependent both *within* and *between* pixels (Atkinson, 1997; Verhoeye and Wulf, 2000).

Conventional classifiers such as the maximum likelihood classifier (MLC) are based on the spectral signatures of training pixels and do not recognize spatial patterns in the same way that a human interpreter does (Gong *et al.*, 1990). However, the MLC generates a substantial amount of information on the class membership properties of a pixel, which provides valuable information on the relative similarity of a pixel to the defined classes (Foody *et al.*, 1992). The objective of the proposed approach was to use such probability measures derived from the MLC together with spatial information at the pixel scale to produce a fine spatial resolution land cover classification (i.e. at the sub-pixel scale), potentially with increased accuracy. Therefore, in the proposed approach for sub-pixel mapping, two stages (spectral–spatial) were employed. In the first stage, a MLC is implemented and *a posteriori* probabilities are predicted. In the second stage, an inverse distance weighting predictor is used to interpolate a probability surface at the sub-pixel scale from the probabilities of the central pixel and its neighbourhood. Subsequently, a sub-pixel hard classification is produced based on the interpolated probabilities at the sub-pixel scale.

Among the four causes of mixed pixels described by Fisher, and shown in Figure 5.1 (Fisher, 1997), the 'boundary pixel' and 'intergrade' cases are investigated in this chapter because they are the most amenable to sub-pixel analysis. The 'sub-pixel' and 'linear sub-pixel' cases are beyond the present scope because of the lack of information on the existence and spatial extent of objects smaller than the pixel size. Nevertheless, the proposed sub-pixel approach aims at solutions to the 'boundary pixel' and 'intergrade' cases without neglecting the potential existence of the other two cases.

Thus, the objective was to present a new method to produce a land cover classification map with finer spatial resolution than that of the original imagery, with potentially increased accuracy, especially for boundary pixels. The results will be used in later research to infer urban land use based on a land cover classification.

2 Methods

2.1 Sub-pixel classification

After implementing the standard MLC, the proportion of each class in a pixel was related to the pixel's probability vector. Each pixel was split into $5 \times 5 = 25$ sub-pixels

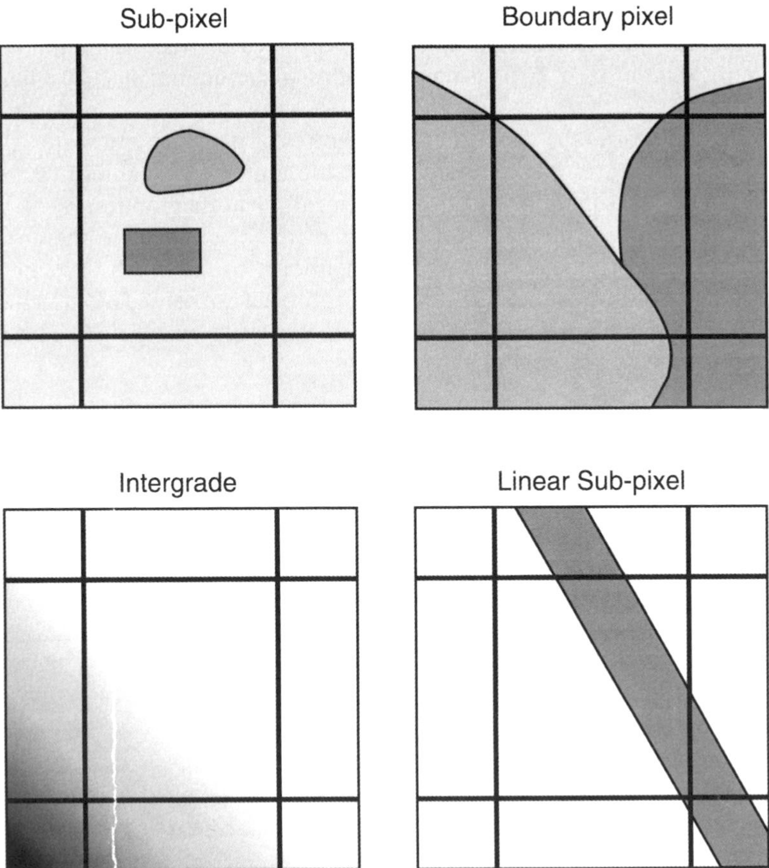

Figure 5.1 *Four causes of mixed pixels (from Fisher, 1997; reproduced from the* International Journal of Remote Sensing *by permission of Taylor & Francis Ltd, http://www.tandf.co.uk/journals)*

with a zoom factor of 5. This zoom factor was chosen to match the difference in the spatial resolutions of Système Pour L'Observation de la Terre (SPOT) High Resolution Visible (HRV) and IKONOS multispectral (MS) imagery, which are a focus of this chapter. To determine the class probability of a sub-pixel for each class, a new probability vector $\hat{\mathbf{z}}$ was calculated based on the probability vectors of the central pixel and its eight neighbouring pixels. The inverse distance weighting predictor was used in computing a new probability value for each sub-pixel. The assumption is that the value of an attribute z at an unvisited point is a distance-weighted average of data points occurring within a neighbourhood or window surrounding the unvisited point (Burrough and McDonnell, 1998).

$$\hat{z}(\mathbf{x}_0) = \frac{\sum_{i=1}^{n} z(\mathbf{x}_i) \cdot d_{ij}^{-r}}{\sum_{i=1}^{n} d_{ij}^{-r}} \tag{1}$$

where, $\hat{z}$ is the value of the attribute at an unvisited location $\mathbf{x}_0$, z is the known value of the attribute at location $\mathbf{x}_i$, d is the distance between the unknown point $\mathbf{x}_j$ and a neighbour $\mathbf{x}_i$, r is a distance weighting factor and n is the number of neighbours.

For a given sub-pixel the distances to the centres of the neighbours are calculated. These distance measures d_{ij} are used to calculate the new probability vector of the sub-pixel $z(\mathbf{x}_0)$, see Figure 5.2 (sub-pixel-central). Another option is to take the distances from a given sub-pixel to the edges of the neighbours to the north, east, south and west and to the corners of the neighbours to the north-west, north-east, south-east and south-west (see Figure 5.3, sub-pixel-corner). The effect of these different distance measures on the interpolation result was tested. The distance weight exponent r was set to 1.0.

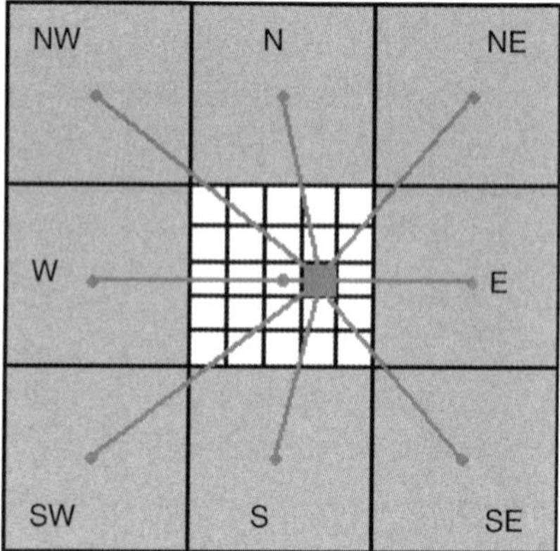

Figure 5.2 Inverse distance interpolation used to compute sub-pixel probability vectors. Distances are taken from each sub-pixel to the centres of neighbouring pixels

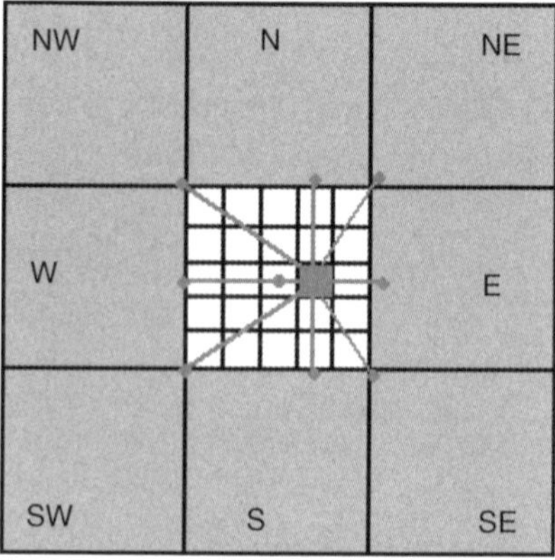

Figure 5.3 Inverse distance interpolation used to compute sub-pixel probability vectors. Distances are taken from each sub-pixel to the corners or edges of neighbouring pixels

An important factor to consider is how to incorporate the probability vector of the current pixel itself into the interpolation. One option is to leave the centre probability vector out. In this case, only the neighbouring probability vectors would be used (to neglect the existence of the 'sub-pixel' and 'linear sub-pixel' cases in Figure 5.1). Another option is to choose a distance value for the centre probability pixel in the interpolation (e.g. to consider the potential existence of the 'sub-pixel' and 'linear sub-pixel' cases in Figure 5.1). For example, the distance from each sub-pixel to the centre pixel could be set to 1.0 to give this pixel a large weight.

2.2 Experimental testing

To assess the accuracy of the proposed sub-pixel approach, several controlled tests were implemented. Four binary testing images were created manually at the sub-pixel scale (200 × 200 pixels) with values of 1 and 0 (white and black respectively in Figure 5.4). Simulated probability images (40 × 40 pixels) were generated from four testing images by aggregating sub-pixels into pixels. Equal-weighted aggregation was used to maintain the statistical and spatial properties of the simulated data (Bian and Butler, 1999). Simulated images can be found in Figures 5.5(a), 5.6(a), 5.7(a) and 5.8(a) (40 × 40 pixels). Each pixel of a simulated image covers 5 × 5 sub-pixels corresponding to the same spatial aggregation scale of SPOT HRV imagery (20 m) relative to IKONOS MS imagery (4 m).

The above sub-pixel approaches were applied to the simulated images. Probability images were generated by applying the proposed interpolation method using the central points of neighbouring pixels (Figures 5.5(d), 5.6(d) 5.7(d) and 5.8(d)) and using the corners or edges of neighbouring pixels (Figures 5.5(g), 5.6(g), 5.7(g) and 5.8(g)) respectively. Corresponding outputs at the sub-pixel scale are shown in Figures 5.5(e), 5.6(e), 5.7(e) 5.8(e) and Figures 5.5(h), 5.6(h), 5.7(h) and 5.8(h) respectively.

2.3 Accuracy assessment

Accuracy assessment was carried out in the following three ways:

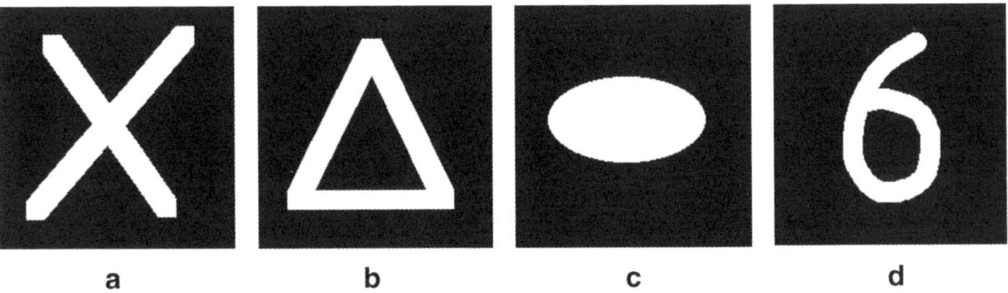

Figure 5.4 Experimental 'truth' images (200 × 200 pixels)

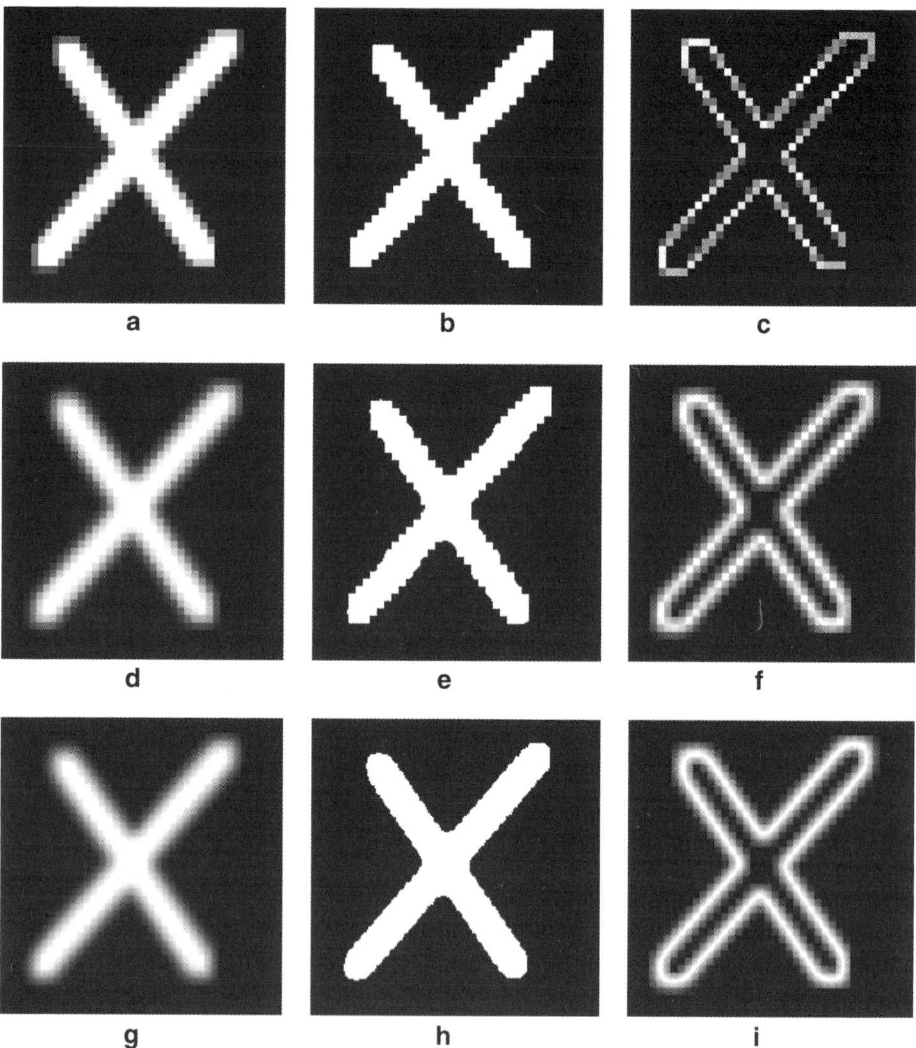

Figure 5.5 Experimental testing results based on the proposed sub-pixel interpolation method: (a) resampled image from experimental truth image (Figure 5.4a); (b) classification results based on resampled image; (c) entropy results of (b) (40 × 40 pixels); (d) probability image at the sub-pixel scale using central points of neighbouring pixels; (e) classification results at the sub-pixel scale based on (d); (f) entropy results of (e) (200 × 200 pixels); (g) probability image at the sub-pixel scale using the corners or edges of neighbouring pixels; (h) classification results at the sub-pixel scale based on (g); (i) entropy results of (h) (200 × 200 pixels)

2.3.1 Overall accuracy and producer's accuracy

The accuracy of classification was estimated by comparing the classified data with the testing data on a pixel-by-pixel basis at the sub-pixel scale (each pixel of results at the pixel scale (b) in Figure 5.5 was split into 5 × 5 pixels at the sub-pixel scale). Overall accuracy was calculated as

Figure 5.6 Experimental testing results based on the proposed sub-pixel interpolation method: (a) resampled image from experimental truth image (Figure 5.4b); (b) classification results based on resampled image; (c) entropy results of (b) (40 × 40 pixels); (d) probability image at the sub-pixel scale using central points of neighbouring pixels; (e) classification results at the sub-pixel scale based on (d); (f) entropy results of (e) (200 × 200 pixels); (g) probability image at the sub-pixel scale using the corners or edges of neighbouring pixels; (h) classification results at the sub-pixel scale based on (g); (i) entropy results of (h) (200 × 200 pixels)

$$Overall\ accuracy = \frac{\sum_{i=1}^{k} n_{ii}}{N} \tag{2}$$

where, n_{ii} is the number of pixels classified correctly, N is the total number of pixels and k is the number of classes. Producer's accuracy was used to represent classification accuracy for individual classes:

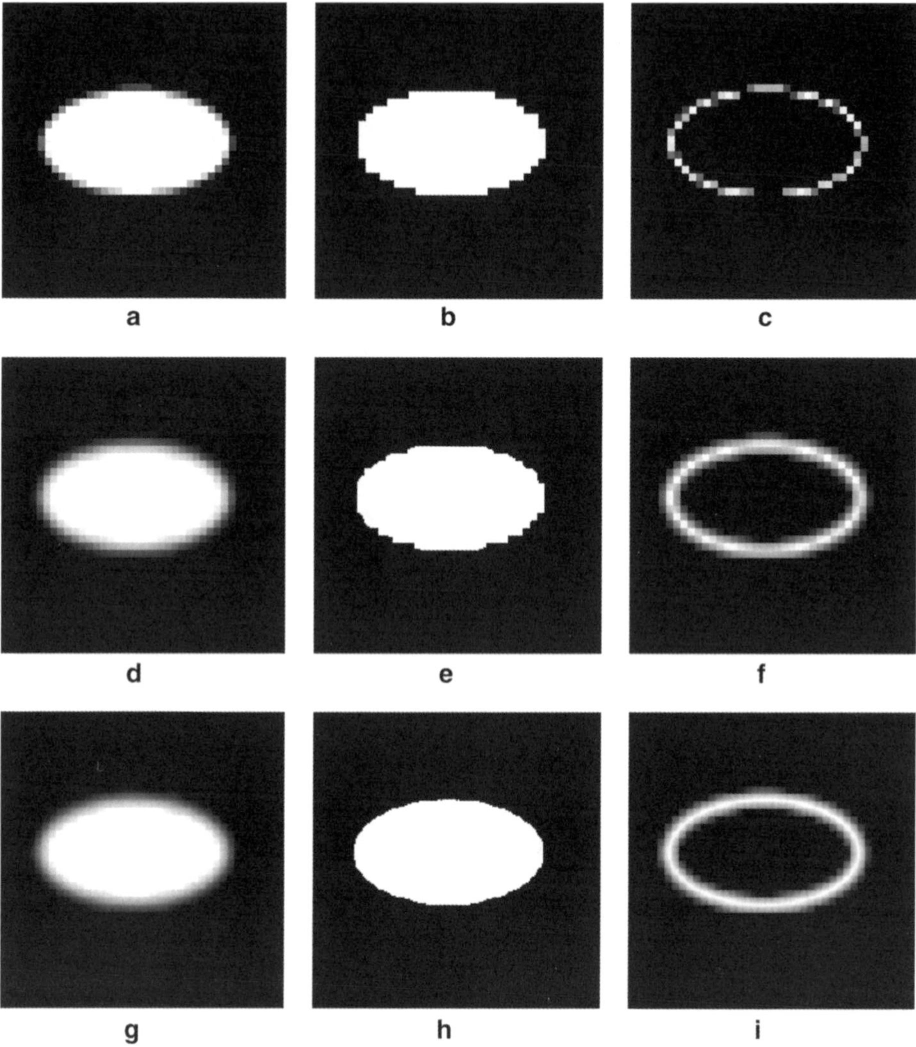

Figure 5.7 *Experimental testing results based on the proposed sub-pixel interpolation method: (a) resampled image from experimental truth image (Figure 5.4c); (b) classification results based on resampled image; (c) entropy results of (b) (40 × 40 pixels); (d) probability image at the sub-pixel scale using central points of neighbouring pixels; (e) classification results at the sub-pixel scale based on (d); (f) entropy results of (e); (200 × 200 pixels); (g) probability image at the sub-pixel scale using the corners or edges of neighbouring pixels; (h) classification results at the sub-pixel scale based on (g); (i) entropy results of (h) (200 × 200 pixels)*

$$Producer's\ accuracy = \frac{\sum_{i=1}^{k} n_{ii}}{n_{+i}} \tag{3}$$

where, n_{ii} is the number of pixels classified correctly, n_{+i} is the number of pixels classified into class i in the testing data set and k is the number of classes.

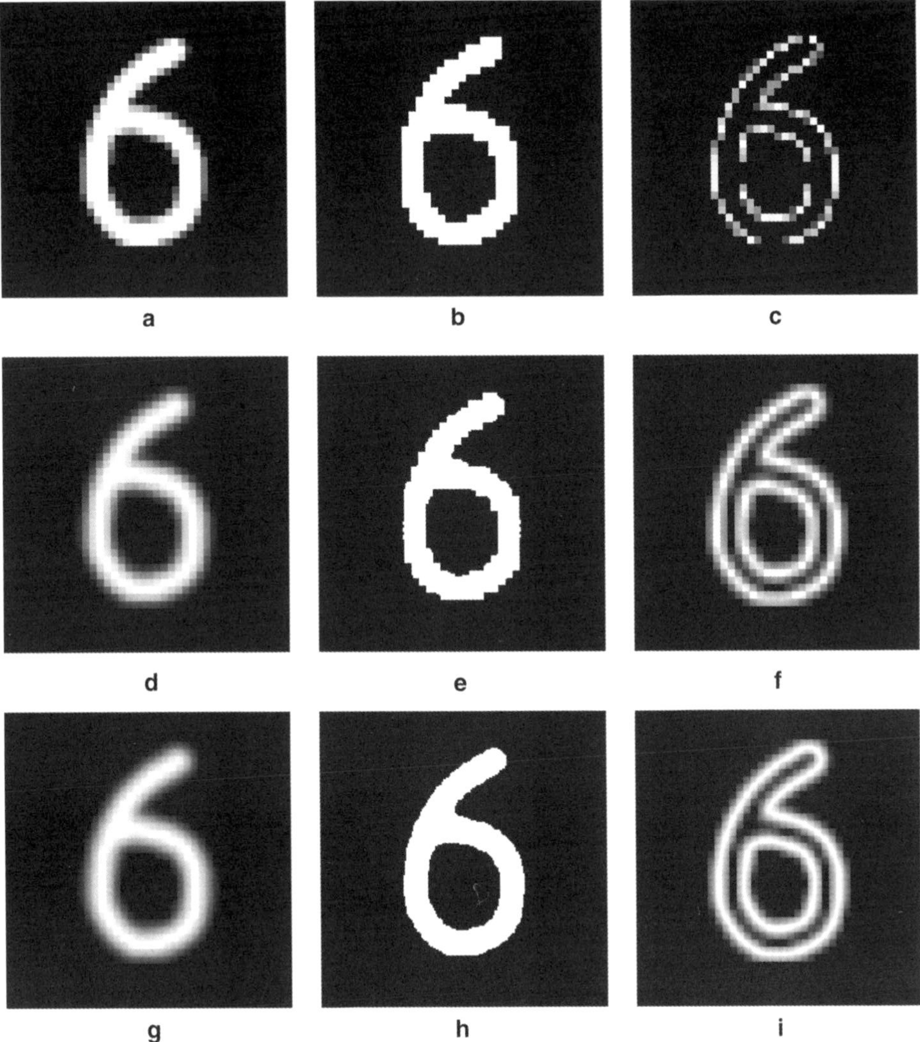

Figure 5.8 *Experimental testing results based on the proposed sub-pixel interpolation method: (a) resampled image from experimental truth image (Figure 5.4d); (b) classification results based on resampled image; (c) entropy results of (b) (40 × 40 pixels); (d) probability image at the sub-pixel scale using central points of neighbouring pixels; (e) classification results at the sub-pixel scale based on (d); (f) entropy results of (e) (200 × 200 pixels); (g) probability image at the sub-pixel scale using the corners or edges of neighbouring pixels; (h) classification results at the sub-pixel scale based on (g); (i) entropy results of (h) (200 × 200 pixels)*

2.3.2 *Kappa coefficient*

Kappa analysis is a discrete multivariate technique used in accuracy assessment for determining statistically if one error matrix is significantly different from another (Bishop *et al.*, 1975). The Kappa analysis has become a popular component of accuracy assessment (Hudson and Ramm, 1987; Congalton, 1991; Richards, 1993; Foody, 1995; Congalton and Green, 1999). The Kappa statistic is given by

$$K = \frac{N \sum_{i=1}^{k} n_{ii} - \sum_{i=1}^{k} n_{i+} n_{+i}}{N^2 - \sum_{i=1}^{k} n_{i+} n_{+i}} \tag{4}$$

where, n_{ii} is the number of pixels classified correctly, n_{i+} is the number of pixels classified into class i, n_{+i} is the number of pixels classified into class i in the testing data set, N is the total number of pixels and k is the number of classes.

2.3.3 Entropy

The accuracy of a classification may be indicated by the way in which the probability of class membership is partitioned between the classes and this may be expressed by entropy measures. Entropy, given by

$$H = -\sum_{i=1}^{k} P_i \log_2 P_i \tag{5}$$

where, P_i is the probability of belonging to class i, was used to further assess uncertainty in this chapter.

The results are indicated (see also Table 5.1) for each case as overall accuracy, Kappa coefficient and entropy images in Figure 5.5(c), (f) and (i), respectively. Figure 5.5 shows that the results are promising for each case. The results show that both interpolation methods have the ability to increase classification accuracy and classify at a spatial resolution that is finer than that of the original data. It also shows that more accurate results have been produced by using the corner or edge points of neighbouring pixels than by using the central point of neighbouring pixels. Therefore, only the interpolation method using the corner or edge points of neighbouring pixels was applied in the following case study based on real imagery.

3 Study Area

A 9 km^2 (3 km $\times$ 3 km) area, south-east of Amsterdam, was selected for the experiment. Approximately 200 000 people live in this sub-urban district. Several types of

Table 5.1 *Testing results using overall accuracy and Kappa coefficient*

Test Image	Pixel Scale		Sub-pixel Scale using corners or edges		Sub-pixel Scale using central points	
	Overall Accuracy	Kappa Coefficient	Overall Accuracy	Kappa Coefficient	Overall Accuracy	Kappa Coefficient
'Cross'	97.85%	93.12%	98.22%	94.32%	99.24%	97.60%
'Triangle'	97.26%	91.35%	97.98%	93.57%	99.39%	98.03%
'Ellipse'	99.11%	97.06%	99.34%	97.80%	99.75%	99.16%
'Six'	98.30%	92.49%	98.62%	93.93%	99.46%	97.61%

residential as well as commercial areas, parks, lakes and canals can be found in the study area. Built-up areas, green space and water are three constituent land cover classes in this study.

SPOT HRV imagery was used for urban land cover classification. A detailed land cover map was prepared for training and testing purposes based on detailed topographic maps (scale 1:1000) together with visual interpretation of IKONOS MS imagery (same 4 m pixel size as the final sub-pixel outputs) of the same area. The final results were analysed to determine the effectiveness of the use of posterior probabilities and the proposed sub-pixel approach to increase the accuracy of land-cover classification and to produce finer spatial resolution classification maps.

4 Applications

4.1 Data and sample selection

The data used in the case study are listed in Table 5.2.

To minimize the potential bias in the MLC, the single-pixel training approach was used in the sample selection. About 100 data for each class were selected manually from the SPOT HRV image following the rules discussed in the literature (Gong and Howarth, 1990). To make the data more representative for MLC training and to allow a certain degree of fuzziness in the training and testing stages, as recommended by Foody (1999), about one-third of the data for each class were selected from mixed pixels.

4.2 Land cover classification using MLC

Land cover classification was implemented using a standard MLC algorithm (Richards, 1993). The classification results consisted of 'hard' classification output (Figure 5.9) and the probability map for each class.

4.3 Preparation of testing data

The training and testing data were prepared based on visual interpretation of IKONOS MS imagery and large-scale topographical maps (1:1000). Colour aerial photographs taken in 1996 and 1997 were used to simulate the situation in 1996 when SPOT HRV imagery was captured. The final map was produced by vector-to-raster conversion with a spatial resolution of 4 m, the same as that of IKONOS MS imagery. Figure 5.10 shows the testing data used in the accuracy assessment.

4.4 Pixel unmixing

Pixel unmixing was implemented by a program developed by the authors in Java using the inverse distance weighting interpolation method based on corner or edge

Table 5.2 List of data used in the experiments

Type of data	Sensor	Date	Band/Colour	Spatial resolution	Scale
Satellite sensor imagery	SPOT	13 Oct. 1996	Multi-spectral	20 m	N/A
	IKONOS	8 June 2000	Multi-spectral	4 m	N/A
Aerial photographs	Optical camera	19 May 1997 30 May 1999	True colour	N/A	1:10 000
Topographic maps		1996, 1997, 1999	Black and white	N/A	1:1000

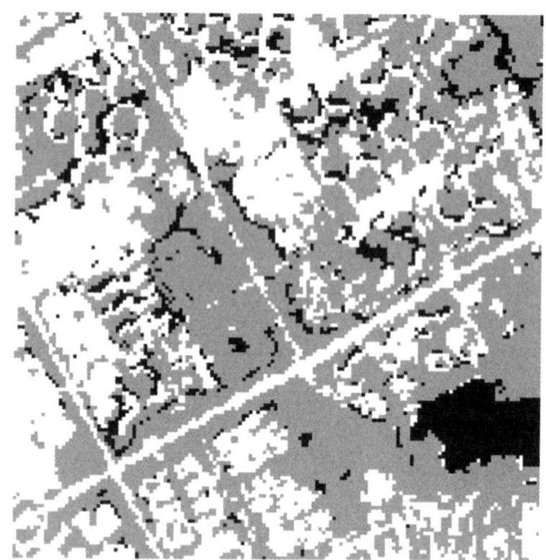

Figure 5.9 Classification map produced by applying the standard MLC approach at the pixel scale (150 × 150 pixels)

Figure 5.10 Image prepared at the sub-pixel scale (750 × 750 pixels), used for both training and testing

points of neighbouring pixels (section 2). The sub-pixel classification results consist of 'hard' classification output (Figure 5.11) and the probability map for each class at the sub-pixel scale as well as the entropy map (Figure 5.12) based on the interpolated probability map for uncertainty assessment.

4.4.1 Accuracy assessment

Accuracy assessment was based on pixel-to-pixel comparison between produced classification results achieved by different approaches with the testing map at the

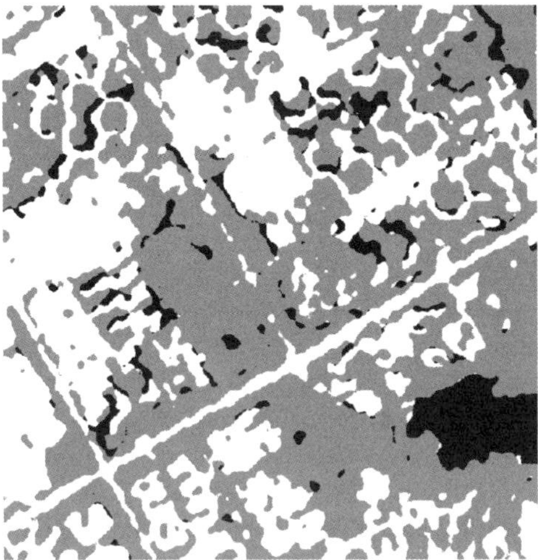

Figure 5.11 *Classification map produced by applying the sub-pixel approach at the sub-pixel scale (750 × 750 pixels)*

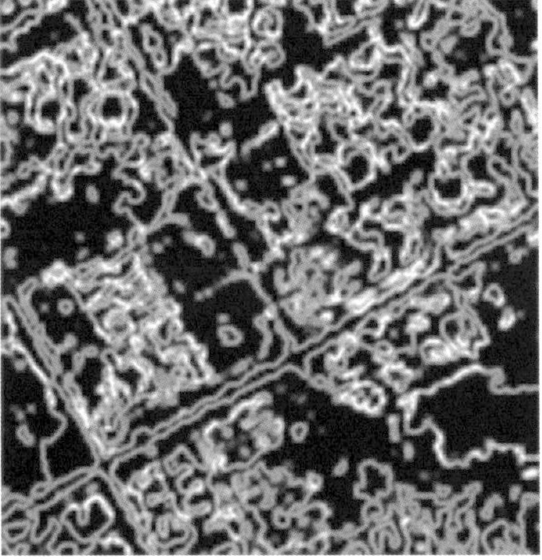

Figure 5.12 *Entropy map produced based on the interpolated probability map at the sub-pixel scale (750 × 750 pixels)*

Table 5.3 Comparison of accuracies obtained by applying the standard MLC and the sub-pixel approach

Accuracy assessment	Standard MLC	Sub-pixel
Overall accuracy	73.66%	74.57%
Kappa	53.15%	54.00%

sub-pixel scale. Each pixel of results (Figure 5.9) was split into 5 × 5 sub-pixels to enable a pixel-to-pixel comparison with the testing map at the sub-pixel scale. Overall accuracy and the Kappa coefficient are presented in Table 5.3. Producer's accuracy was used to represent classification accuracy for individual land cover classes Table 5.4). An entropy map was produced based on the interpolated probability map at the sub-pixel scale (Figure 5.12).

4.5 Impacts of applying the sub-pixel approach to boundary pixels

To assess the impacts of the proposed approach on boundary pixels (the main goal of this chapter) in detail, the entropy map was produced based on the probability values to each class at the pixel scale (Figure 5.13). About 14% of the total number of pixels were extracted as boundary pixels (Figure 5.14) based on the entropy map (Figure 5.13) using a threshold value of 0.469. A threshold value of 0.469 means that no one class is dominant (maximum likelihood > 0.9) in comparing probabilities to each class (with three classes). Non-boundary pixels can be interpreted as having at least one class dominant among all classes.

Classification maps produced by applying the standard MLC and the sub-pixel approach were compared with the testing map using boundary pixels and non-boundary pixels as masks. The comparison results are shown in Tables 5.4 and 5.5.

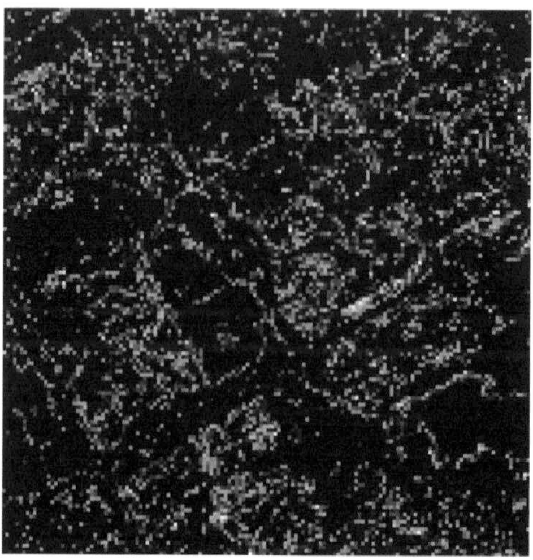

Figure 5.13 Entropy map produced based on MLC probability at the pixel scale (150 × 150 pixels)

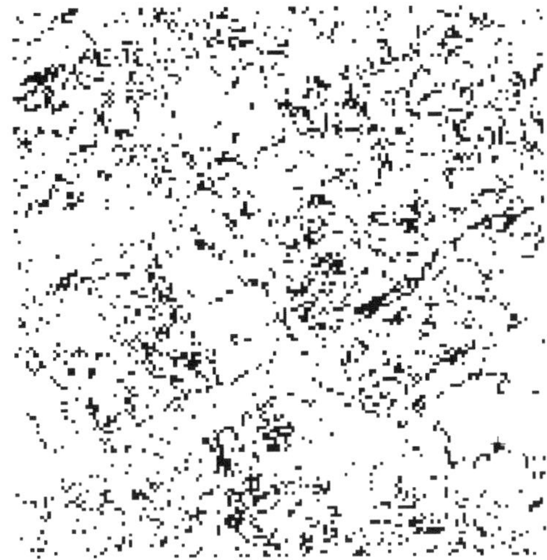

Figure 5.14 Boundary pixels (black) extracted from entropy map in Figure 5.13 (150 × 150 pixels)

5 Discussion

From Table 5.2, both overall accuracy and Kappa increased slightly (about 1%) using the proposed sub-pixel approach. Table 5.4 shows that the overall accuracy also increased by 4.58% for boundary cases for the sub-pixel approach.

The producer's accuracy increased for the built-up area and green space, but decreased for water (Table 5.4). For boundary cases (Table 5.6), the producer's accuracy increased for the built-up area and green space (6.09% and 7.36%). However, the producer's accuracy decreased for water again (13%). When the classification map (Figure 5.11) is compared with the testing data map (Figure 5.10), it can be seen that some pixels along narrow canals had been eliminated by the sub-pixel

Table 5.4 Comparison of producer's accuracies obtained for each land cover class by applying the standard MLC and the sub-pixel approach

Land cover classes	Standard MLC	Sub-pixel
Built-up area	80.19%	80.99%
Green space	72.29%	74.01%
Water body	58.02%	53.61%

Table 5.5 Comparison of accuracies obtained by applying the standard MLC and the sub-pixel approach to boundary and non-boundary cases

Overall accuracy	Standard MLC	Sub-pixel
Boundary pixels	48.65%	53.23%
Non-boundary pixels	77.71%	78.03%

Table 5.6 *Comparison of producer's accuracies for land cover classes by applying standard MLC and the sub-pixel approach in boundary and non-boundary cases*

Producer's accuracy for land cover classes		Standard MLC	Sub-pixel
Boundary pixels	Built-up area	44.98%	51.07%
	Green space	50.22%	57.58%
	Water body	48.01%	35.01%
Non-boundary pixels	Built-up area	84.44%	84.59%
	Green space	76.13%	76.87%
	Water body	60.58%	58.36%

approach. From the entropy map (Figure 5.12), it can be seen that these pixels along canals gained greater uncertainty as well. The main reason is that the proposed approach fails to tackle the 'sub-pixel' and 'linear sub-pixel' cases. Further, pixels in shadow areas are often mixed up with water pixels in spectral-feature space. However, it increases classification accuracy in the 'boundary' and 'intergrade' cases, which are the main objectives of this chapter.

Based on the above results, it can be seen that the proposed sub-pixel approach has the ability to increase classification accuracy and produce a map at a finer spatial resolution than that of the original imagery, with objects consisting of at least four pixels or narrow linear objects at least two pixels wide, but certainly not with small objects of a single pixel or smaller. It may be possible to apply different interpolation settings to preserve small objects being eliminated in the proposed spatial allocation stage.

6 Conclusions

Pixel unmixing is an important issue especially for urban applications due to the fragmentary nature of urban features. The experiment shows that the posterior probability is useful in detecting mixed pixels as well as for spatial sub-pixel analysis. Additional efforts have to be made in dealing with pixels in shadow areas.

The results indicate that the proposed sub-pixel approach can be useful in increasing the spatial resolution and increasing overall classification accuracy, especially for class boundaries. Since the proposed sub-pixel approach consists of two independent stages, class components analysis and spatial allocation respectively, it is also suitable for pixel unmixing based on class components derived from other approaches.

The proposed sub-pixel approach tries to recognize the potential existence of features smaller than a pixel by using the value of the central pixel when spatially allocating class components according to their neighbouring pixels. However, the factors that contribute to producing the digital number (DN) of a remotely sensed image are complicated and interrelated (Fisher, 1997; Cracknell, 1998). Further improvements are needed to deal with such difficult pixel unmixing issues.

References

Aplin, P. and Atkinson, P. M., 2001, Sub-pixel land cover mapping for per-field classification, *International Journal of Remote Sensing*, **22**(14), 2853–8.

Aplin, P., Atkinson P. M. and Curran P. J., 1999a, Fine spatial resolution simulated sensor imagery for land cover mapping in the United Kingdom, *Remote Sensing of Environment*, **68**, 206–16.

Aplin, P., Atkinson P. M. and Curran P. J., 1999b, Per-field classification of land use using the forthcoming very fine spatial resolution satellite sensors: problems and potential solutions, in P. M. Atkinson and N. J. Tate (eds), *Advances in Remote Sensing and GIS Analysis*, (Chichester: Wiley), pp. 219–39.

Atkinson, P. M., 1997, Mapping sub-pixel boundaries from remotely sensed images, in Z. Kemp (ed.) *Innovations in GIS 4* (London: Taylor and Francis) pp. 166–80.

Atkinson, P. M., Cutler, M. E. J. and Lewis, H. G., 1997, Mapping sub-pixel proportional land cover with AVHRR imagery, *International Journal of Remote Sensing*, **18**(4), 917–35.

Bastin, L., 1997, Comparison of fuzzy c-means classification, linear mixture modelling and MLC probabilities for unmixing coarse pixels, *International Journal of Remote Sensing*, **18**(17), 3629–48.

Bian, L. and Butler, R., 1999, Comparing effects of aggregation methods on statistical and spatial properties of simulated spatial data, *Photogrammetric Engineering and Remote Sensing*, **65**(1), 73–84.

Bishop, Y., Fienberg S. and Holland, P., 1975, *Discrete Multivariate Analysis: Theory and Proactice* (Cambridge, MA: MIT Press).

Burrough, P. A. and McDonnell, R. A., 1998, *Principles of Geographical Information Systems – Spatial Information Systems and Geostatistics* (Oxford: Oxford University Press), p. 333.

Congalton, R., 1991, A review of assessing the accuracy of classification of remotely sensed data, *Remote Sensing of Environment*, **37**, 35–46.

Congalton, R. G. and Green, K., 1999, *Assessing the Accuracy of Remotely Sensed Data: Principles and Practices* (Boca Raton: Lewis).

Cracknell, A. P., 1998, Synergy in remote sensing – what's in a pixel? *International Journal of Remote Sensing*, **19**, 2025–47.

Duda, R. O., Hart P. E. and Stork, D. G., 2001, *Pattern Classification* (New York: Wiley).

Fisher, P., 1997, The pixel: a snare and a delusion, *International Journal of Remote Sensing*, **18**(3), 679–85.

Foody, G. M., 1995, Cross-entropy for the evaluation of the accuracy of a fuzzy land cover classification with fuzzy ground data, *ISPRS Journal of Photogrammetry and Remote Sensing*, **50**(5), 2–12.

Foody, G. M., 1996, Approaches for the production and evaluation of fuzzy land cover classification from remotely-sensed data, *International Journal of Remote Sensing*, **17**(17), 1317–40.

Foody, G. M., 1998, Sharpening fuzzy classification output to refine the representation of sub-pixel land cover distribution, *International Journal of Remote Sensing*, **19**(13), 2593–9.

Foody, G. M., 1999, The continuum of classification fuzziness in thematic mapping, *Photogrammetric Engineering and Remote Sensing*, **65**(4), 443–51.

Foody, G. M., Campbell, N. A., Trodd, N. M. and Wood, T. F., 1992, Derivation and applications of probabilistic measures of class membership from the maximum-likelihood classification, *Photogrammetric Engineering and Remote Sensing*, **58**(9), 1335–41.

Foody, G. M., Lucas R. M., Curran, P. J. and Honzak, M., 1997, Non-linear mixture modelling without end-members using an artificial neural network, *International Journal of Remote Sensing*, **18**(4), 937–53.

Gong, P. and Howarth, P. J., 1990, An assessment of some factors influencing multi-spectral land-cover classification, *Photogrammetric Engineering and Remote Sensing*, **56**(5), 597–603.

Hudson, W. D. and Ramm C. W., 1987, Correct formulation of the Kappa coefficient of agreement, *Photogrammetric Engineering and Remote Sensing*, **53**, 421–2.

Richards, J. A., 1993, *Remote Sensing Digital Image Analysis: An Introduction* (Berlin: Springer-Verlag).

Schowengerdt, R. A., 1997, *Remote Sensing: Model and Methods for Image Processing* (San Diego: Academic Press).

Steinwendner, J., 1999, From satellite images to scene description using advanced image processing techniques, *RSS'99* (Nottingham: Remote Sensing Society).

Tatem, A. J., Lewis, H. G., Atkinson, P. M. and Nixon, M. S., 2001a, Super-resolution target identification from remotely sensed images using a Hopfield neural network, *IEEE Transactions on Geoscience and Remote Sensing*, **39**, 781–96.

Tatem, A. J., Lewis H. G., Atkinson, P. M. and Nixon, M. S., 2001b, Multi-class land-cover mapping at the sub-pixel scale using a Hopfield neural network, *International Journal of Applied Earth Observation and Geoinformation*, **3**(2), 184–90.

Verhoeye, J. and De Wulf, R., 2000, Sub-pixel mapping of Sahelian wetlands using multi-temporal SPOT VEGETATION images, in *28th International Symposium on Remote Sensing of Environment, Information for Sustainable Development* (Cape Town: CSIR Satellite Applications Centre).

6

Super-resolution Land Cover Mapping from Remotely Sensed Imagery using a Hopfield Neural Network

A. J. Tatem, H. G. Lewis, P. M. Atkinson and M. S. Nixon

1 Introduction

Accurate information on land cover is required to aid the understanding and management of the environment. Land cover represents a critical biophysical variable in determining the functioning of terrestrial ecosystems in bio-geochemical cycling, hydrological processes and the interaction between surface and atmosphere (Cihlar *et al.*, 2000). Information on land cover is central to all scientific studies that aim to understand terrestrial dynamics and is required from local to global scales to aid planning, while safeguarding environmental concerns. It is, therefore, central to all scientific studies that aim to understand terrestrial dynamics at any scale. In addition to this, the identification, extraction and mapping of target land cover features with low levels of uncertainty is vital in many areas of land management. Such information has many uses including those related to inventories of land resources, monitoring environmental change, and predicting future environmental scenarios.

Digital remotely sensed imagery derived from aircraft and satellite mounted sensors can provide the information required for land cover mapping and, since the 1970s, automated techniques for thematic map production have been used widely. The dominant approach, still used widely today, is hard classification (Campbell, 1996).

Uncertainty in Remote Sensing and GIS. Edited by G.M. Foody and P.M. Atkinson.
 ISBN: 0–470–84408–6

In a hard classification, land cover is represented as a series of discrete units, whereby each pixel is associated with a single land cover class. Among the most frequently used hard classification algorithms are the parallelepiped, minimum distance and maximum likelihood decision rules (Campbell, 1996). However, there exist practical limitations with the single class per-pixel assumption underlying hard classification.

The use of a hard classifier corresponds to the partitioning of feature space into mutually exclusive decision regions. These algorithms ignore the fact that many pixels in a remotely sensed image represent an average of spectral signatures from two or more surface categories. For example, a Landsat Thematic Mapper (TM) image (with a spatial resolution of 30 m by 30 m (120 m by 120 m in band 6)) of an urban scene may contain many pixels that represent more than one land cover class. Even in rural areas a Landsat TM scene may contain mixed pixels at the boundaries between land cover types, for example, between agricultural fields. This mixing of signatures occurs as a function of:

(i) Frequency of land cover – the physical continuum that exists in many cases between discrete category labels, combined with the spatially mixed nature of most natural land cover classes.
(ii) Frequency of sampling – the spatial integration within the pixel of land cover classes due to factors such as the sensor spatial resolution, point spread function (PSF) and resampling for geometric rectification.

The mixed pixel problem affects data acquired by satellite sensors of all spatial resolutions. Distinguishing the sources of mixing within an image is only possible in controlled and limited situations (Schowengerdt, 1996) meaning that classifications produced via hard classification may contain much uncertainty.

The problems above prompted the development and usage of classifiers that attempt to reduce error and uncertainty by accounting for mixed pixels. A hard classification fails to recognize or represent the existence of classes and objects which grade into one another and class boundaries at sub-pixel scales, leaving the classification user uncertain of the accuracy of the prediction. Such simplification can be seen as a waste of the available multispectral information, which could be interpreted more efficiently (Wang, 1990; Foody, 1995). Therefore, since an initial approach by Horowitz *et al.* (1971), many studies have used various techniques to attempt to 'unmix' the information from pixels. Typically, these produce a set of images, one for each land cover class, in which each pixel has a membership value between 0 and 1 to the relevant class (Foody, 1992; Bastin, 1997). In general, two different types of soft classification technique exist. The most commonly used methods predict posterior probabilities of class membership using statistical pattern recognition methods, and correlate these with area proportions. However, as Lewis *et al.* (1999) state, posterior probabilities are measures of statistical uncertainty, and there is no causal relationship with proportions of pixels containing the class, despite their correlation. Thus, as a consequence, posterior probabilities cannot represent optimum predictions of area (Manslow and Nixon, 2000). The second type of soft classification technique predicts directly class area proportions using regression models. Lewis *et al.* (1999) demonstrate that posterior probabilities do not represent optimum area predictions, and direct area proportion models achieve more accurate predictions of true land cover proportions.

Soft classification involves the prediction of sub-pixel class composition through the use of techniques such as spectral mixture modelling (Garcia-Haro *et al.*, 1996), multi-layer perceptrons (Atkinson *et al.*, 1997), nearest neighbour classifiers (Schowengerdt, 1997) and support vector machines (Brown *et al.*, 1999). The output of these techniques generally takes the form of a set of proportion images, each displaying the proportion of a certain class within each pixel. In most cases, this results in a more appropriate and informative representation of land cover than that produced using a hard, one class per-pixel classification. However, while the class composition of every pixel is predicted, the spatial distribution of these class components within the pixel remains unknown. Therefore, while soft classification conveys more information than hard classification, the resultant predictions still contain a large degree of uncertainty.

The overview below of work previously undertaken by the authors of this chapter, along with other work in the literature, demonstrates that it is possible to map land cover at the sub-pixel scale, thus, reducing the uncertainty inherent in maps produced solely by hard or soft classification.

2 Previous Image Mapping at the Sub-pixel Scale

Previous work on mapping or reconstructing images at the sub-pixel scale to reduce uncertainty has taken the form of three differing approaches, namely (i) image fusion and sharpening, (ii) image reconstruction and restoration and (iii) super-resolution.

2.1 Image fusion and sharpening

The basis for image fusion and sharpening stems from the fact that there will always be some trade-off within the field of remote sensing between spatial and spectral resolution. Images with fine spatial resolution can be used to locate objects accurately, whereas images with fine spectral resolution can be used to identify and quantify variables accurately. With different sensors acquiring information over the same area, it is often useful to merge the data into a hybrid product containing the useful information from each. Such a hybrid image can be used to create detailed 'sharpened' images that map the abundance of various materials within a scene. Therefore, image fusion techniques may be used to merge images of different spatial and spectral resolutions to create a fine spatial resolution multi-spectral combination (Gross and Schott, 1998).

Generally, two approaches to fusion and sharpening exist. The first uses a two-step approach whereby, firstly, the coarse spatial resolution multi-spectral image is used to identify the materials in the scene via soft classification. Second, the proportion maps are combined with a fine spatial resolution panchromatic image of the same area, which serves to constrain the proportions to produce a set of 'sharpened' fine spatial resolution maps. This has been carried out for synthetic images (Gross and Schott, 1998), agricultural fields (Li *et al.*, 1998) and for a lake (Foody, 1998). Foody

(1998) used a finer spatial resolution image and a simple regression based approach to sharpen the output of a soft classification of a coarser spatial resolution image, producing a sub-pixel land cover map. The results produced a visually realistic representation of the lake being studied, and this was further refined by fitting class membership contours, lessening the blocky nature of the representation. However, the areal extent of the lake was not maintained, and generally, obtaining two coincident images of differing spatial resolution is difficult.

The second approach to fusion and sharpening should conceptually produce identical results. Image fusion is undertaken to produce a fine spatial resolution multi-spectral image that is then unmixed into fine spatial resolution material maps. There is little published work of this approach, although Robinson *et al.* (2000) used it with synthetic imagery.

2.2 Image reconstruction and restoration

Digital image reconstruction refers to the process of predicting a continuous image from its samples and has fundamental importance in digital image processing, particularly in applications requiring image resampling. Its role in such applications is to provide a spatial continuum of image values from discrete pixels so that the input image may be resampled at any arbitrary position, even those at which no data were originally supplied (Boult and Wolberg, 1993).

Whereas reconstruction simply derives a continuous image from its samples, restoration attempts to go one step further. It assumes that the underlying image has undergone some degradation before sampling, and so attempts to predict the original continuous image from its corrupted samples. Restoration techniques must, therefore, model the degradation and invert its effects on the observed image samples, a process which itself contains a large degree of uncertainty when dealing with satellite sensor imagery.

2.3 Super-resolution

Super-resolution techniques are similar to the reconstruction and restoration approaches in that they predict image values at points where no data were originally supplied. However, where the reconstruction and restoration techniques attempt the difficult and uncertain task of recovering a continuous image, the super-resolution approaches merely attempt to increase the spatial resolution of an image to a desired level. This solves the problem of the computational complexity associated with the reconstruction and restoration techniques, making the super-resolution techniques applicable to remotely sensed data. The problem of 'scaling-up' ground data to match an image pixel support, as described by Atkinson and Foody (this volume, Chapter 1), can also be solved by 'scaling-down' the land cover information obtained from satellite sensor imagery using this technique. Various differing approaches to super-resolution mapping have been attempted.

Schneider (1993) introduced a knowledge-based analysis technique for the automatic localization of field boundaries with sub-pixel accuracy. The technique relies on knowledge of straight boundary features within Landsat TM scenes, and serves as a pre-processing step prior to automatic pixel-by-pixel land cover classification. With knowledge of pure pixel values either side of a boundary, a model can be defined for each 3 by 3 block of pixels. The model uses variables such as pure pixel values, boundary angle, and distance of boundary from the centre pixel. Using least squares adjustment, the most appropriate model parameters are chosen for location of a sub-pixel boundary, dividing mixed pixels into their respective pure components. Improvements on this technique were described by Steinwendner and Schneider (1997), who used a neural network to speed up processing. In addition Steinwendner and Schneider (1998), Steinwendner *et al.* (1998) and Steinwendner (1999) suggested algorithmic improvements, along with the addition of a vector segmentation step. The technique represents a successful, automated and simplistic pre-processing step for increasing the spatial resolution of satellite sensor imagery. However, its application is limited to imagery containing large features with straight boundaries at a certain spatial resolution, and the models used still have problems resolving image pixels containing more than two classes (Schneider, 1999).

Flack *et al.* (1994) concentrated on sub-pixel mapping at the borders of agricultural fields, where pixels of mixed class composition occur. Edge detection and segmentation techniques were used to identify field boundaries and the Hough transform (Leavers, 1993) was applied to identify the straight, sub-pixel boundaries. These vector boundaries were superimposed on a sub-sampled version of the image, and the mixed pixels were reassigned each side of the boundaries. By altering the image sub-sampling, the degree to which the spatial resolution was increased could be controlled. No validation was carried out, and the work was not followed up, and so the appropriateness of the technique remains unclear.

Aplin (1999, 2000) made use of sub-pixel scale vector boundary information, along with fine spatial resolution satellite sensor imagery to map land cover. By utilizing Ordnance Survey land line vector data fused with CASI imagery, and undertaking per-field rather than the traditional per-pixel land cover classification, mapping at a sub-pixel scale was demonstrated. Assessments suggested that the per-field classification technique was generally more accurate than the per-pixel classification (Aplin *et al.*, 1999). However, in most cases, availability of accurate vector data sets to apply the approach will be rare, and the technique is limited to features large enough to appear on such data sets.

The three techniques described so far are based on direct processing of the remotely sensed imagery. Atkinson (1997) suggested using super-resolution mapping to attempt to reduce the uncertainty inherent in soft classification output, by mapping the resulting class proportions within the pixels. An assumption of spatial dependence within and between pixels was used, to map the location within each pixel of the proportions output from a soft classification. The assumption proved to be valid for recreating the layout and areal coverage of the land cover. However, the algorithm attempted to cluster similar sub-pixels spatially by comparing sub-pixels to neighbouring pixels, thus mixing scales and producing linear artefacts in the output imagery.

Verhoeye and De Wulf (2001) extended the work of Atkinson (1997), by formulating a solution using linear optimization techniques. The algorithm was applied to synthetic imagery and a SPOT HRV image of Sahelian wetlands. However, since the algorithm also compared sub-pixels to pixels, linear artefacts resulted in the super-resolution map.

This chapter provides an overview of work done by the authors to develop a super-resolution land cover mapping approach that uses the output from a soft classification technique to constrain a Hopfield neural network formulated as an energy minimization tool. By actually mapping the location of land cover class components within each pixel, the technique manages to reduce the uncertainty of predictions produced through soft classification by converting them to single, hard super-resolution predictions. The majority of the remainder of this chapter will focus on the basic mapping approach of the network, but an extension of the technique is also described.

3 Using the Hopfield Network for Land Cover Mapping at the Sub-pixel Scale

For the task of super-resolution land cover mapping, the basic design of the Hopfield neural network described in Tatem *et al.* (2001a) was extended to cope with multiple classes. The class proportion images output from soft classification are represented by h inter-connected layers in the network (where h is the number of land cover classes). Each neuron of largest output value within these layers corresponds to a pixel in the finer spatial resolution map produced after the network has converged. Therefore, neurons will be referred to by co-ordinate notation, for example, neuron (h,i,j) refers to a neuron in row i and column j of the network layer representing land cover class h, and has an input value of u_{hij} and an output value of ν_{hij}. The zoom factor, z, determines the increase in spatial resolution from the original satellite sensor image to the new fine spatial resolution image. After convergence to a stable state, the output values, ν, of all neurons are either 0 or 1, representing a binary classification of the land cover *at the finer spatial resolution*. The specific goals and constraints of the Hopfield neural network energy function determine the final distribution of neuron output values.

The input, u, to neuron with co-ordinates (h,i,j) is made up of a weighted sum of the outputs from every other neuron,

$$u_{hij}(t) = \sum_{a=0}^{N} \sum_{\substack{b=0 \\ b \neq i}}^{pxz} \sum_{\substack{c=0 \\ c \neq j}}^{qxz} w_{abc,\,hij}\, \nu_{abc} \tag{1}$$

where N is the number of land cover classes, $w_{abc,\,hij}$ represents the weight between neuron (h,i,j) and (a,b,c), and ν_{abc} is the output of neuron (a,b,c). The function describing the neural output at time t, $\nu_{hij}(t)$, as a function of the input is,

$$\nu_{hij}(t) = \frac{1}{2}(1 + \tanh u_{hij}\, \lambda) \tag{2}$$

where λ is the gain, which determines the steepness of the function and $\nu_{kij}(t)$ lies in the range [0,1].

The dynamics of the neurons are simulated numerically by the Euler method. After a user-defined time step, dt, the input to neuron (h,i,j), u_{hij}, becomes,

$$u_{hij}(t + dt) = u_{hij}(t) + \frac{du_{hij}(t)}{dt} dt \tag{3}$$

where $\frac{du_{hij}(t)}{dt}$ represents the energy change of neuron (h,i,j) at time t. The total energy change of the network is then,

$$\frac{dE}{d\nu} = -\sum_{h=0}^{N} \sum_{i=0}^{p\times z} \sum_{j=0}^{q\times z} \left| \frac{du_{hij}(t)}{dt} \right| \tag{4}$$

and when this reaches zero, or the change in energy from E_t to E_{t+1} is very small, the network has converged on a stable solution.

The goal and constraints of the sub-pixel mapping task are defined such that the network energy function is,

$$E = -\sum_{h} \sum_{i} \sum_{j} (k_1 G1_{hij} + k_2 G2_{hij} + k_3 P_{hij} + k_4 M_{hij}) \tag{5}$$

For neuron (h,i,j), $G1_{hij}$ and $G2_{hij}$ are the values of spatial clustering (goal) functions, P_{hij} is the value of a proportion constraint, and M_{hij} is the value of a multi-class constraint. The constants k_1, k_2, k_3 and k_4 are used to decide which constraint weightings to apply to solve the problem.

3.1 The goal functions

The spatial clustering (goal) functions, $G1_{hij}$ and $G2_{hij}$, are based upon an assumption of spatial dependence. For neuron (h,i,j), an average output of its neighbours for each land cover class h, $\bar{\nu}_n$, is calculated, which represents the target output for that neuron,

$$\bar{\nu}_n = \frac{1}{8} \sum_{\substack{b=i-1 \\ b\neq i}}^{i+1} \sum_{\substack{c=j-1 \\ c\neq j}}^{j+1} \nu_{bc} \tag{6}$$

The first function aimed to increase the output for each layer, h, of the centre neuron, ν_{ij}, to 1, if the average output of the surrounding eight neurons, $\frac{1}{8} \sum_{\substack{b=i-1 \\ b\neq i}}^{i+1} \sum_{\substack{c=j-1 \\ c\neq j}}^{j+1} \nu_{bc}$, was greater than 0.5,

$$\frac{dG1_{hij}}{d\nu_{hij}} = \frac{1}{2} \left(1 + \tanh \left(\frac{1}{8} \sum_{\substack{b=i-1 \\ b\neq i}}^{i+1} \sum_{\substack{c=j-1 \\ c\neq j}}^{j+1} \nu_{bc} - 0.55 \right) \lambda \right) (\nu_{hij} - 1), \tag{7}$$

where λ is a gain which controls the steepness of the tanh function. The tanh function controls the effect of the neighbouring neurons. If the averaged output of the neighbouring neurons is less than 0.5, then equation (7) evaluates to 0, and the function has no effect on the energy function (equation (5)). If the averaged output is greater than 0.5, equation (7) evaluates to 1, and the $(\nu_{hij} - 1)$ function controls the magnitude of the negative gradient output, with only $\nu_{hij} = 1$ producing a zero gradient. A negative gradient is required to increase neuron output.

The second clustering function aimed to decrease the output for each layer, h, of the centre neuron, ν_{ij}, to 0, given that the average output of the surrounding eight neurons, $\frac{1}{8}\sum_{\substack{b=i-1\\b\neq i}}^{i+1}\sum_{\substack{c=j-1\\c\neq j}}^{j+1}\nu_{bc}$, was less than 0.5,

$$\frac{dG2_{hij}}{d\nu_{hij}} = \frac{1}{2}\left(1 + \left(-\tanh\left(\frac{1}{8}\sum_{\substack{b=i-1\\b\neq i}}^{i+1}\sum_{\substack{c=j-1\\c\neq j}}^{j+1}\nu_{bc}\right)\lambda\right)\right)\nu_{hij}. \tag{8}$$

This time the tanh function evaluates to 0 if the averaged output of the neighbouring neurons is more than 0.5. If it is less than 0.5, the function evaluates to 1, and the centre neuron output, ν_{hij}, controls the magnitude of the positive gradient output, with only $\nu_{hij} = 0$ producing a zero gradient. A positive gradient is required to decrease neuron output and only when $\nu_{hij} = 1$ and $\frac{1}{8}\sum_{\substack{b=i-1\\b\neq i}}^{i+1}\sum_{\substack{c=j-1\\c\neq j}}^{j+1}\nu_{bc} > 0.5$, or $\nu_{hij} = 0$ and $\frac{1}{8}\sum_{\substack{b=i-1\\b\neq i}}^{i+1}\sum_{\substack{c=j-1\\c\neq j}}^{j+1}\nu_{bc} < 0.5$, is the energy gradient equal to zero, and $G1_{hij} + G2_{hij} = 0$. This satisfies the objective of recreating spatial order, while also forcing neuron output to either 1 or 0 to produce a bipolar image.

3.2 The proportion constraint

The proportion constraint, P_{hij}, aims to retain the pixel class proportions output from the soft classification. This is achieved by adding in the constraint that for each land cover class layer, h, the total output from each pixel should be equal to the predicted class proportion for that pixel. An area proportion prediction, $\hat{a}_{hxy}$, is calculated for all the neurons representing pixel (h,x,y),

$$\text{Area Proportion Prediction} = \frac{1}{2z^2}\sum_{b=xz}^{xz+z}\sum_{c=yz}^{yz+z}(1 + \tanh(\nu_{bc} - 0.55)\lambda). \tag{9}$$

The tanh function ensures that if a neuron output is above 0.55, it is counted as having an output of 1 within the prediction of class area per pixel. Below an output of 0.55, the neuron is not counted within the prediction, which simplifies the area proportion prediction procedure, and ensures that neuron output must exceed the random initial assignment output of 0.55 to be counted within the calculations.

To ensure that the class proportions per pixel output from the soft classification were maintained, the proportion target per pixel, a_{hxy}, was subtracted from the area proportion prediction (equation (9)),

$$\frac{dP_{hij}}{d\nu_{hij}} = \frac{1}{2z^2}\sum_{b=xz}^{xz+z}\sum_{c=yz}^{yz+z}(1 + \tanh(\nu_{bc} - 0.55)\lambda) - a_{hxy}. \tag{10}$$

If the area proportion prediction for pixel (h,x,y) is lower than the target area, a negative gradient is produced which corresponds to an increase in neuron output to counteract this problem. An over-prediction of class area results in a positive gradient, producing a decrease in neuron output. Only when the area proportion prediction is identical to the target area proportion for each pixel does a zero gradient occur, corresponding to $P_{hij} = 0$ in the energy function (equation (5)).

3.3 The multi-class constraint

The multi-class constraint, M_{hij}, aims to ensure that the outputs from each class layer fit together with no gaps or overlap between land cover classes in the final prediction map. This is achieved by ensuring that the sum of the outputs of each set of neurons with position (i,j) equals one.

$$\frac{dM_{hij}}{d\nu_{hij}} = \left(\sum_{k=0}^{N} \nu_{hij}\right) - 1 \tag{11}$$

If the sum of the outputs of each set of neurons representing pixel (i,j) in the final prediction image is less than one, a negative gradient is produced which corresponds to an increase in neuron output to counteract this problem. A sum greater than one produces a positive gradient, leading to a decrease in neuron output. Only when the output sum for the set of neurons in question equals one, does a zero gradient occur, corresponding to $M_{hij} = 0$ in the energy function (equation (5)).

4 Simulated Remotely Sensed Imagery

In this section, simulated remotely sensed imagery from two sensors, Landsat TM and SPOT HRV, are used to enable refinement of the technique and clear demonstration of the workings of the network. The use of simulated imagery avoids the uncertainty inherent in real imagery caused by the sensor's PSF, atmospheric and geometric effects and classification error. These effects are explored in Tatem *et al.* (2001b). Figure 6.1 shows the area near Bath, UK, used for this study, the ground or reference data and the simulated satellite sensor images. The corresponding class proportion images are shown in Figures 6.2 and 6.3. All the proportion images were derived by degrading the reference data using a square mean filter, to avoid the potential problems of incorporating error from the process of soft classification. For the simulated Landsat TM imagery, each class within the reference data was degraded to produce pixels with a spatial resolution of 30 m. For the simulated SPOT HRV imagery, the reference data were degraded by a smaller amount to produce pixels of 20 m.

4.1 Network settings

The class proportion images shown in Figures 6.2 and 6.3 provided the inputs to the Hopfield neural network. These proportions were used to initialize the network and

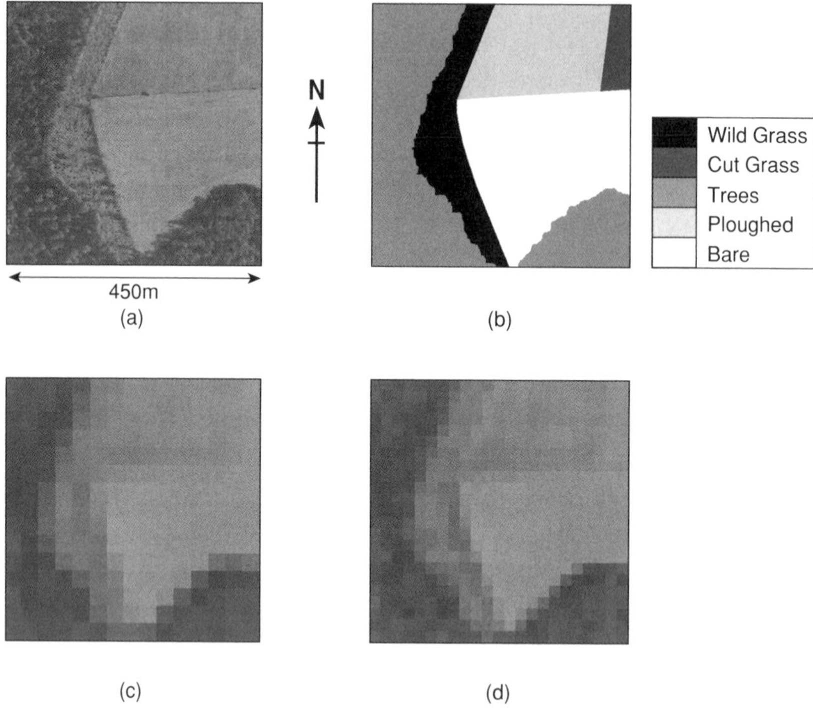

Figure 6.1 *(a) Aerial photograph of an agricultural area near Bath, UK. (b) Reference data map derived from aerial photographs. (c) Image with a spatial resolution equivalent to Landsat TM (30 m). (d) Image with a spatial resolution equivalent to SPOT HRV (20 m)*

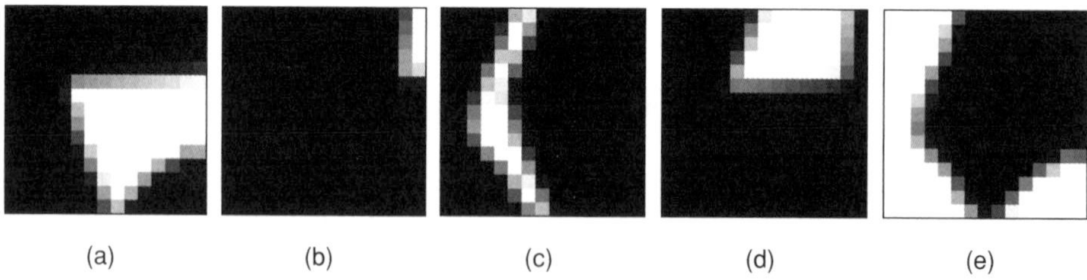

Figure 6.2 *Class area proportions for the simulated Landsat TM imagery: (a) Bare, (b) Cut Grass, (c) Wild Grass, (d) Ploughed, (e) Trees. Pixel grey level represents class area proportion, ranging from white (100%) to black (0%)*

provide area predictions for the proportion constraint. For both sets of proportions, a zoom factor of $z = 7$ was used. This meant that from the simulated Landsat TM input images (with a spatial resolution of 30 m), a prediction map with a spatial resolution of 4.3 m was produced. From the simulated SPOT HRV input images (with a spatial resolution of 20 m), a prediction map with a spatial resolution of 2.9 m was produced. In both cases, the constraint weightings k_1, k_2, k_3 and k_4 were set to 1.0, to ensure that no single function had a dominant effect on the energy function.

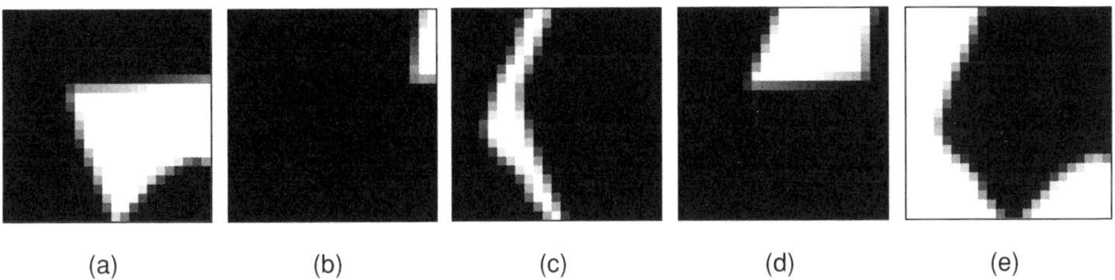

Figure 6.3 *Class area proportions for the simulated SPOT HRV imagery: (a) Bare, (b) Cut Grass, (c) Wild Grass, (d) Ploughed, (e) Trees. Pixel grey level represents class area proportion, ranging from white (100%) to black (0%)*

4.2 Hard classification

To evaluate the success of the Hopfield neural network technique, the traditional method of producing a land cover map from a satellite sensor image was also carried out for comparison. This involved undertaking a 'hard' classification of the imagery using a maximum likelihood classifier (Campbell, 1996). Comparison of the results of this classification to the Hopfield neural network approach should highlight the error and uncertainty issues which arise from using hard classification on imagery where the support is large relative to the land cover features (Atkinson and Foody, this volume, Chapter 1).

4.3 Accuracy assessment

Four measures of accuracy were estimated to assess the difference between each network prediction and the reference images. These were:

(i) Area Error Proportion (AEP)
One of the simplest measures of agreement between a set of known proportions in matrix **y**, and a set of predicted proportions in matrix **a**, is the area error proportion (AEP) per class,

$$AEP_q = \frac{\sum_{i=1}^{n} (y_{iq} - a_{iq})}{\sum_{i=1}^{n} a_{iq}}, \tag{12}$$

where, q is the class and n is the total number of pixels. This statistic informs about bias in the prediction image.

(ii) Correlation Coefficient (CC)
The correlation coefficient, r, measures the amount of association between a target, **y**, and predicted, **a**, set of proportions,

$$r_q = \frac{c_{yq \cdot aq}}{s_{yq} \cdot s_{aq}}, \quad c_{yq \cdot aq} = \frac{\sum_{i=1}^{n} (\bar{y}_{iq} - y_{iq}) \cdot (\hat{a}_{iq} - a_{iq})}{n - 1}, \tag{13}$$

where, $c_{yq \cdot aq}$ is the covariance between **y** and **a** for class q and s_{yq} and s_{aq} are the standard deviations of **y** and **a** for class q. This statistic informs about the prediction variance.

(iii) Closeness (S)
Foody (1996) suggests a measure related to the Euclidean distance between the land cover proportions predicted by the classification, and those of the reference data. This measures the separation of the two data sets, per pixel, based on the relative proportion of each class in the pixel. It is calculated as:

$$S_i = \sum_{q=1}^{c} (y_{iq} - a_{iq})^2 / c \tag{14}$$

where y_{iq} is the proportion of class i in a pixel from reference data, a_{iq} is the measure of the strength of membership to class q, taken to represent the proportion of the class in the pixel from the soft classification, and c is the total number of classes.

(iv) Root Mean Square Error (RMSE)
The root mean square error (RMSE) per class,

$$RMSE_q = \sqrt{\frac{\sum_{i=1}^{n} (y_{iq} - a_{iq})^2}{n}}, \tag{15}$$

informs about the inaccuracy of the prediction (bias and variance).

4.4 Results

Illustrative results were produced using the Hopfield network run on the class proportions derived from the simulated satellite sensor imagery.

4.4.1 Landsat TM

After convergence of the network with a zoom factor of $z = 7$, a prediction image was produced (Figure 6.4(c)) with spatial resolution seven times finer than that of the input class proportions in Figure 6.2. In addition, a maximum likelihood classification was produced for comparison (Figure 6.4(b)). Per-class and overall accuracy

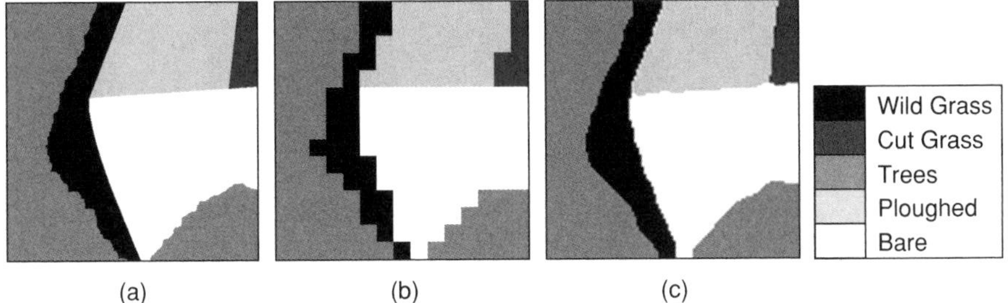

Figure 6.4 Result of super-resolution mapping. (a) Reference image. (b) Hard classification of the simulated Landsat TM imagery (spatial resolution 30 m). (c) Hopfield neural network prediction map (spatial resolution 4.3 m)

Table 6.1 Accuracy assessment results for the Landsat TM hard classification map shown in Figure 6.4(b)

Class	CC	AEP	S	RMSE
Bare	0.934	−0.055	0.029	0.169
Cut Grass	0.871	0.037	0.008	0.089
Wild Grass	0.832	0.058	0.038	0.194
Ploughed	0.917	0.059	0.024	0.153
Trees	0.931	0.002	0.032	0.178
Entire Image		0.0072	0.026	0.161

Table 6.2 Accuracy assessment results for the SPOT HRV hard classification map shown in Figure 6.5(b)

Class	CC	AEP	S	RMSE
Bare	0.954	0.019	0.019	0.139
Cut Grass	0.907	0.023	0.006	0.076
Wild Grass	0.899	−0.011	0.023	0.153
Ploughed	0.956	0.008	0.013	0.113
Trees	0.963	−0.009	0.017	0.131
Entire Image		0.0025	0.016	0.125

statistics were calculated to assess the differences between each map and the reference data (Tables 6.1 and 6.3).

4.4.2 SPOT HRV

After convergence of the network with a zoom factor of $z = 7$, a prediction image was produced (Figure 6.5(c)) with spatial resolution seven times finer than that of the input class proportions in Figure 6.3. In addition, a maximum likelihood classification was produced for comparison (Figure 6.5(b)). Per-class and overall accuracy statistics were calculated to assess the differences between each map and the reference data (Tables 6.2 and 6.4).

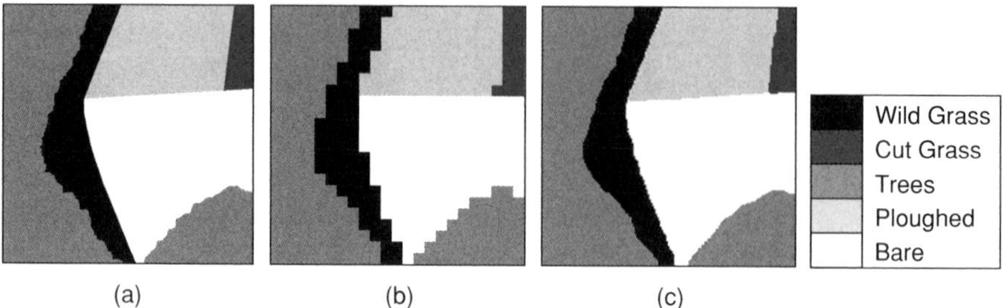

Figure 6.5 Result of super-resolution mapping. (a) Reference image. (b) Hard classification of the simulated SPOT HRV imagery (spatial resolution 30 m). (c) Hopfield neural network prediction map (spatial resolution 2.9 m)

Table 6.3 Accuracy assessment results for the Landsat TM Hopfield neural network prediction map shown in Figure 6.4(c)

Class	CC	AEP	S	RMSE
Bare	0.98	0.007	0.008	0.091
Cut Grass	0.976	0.012	0.002	0.038
Wild Grass	0.96	−0.021	0.009	0.096
Ploughed	0.986	0.001	0.004	0.064
Trees	0.983	−0.002	0.008	0.088
Entire Image		0.0012	0.006	0.079

Table 6.4 Accuracy assessment results for the SPOT HRV Hopfield neural network prediction map shown in Figure 6.5(c)

Class	CC	AEP	S	RMSE
Bare	0.988	−0.001	0.005	0.07
Cut Grass	0.98	0.026	0.001	0.035
Wild Grass	0.968	−0.012	0.007	0.086
Ploughed	0.991	−0.004	0.002	0.051
Trees	0.985	−0.004	0.007	0.083
Entire Image		0.001	0.005	0.069

4.5 Discussion

The high accuracy shown for the results in Figures 6.4(c) and 6.5(c) indicates that the Hopfield neural network has considerable potential for the accurate mapping of land cover class proportions within pixels and, consequently reducing the uncertainty of the soft classification it was derived from. Spatial resolution and accuracy increases over hard classification also demonstrate that the Hopfield neural network predictions have reduced classification uncertainty by mapping class proportions at the sub-pixel scale. While the soft classifications retain more information than hard classification, the resultant predictions still display a large degree of uncertainty. The results in Figures 6.4(c) and 6.5(c) demonstrate that it is possible to map land cover at the sub-pixel scale, thus, reducing the error and uncertainty inherent in maps produced solely by hard or soft classification.

4.5.1 Landsat TM

Visual assessment of Figure 6.4 suggests immediately that the Hopfield neural network has mapped the land cover more accurately than the hard classifier. By forcing each 30×30 m pixel to represent just one land cover class, Figure 6.4(b) appears blocky, and does not recreate the boundaries between classes well. The use of such large classification units, in comparison to the land cover features of interest, resulted in uncertainty as to whether the classification represented correctly the land cover. In contrast, the Hopfield neural network prediction map appeared to have recreated each land cover feature more precisely.

The statistics shown in Tables 6.1 and 6.3 confirm the above visual assessments. For all classes, the Hopfield neural network produced an increase in correlation coefficient over the hard classification, particularly for the two grass classes, with increases of greater than 0.1 in both cases. These increases in precision were mirrored by the closeness and RMSE values, which were less for the Hopfield neural network approach than for the hard classification for all classes, and overall. These statistics indicate that the Hopfield neural network approach produced a reduction in classification uncertainty over the hard classification-derived prediction.

4.5.2 SPOT HRV

Visual assessment of Figure 6.5 suggests that the Hopfield neural network approach was able to map land cover more accurately than the hard classifier. The 20×20 m pixels of the hard classification appeared insufficient to recreate accurately the boundaries between classes, whereas the Hopfield neural network approach appears to have achieved this accurately.

Tables 6.2 and 6.4 confirm the above assessments. Again, the Hopfield neural network approach maintained class area proportions more accurately than the hard classification, as shown by the AEP statistics. The Hopfield neural network approach also recreated each land cover class more precisely with correlation coefficient values of above 0.96 for all classes, and as high as 0.991 for the ploughed class. Such increases were also apparent in the closeness and RMSE values, indicating that the Hopfield neural network reduced the uncertainty shown in the hard-classified prediction.

Comparisons between the two Hopfield neural network predictions in Figures 6.4(c) and 6.5(c) show that, as expected, the prediction map derived from simulated SPOT HRV imagery produced the more accurate results. The fact that the SPOT HRV image had a finer spatial resolution meant that by using the same zoom factor, its prediction map was also of a finer spatial resolution. This enabled the Hopfield neural network to recreate each land cover feature more accurately and with less uncertainty about pixel composition.

Further work on the application of this technique to simulated and real Landsat TM imagery can be found in Tatem *et al.* (2002a) and Tatem *et al.* (2001b), respectively.

5 Super-resolution Land Cover Pattern Prediction

The focus of each of the techniques described so far on land cover features larger than a pixel (e.g. agricultural fields), enables the utilization of information contained in surrounding pixels to produce super-resolution maps and, consequently, reduce uncertainty. However, this source of information is unavailable when examining imagery of land cover features that are smaller than a pixel (e.g. trees in a forest). Consequently, while these features can be detected within a pixel by soft classification techniques, surrounding pixels hold no information for inference of sub-pixel class location. Attempting to map such features is, therefore, an under-constrained task fraught with uncertainty. This section describes briefly a novel and effective approach introduced in Tatem *et al.* (2001c, 2002b), which forms an extension to the Hopfield neural network approach described previously. The technique is based on prior information on the spatial arrangement of land cover. Wherever prior information is available in a land cover classification situation, it should be utilized to further constrain the problem at hand, therefore, reducing uncertainty. This section will show that the utilization of prior information can aid the derivation of realistic prediction maps. For this, a set of functions to match land cover distribution within each pixel to this prior information was built into the Hopfield neural network.

5.1 The new energy function

The goal and constraints of the sub-pixel mapping task were defined such that the network energy function was,

$$E = -\sum_h \sum_i \sum_j \left(\left(\sum_{n=1}^{z} k_n Sn_{hij} \right) + k_{z+1} G1_{hij} + k_{z+2} G2_{hij} + k_{z+3} P_{hij} + k_{z+4} M_{hij} \right) \tag{16}$$

where Sn_{hij} represents the output values for neuron (h,i,j) of the z semivariance functions.

5.2 The semivariance functions

The semivariance functions aim to model the spatial pattern of each land cover at the sub-pixel scale. Prior knowledge about the spatial arrangement of the land cover in question is utilized, in the form of semivariance values, calculated by:

$$\gamma(l) = \frac{1}{2N(l)} \sum_{i=1, j=1}^{N(l)} [\nu_{hij} - \nu_{h,\ i \pm l,\ j \pm l}]^2 \tag{17}$$

where $\gamma(l)$ is the semivariance at lag l, and $N(l)$ is the number of pixels at lag l from the centre pixel (h,i,j). The semivariance is calculated for n lags, where $n =$ the zoom factor, z, of the network, from, for example, an aerial photograph. This provides information on the typical spatial distribution of the land cover under study, which can then be used to reduce uncertainty in land cover simulation from remotely sensed imagery at the sub-pixel scale. By using the values of $\gamma(l)$ from equation (17), the output of the centre neuron, $\nu(c)_{ij}$, which produces a semivariance of $\gamma(l)$ can be calculated using:

$$\nu(c)_{hij} = \frac{1}{2a}\left[-b \pm \sqrt{b^2 - 4ac}\right] \tag{18}$$

where $a = 2N(l)$, $b = \sum_{i=1,j=1}^{N(l)} \nu_{h,\ i\pm l,\ j\pm l}$, $c = \sum_{i=1,j=1}^{N(l)} (\nu_{h,\ i\pm l,\ j\pm l})^2 - 2N(l)\gamma(l)$. The semivariance function value for lag 1 ($l = 1$), $S1_{hij}$, is then given by:

$$\frac{dS1_{hij}}{d\nu_{hij}} = \nu_{hij} - \nu(c)_{hij} \tag{19}$$

If the output of neuron (h,i,j), ν_{hij}, is lower than the target value, $\nu(c)_{hij}$, calculated in equation (18), a negative gradient is produced that corresponds to an increase in neuron output to counteract this problem. An overestimation of neuron output results in a positive gradient, producing a decrease in neuron output. Only when the neuron output is identical to the target output, does a zero gradient occur, corresponding to $S1_{hij} = 0$ in the energy function (equation (16)). The same calculations are carried out for lags 2 to z.

5.3 Results

The new network set up was tested on simulated SPOT HRV imagery. Figure 6.6(a) shows an aerial photograph of the chosen test area, which contained both a large area of woodland (feature larger than the image support), and lone trees (smaller than the image support) amongst grassland. The reference image in Figure 6.6(c) was degraded, using a square mean filter, to produce three class proportion images that

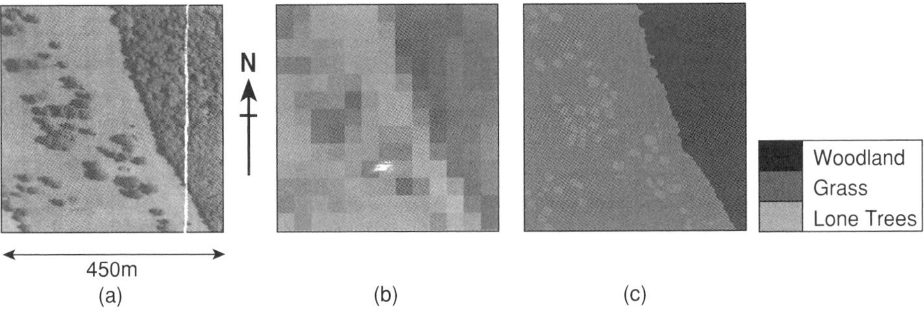

Figure 6.6 Construction of simulated SPOT HRV image. (a) Aerial photograph of an agricultural area near Bath, UK. (b) Image with a spatial resolution equivalent to SPOT HRV (20 m). (c) Reference image derived from aerial photographs (reproduced, with permission, from Tatem et al., 'Super-resolution mapping of multiple scale land cover features using a Hopfield neural network', Proceedings of the International Geoscience and Remote Sensing Symposium, Sydney) ©2001 IEEE.

Table 6.5 Accuracy assessment results for the hard classification map shown in Figure 6.7(b) (reproduced, with permission, from Tatem et al., 'Super-resolution mapping of multiple scale land cover features using a Hopfield neural network,' Proceedings of the International Geoscience and Remote Sensing Symposium, Sydney) © 2001 IEEE

Class	CC	AEP	S	RMSE
Lone Trees	N/A	1.0	0.0627	0.251
Woodland	0.887	0.0095	0.0476	0.218
Grass	0.761	−0.101	0.11	0.331
Entire Image		0.00434	0.073	0.271

Table 6.6 Accuracy assessment results for the Hopfield neural network prediction map shown in Figure 6.7(c) (reproduced, with permission, from Tatem et al., 'Super-resolution mapping of multiple scale land cover features using a Hopfield neural network,' Proceedings of the International Geoscience and Remote Sensing Symposium, Sydney) © 2001 IEEE

Class	CC	AEP	S	RMSE
Lone Trees	0.43	−0.161	0.0721	0.269
Woodland	0.985	0.0085	0.0066	0.0809
Grass	0.831	0.0136	0.0786	0.28
Entire Image		0.00711	0.052	0.229

provided input to the Hopfield neural network. In addition, a variogram (Figure 6.8(a)) was calculated from a small section of Figure 6.6(c) to provide the prior spatial information on the lone tree class, required by the SV function. After convergence of the network with $z = 7$, the map shown in Figure 6.7(c) was produced, and a traditional hard classification was undertaken for comparison. Both images were compared to the reference data, and accuracy statistics calculated. These included correlation coefficients between classes, area error proportion, closeness and RMSE, all shown in Tables 6.5 and 6.6.

5.4 Discussion

The results show clearly that the super-resolution technique provided an increase in mapping accuracy over that derived with a traditional hard classification. Visual inspection of Figure 6.7 revealed that the hard classification failed to identify the lone tree class, and produced an uneven woodland boundary. The hard classifier has attempted to minimize the probability of mis-classification and, as each lone tree covers less than half a pixel, no pixels were assigned to this class. In contrast, the Hopfield network prediction appeared to have identified and mapped both features correctly. This was confirmed after inspection of the accuracy statistics and variograms. While there was little difference between the AEP values in Tables 6.5 and 6.6, showing that both techniques maintained class area to a similar degree, the other statistics show how accurate the Hopfield network was. The woodland class was mapped accurately, with a correlation coefficient of 0.985, compared to just 0.887 using hard classification.

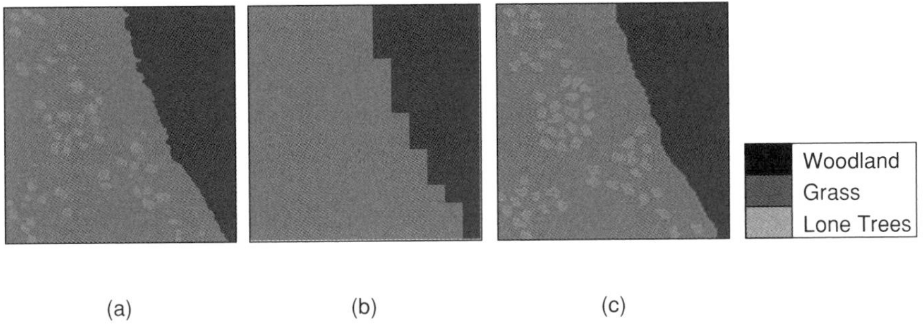

Figure 6.7 Result of super-resolution mapping. (a) Reference image. (b) Hard classification of the image shown in Figure 6.6(b) (spatial resolution 20 m). (c) Hopfield neural network prediction map (spatial resolution 2.9 m) (reproduced, with permission, from Tatem et al., 'Super-resolution mapping of multiple scale land cover features using a Hopfield neural network', Proceedings of the International Geoscience and Remote Sensing Symposium, Sydney) ©2001 IEEE.

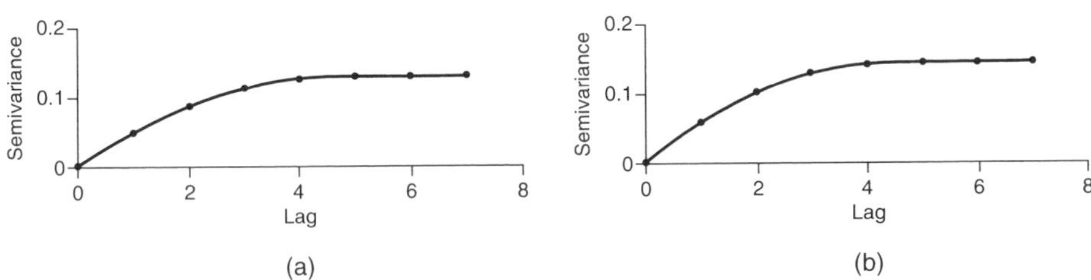

Figure 6.8 (a) Variogram of lone tree class in reference image (Figure 6.6(c)). (b) Variogram of lone tree class in Hopfield network prediction image (Figure 6.7(c)) (reproduced, with permission, from Tatem et al., 'Super-resolution mapping of multiple scale land cover features using a Hopfield neural network', Proceedings of the International Geoscience and Remote Sensing Symposium, Sydney) ©2001 IEEE.

Overall image results also show an increase in accuracy, with closeness and RMSE values of just 0.052 and 0.229, respectively. The only low accuracy was for the lone tree class, with a correlation coefficient of just 0.43. However, as the aim of the SV function is to recreate the spatial arrangement of sub-pixel scale features, rather than accurately map their locations, this was expected. The most appropriate way to test the performance of this function was, therefore, to compare the shape of the variograms, which confirms that a similar spatial arrangement of trees to that of the reference image had been derived (Figure 6.8). Such results demonstrate the importance of utilizing prior information to constrain inherently uncertain problems in order to increase the accuracy and reduce the uncertainty in predictions.

6 Summary

Accurate land cover maps with low levels of uncertainty are required for both scientific research and management. Remote sensing has the potential to provide the information needed to produce these maps. Traditionally, hard classification has been used to assign each pixel in a remotely sensed image to a single class to produce a thematic map. However, where mixed pixels are present, hard classification has been shown to be inappropriate, inaccurate and the source of uncertainty.

Soft classification techniques have been developed to predict class composition of image pixels, but their output are images of great uncertainty that provide no indication of how these classes are distributed spatially within each pixel. This chapter has described techniques for locating these class proportions and, consequently, reducing this uncertainty to produce super-resolution maps from remotely sensed imagery.

Pixels of mixed land cover composition have long been a source of inaccuracy and uncertainty within land cover classification from remotely sensed imagery. By attempting to match the spatial resolution of land cover classification predictions to the spatial frequency of the features of interest, it has been shown that this source of uncertainty can be reduced. Additionally, extending such approaches to include extra sources of prior information can further reduce uncertainty to produce accurate land cover predictions. This chapter demonstrated advancement in the field of land cover classification from remotely sensed imagery and it is hoped that it stimulates further research to continue this advancement.

Acknowledgements

This work was supported by an EPSRC studentship awarded to AJT (98321498). Imagery was provided by the E.U. FLIERS Project (ENV4-CT96-0305).

References

Aplin, P., 1999, Fine spatial resolution satellite sensor imagery for per-field land cover classification, Unpublished PhD Thesis (Southampton: University of Southampton).

Aplin, P., 2000, Comparison of simulated IKONOS and SPOT HRV imagery for land cover classification, *Proceedings of the 26th Annual Conference of the Remote Sensing Society* (Nottingham: Remote Sensing Society), CD-ROM.

Aplin, P., Atkinson, P. M. and Curran, P. J., 1999, Fine spatial resolution simulated satellite imagery for land cover mapping in the United Kingdom, *Remote Sensing of Environment*, **68**, 206–16.

Atkinson, P. M., 1997, Mapping sub-pixel boundaries from remotely sensed images, in Z. Kemp (ed.), *Innovations in G.I.S. 4* (London: Taylor and Francis), pp. 166–80.

Atkinson, P. M., Cutler, M. E. J. and Lewis, H. G., 1997, Mapping sub-pixel proportional land cover with A.V.H.R.R. imagery, *International Journal of Remote Sensing*, **18**, 917–35.

Bastin, L., 1997, Comparison of fuzzy c-means classification, linear mixture modelling and M.L.C. probabilities as tools for unmixing coarse pixels, *International Journal of Remote Sensing*, **18**, 3629–48.

Boult, T. E and Wolberg, G., 1993, Local image reconstruction and subpixel restoration algorithms, *CVGIP: Graphical Models and Image Processing*, **55**, 63–77.

Brown, M., Gunn, S. R. and Lewis, H. G., 1999, Support vector machines for optimal classification and spectral unmixing, *Ecological Modelling*, **120**, 167–79.

Campbell, J. B., 1996, *Introduction to Remote Sensing*, 2nd edition (New York: Taylor and Francis).

Cihlar, J., Latifovic, R., Chen, J., Beaubien, J., Li, Z. and Magnussen, S., 2000, Selecting representative high resolution sample images for land cover studies. Part 2: application to estimating land cover composition, *Remote Sensing of Environment*, **72**, 127–38.

Flack, J., Gahegan, M. and West, G., 1994, The use of sub-pixel measures to improve the classification of remotely sensed imagery of agricultural land, *Proceedings of the 7th Australasian Remote Sensing Conference* (Melbourne: Floeat Park) pp. 531–41.

Foody, G. M., 1992, A fuzzy sets approach to the representation of vegetation continua from remotely sensed data: an example from lowland heath, *Photogrammetric Engineering and Remote Sensing*, **58**, 221–5.

Foody, G. M., 1995, Cross-entropy for the evaluation of the accuracy of a fuzzy land cover classification with fuzzy ground data, *ISPRS Journal of Photogrammetry and Remote Sensing*, **61**, 2–12.

Foody, G. M., 1996, Approaches for the production and evaluation of fuzzy land cover classifications from remotely sensed data, *International Journal of Remote Sensing*, **17**, 1317–40.

Foody, G. M., 1998, Sharpening fuzzy classification output to refine the representation of sub-pixel land cover distribution, *International Journal of Remote Sensing*, **19**, 2593–9.

Garcia-Haro, F. J., Gilabert, M. A. and Melia, J., 1996, Linear spectral mixture modelling to estimate vegetation amount from optical spectral data, *International Journal of Remote Sensing*, **17**, 3373–400.

Gross, H. N. and Schott, J. R., 1998, Application of spectral mixture analysis and image fusion techniques for image sharpening, *Remote Sensing of Environment*, **63**, 85–94.

Horowitz, H. M., Nalepka, R. F., Hyde, P. D. and Morgenstern, J. P., 1971, Estimating the proportions of objects within a single resolution element of a multispectral scanner, *Proceedings of the 7th International Symposium on Remote Sensing of the Environment* (Michigan: Ann Arbor) pp. 1307–20.

Leavers, V. F., 1993, Which Hough transform, *Computer Vision and Graphical Image Processing – Image Understanding*, **58**, 250–64.

Lewis, H. G., Nixon, M. S., Tatnall, A. R. L. and Brown, M., 1999, Appropriate strategies for mapping land cover from satellite imagery, In *Proceedings of the 25th Remote Sensing Society Conference* (Nottingham: Remote Sensing Society), pp. 717–24.

Li, X., Kudo, M., Toyama, J. and Shimbo, M., 1998, Knowledge-based enhancement of low spatial resolution images, *IEICE Transactions on Information and Systems*, **81**, 457–63.

Manslow, J. and Nixon, M., 2000, On the representation of fuzzy land cover classifications, *Proceedings of the 26th Annual Conference of the Remote Sensing Society* (Nottingham: Remote Sensing Society).

Robinson, G. D., Gross, H. N. and Schott, J. R., 2000, Evaluation of two applications of spectral mixing models to image fusion, *Remote Sensing of Environment*, **71**, 272–81.

Schneider, W., 1993, Land use mapping with subpixel accuracy from Landsat TM image data, *Proceedings of the 25th International Symposium on Remote Sensing and Global Environmental Change* (Michigan: Ann Arbor), pp. 155–61.

Schneider, W., 1999, Land cover mapping from optical satellite images employing subpixel segmentation and radiometric calibration, in I. Kanellopoulos, G. Wilkinson, and T. Moons (eds), *Machine Vision and Advanced Image Processing in Remote Sensing* (London: Springer), pp. 229–37.

Schowengerdt, R. A., 1996, On the estimation of spatial-spectral mixing with classifier likelihood functions, *Pattern Recognition Letters*, **17**, 1379–87.

Schowengerdt, R. A., 1997, *Remote Sensing: Models and Methods for Image Processing*, 2nd edition (San Diego: Academic Press).

Steinwendner, J., 1999, From satellite images to scene description using advanced image processing techniques, *Proceedings of the 25th Remote Sensing Society Conference* (Nottingham: Remote Sensing Society), pp. 865–72.

Steinwendner, J. and Schneider, W., 1997, A neural net approach to spatial subpixel analysis in remote sensing, in W. Burger and M. Burge (eds), *Pattern Recognition 1997, Proceedings 21st OAGM Workshop* (Halstatt, Austria: OCG-Shriftenreihe).

Steinwendner, J. and Schneider, W., 1998, Algorithmic improvements in spatial subpixel analysis of remote sensing images, *Proceedings of the 22nd Workshop of the Austrian Association of Pattern Recognition* (Illmitz: OAGM), pp. 205–13.

Steinwendner, J., Schneider, W. and Suppan, F., 1998, Vector segmentation using spatial subpixel analysis for object extraction, *International Archives of Photogrammetry and Remote Sensing*, **32**, 265–71.

Tatem, A. J., Lewis, H. G., Atkinson, P. M. and Nixon, M. S., 2001a, Super-resolution target identification from remotely sensed images using a Hopfield neural network, *IEEE Transactions on Geoscience and Remote Sensing*, **39**, 781–96.

Tatem, A. J., Lewis, H. G., Atkinson, P. M. and Nixon, M. S., 2001b, Increasing the spatial resolution of satellite sensor imagery for agricultural land cover mapping, *Proceedings of the 1st Annual Meeting of the Remote Sensing and Photogrammetry Society* (Nottingham: Remote Sensing and Photogrammetry Society), CD-ROM.

Tatem, A. J., Lewis, H. G., Atkinson, P. M. and Nixon, M. S., 2001c, Super-resolution mapping of multiple scale land cover features using a Hopfield neural network, *Proceedings of the International Geoscience and Remote Sensing Symposium*, IEEE, Sydney, CD-ROM.

Tatem, A. J., Lewis, H. G., Atkinson, P. M. and Nixon, M. S., 2002a, Land cover mapping at the sub-pixel scale using a Hopfield neural network, *International Journal of Applied Earth Observation and Geoinformation*, **3**, 184–90.

Tatem, A. J., Lewis, H. G., Atkinson, P. M. and Nixon, M. S., 2002b, Super-resolution land cover pattern prediction using a Hopfield neural network, *Remote Sensing of Environment*, **79**, 1–14.

Verhoeye, J. and De Wulf, R., 2001, Land cover mapping at the sub-pixel scale using linear optimisation techniques, *Remote Sensing of Environment*, **79**, 96–104.

Wang, F., 1990, Improving remote sensing image analysis through fuzzy information representation, *Photogrammetric Engineering and Remote Sensing*, **56**, 1163–9.

7

Uncertainty in Land Cover Mapping from Remotely Sensed Data using Textural Algorithm and Artificial Neural Networks

Anna M. Jakomulska[†] *and Jan P. Radomski*

1 Introduction

Textural information is increasingly recognized as being useful in the classification of remotely sensed imagery. Many algorithms for the quantification of texture have been proposed, such as second-order statistics derived from the grey level co-occurrence matrix (Haralick *et al.*, 1973) and geostatistical measures of spatial variability (Miranda *et al.*, 1992, 1996; Ramstein and Raffy, 1989) and most are derived from a sliding window area passed over the imagery. Although combining spectral with textural information generally results in an increase in classification accuracy over that derived from a standard spectral classification, textural information has been used relatively rarely. A major reason for this situation is that there is no general rule as to which textural measures should be used and perhaps more importantly, what size of a sliding window should be chosen for their derivation. Most applications, therefore, have significant shortcomings with texture derived from a moving kernel of a fixed size, where window size is chosen experimentally in the pre-processing phase and may be appropriate for only the particular data used in a specific study.

In this chapter, issues connected with the uncertainty in land cover mapping from remotely sensed data are addressed. The main focus is on evaluating the potential of

[†]deceased

Uncertainty in Remote Sensing and GIS. Edited by G.M. Foody and P.M. Atkinson.

a variogram-based algorithm for the quantification of image texture. Focusing on the optimal use of variogram-derived texture measures, two sets of variables were used. Firstly, a number of texture bands derived within a kernel of an adapting size (Franklin and McDermid, 1993; Franklin *et al.*, 1996) was used in image classification. Secondly, the size of an adapting window, being a measure of both texture and border proximity, was used as an additional input to the classification. Additionally, attention was focused on areas located in the proximity of boundaries between classes which are particularly problematical in image classification, particularly when using textural measures. The research used medium spatial resolution optical (Landsat-5 TM) and microwave (ERS-1 SAR) imagery of an area characterized by a complex mosaic of small land cover patches. Due to the high density of boundaries in the study area, the size of the kernel for texture quantification was of particular importance in terms of the accuracy of the resulting classification. Finally, the uncertainty in class allocation resulting from the application of a neural network classifier to the derived data is discussed in this chapter. Two criteria, the magnitude of class membership strength and the difference in memberships observed for a pixel (Gopal and Woodcock, 1994), were used to discriminate between pixels allocated to a class with a high degree of confidence, those for which an uncertain allocation was made and those for which it may be most appropriate to leave unclassified.

1.1 The variogram and its applications in image classification

The variogram, is a function of variance with distance that describes the spatial dependence in a data set (Figure 7.1). Generally observations that are close in space are more alike than those that are far apart. In the bounded model of the variogram, semivariance reaches its maximum (sill) at a distance over which the data are correlated (range). The nugget of the variogram indicates the level of spatially independent noise present.

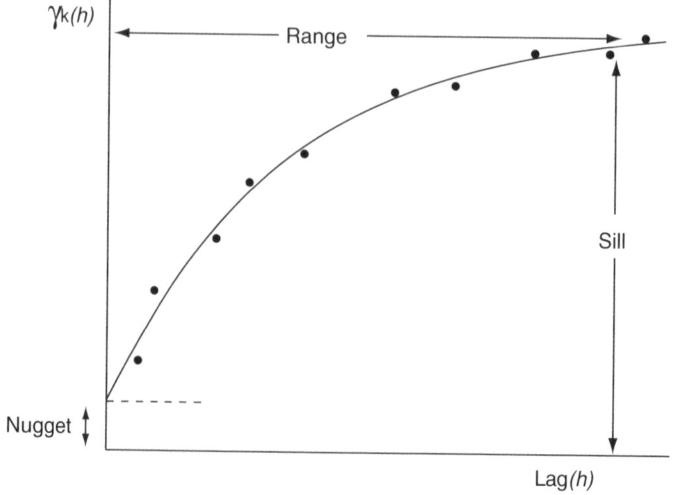

Figure 7.1 *Variogram and its characteristic features: sill, nugget and range*

The experimental variogram can be described mathematically by equation (1) (Cressie, 1993):

$$\gamma_k(h) = \frac{1}{2n(h)} \sum_{i=1}^{n(h)} \{dn_k(x_i) - dn_k(x_i + h)\}^2 \qquad (1)$$

where $n(h)$ is the number of pairs in lag h, $dn()$ represents the image digital number values at location x_i and $x_i + h$; and k denotes a waveband.

Multivariate variograms (Journel and Huijbregts, 1978; Wackernagel, 1995) measure the joint spatial variability between two wavebands (the cross variogram) equation (2):

$$\gamma_{jk}(h) = \frac{1}{2n(h)} \sum_{i=1}^{n(h)} \{[dn_j(x_i) - dn_j(x_i + h)] \times [dn_k(x_i) - dn_k(x_i + h)]\} \qquad (2)$$

and the variance of cross-differences of two wavebands (the pseudo-cross variogram) equation (3):

$$\gamma_{jk}(h) = \frac{1}{2n(h)} \sum_{i=1}^{n(h)} \{dn_j(x_i) - dn_k(x_i + h)\}^2 \qquad (3)$$

where j and k indicate two radiometric wavebands (e.g. spectral wavebands or, as in this study, two principal components images derived from the remotely sensed data).

Geostatistics, and the variogram in particular, are currently well understood and frequently applied in image analysis. It has been shown (Woodcock *et al.*, 1988a, 1988b) that the variogram's range is directly related to texture and/or object size, while its sill is proportional to the global object (class) variance, although it is also affected by external factors, such as sensor gain, image noise, atmospheric conditions etc.

Recently there has been increasing interest in the potential applications of variogram-derived texture measures for use in image classification. Two approaches to the derivation of texture from the variogram have been suggested and used to increase image classification accuracy: the semivariogram textural classifier algorithm (STC, Miranda *et al.*, 1992, 1996) and methods based on the coefficients of the variogram model (Ramstein and Raffy, 1989). In STC, semivariances for the first consecutive lags are used directly as additional input layers to an image classification, while in the second technique, variogram model coefficients are used. However, the general application of both techniques has an important shortcoming in that the computation of the variogram is typically based on a moving window area or kernel of a fixed size.

There is a trade-off between the application of a window that is too large, and one that is too small. The first approach leads to the window straddling class boundaries, and so encompassing different classes in adjacent areas in the same variogram. On the other hand, the sill of a variogram may be not achieved if the calculations are undertaken within a small window. Furthermore, variograms for short lags measure field edges rather than within-field variability that is important texturally (Jakomulska and Stawiecka, 2002). Franklin and McDermid (1993) and Franklin *et al.* (1996) proposed an alternative technique, a geographic window in which the dimensions of a moving kernel are customized and are determined by the variogram's range. The range of a variogram is short for homogeneous objects (e.g. water) and increases for more textured objects. Furthermore, the range contracts at object boundaries while the variance increases very rapidly due to the fact that the window straddles the boundary separating adjacent areas of different classes. These features allow the moving window to remain small when the edge of an object is approached. The geographic window based technique has proved to be promising in the derivation of textural measures (Wulder *et al.*, 1997).

Unfortunately, no consensus has been reached on which textural measures are appropriate to achieve the greatest accuracy in image classification. Although Ramstein and Raffy (1989) showed that some land cover classes can be well differentiated using only the range of the variogram (assuming an exponential variogram model), other reports show that both the sill and range are distinctive for different land cover types (St-Onge and Cavayas, 1995; Wallace *et al.*, 2000; Jakomulska and Clarke, 2001). The range of variogram indicates the coarseness of texture and has been shown to be an important discriminant of a variety of land cover classes. However, the variogram's range is also directly related to the size of objects (and boundary proximity): for small patches (in landscapes characterized by high spatial frequency) the range of a variogram is shorter than for large patches. Hence, it seems inappropriate to use a constant value.

Increases in classification accuracy through the use of textural information have also generally been observed in studies with relatively ideal conditions. For example, studies have used fine spatial resolution data in which the high variance of objects results from the amount of detail present in the image (Wharton, 1982; Wulder *et al.*, 1996, 1998), microwave imagery in which the textural analysis reduces speckle noise and works like a low pass filter (Miranda *et al.*, 1992, 1996), and per-field image classification in which the problematical boundaries are removed (Berberoglu *et al.*, 2000).

In this study, variogram-based measures (sills, ranges, mean semivariances, sum of semivariances up to a range and semivariances at lag = 0) were derived from a kernel of an adapting size, as proposed by Franklin and McDermid (1993). To analyse the 'border effect' of spatial classification, the accuracy of classification based on textural information derived from a moving window was assessed. Later also the range (the window size itself) was used as an additional textural measure. Due to the non-Gaussian distribution of textural data an artificial neural network classifier, which does not make assumptions about the data's distribution was used (Benediktsson *et al.*, 1990; Atkinson and Tatnall, 1997). To gain an understanding of utility and limitations of the approach, attention focuses on the overall accuracy of the classification as well as the spatial distribution of error.

2 Study Area and Data

The study area was the Kolno Upland, located in north-eastern Poland (between 21°70′ and 22°60′E and 53°20′ and 53°60′N). The landscape was agricultural, with arable lands prevailing on the upland and meadows along valley floors. The arable lands were characterized by a complicated pattern of ownership that resulted in a mosaic of small parcels (on average 5 ha). Coniferous and deciduous forests covered a few per cent of the area. Along the valley bottoms, sparsely timbered deciduous forests were common. Urbanized areas were rare, but dispersed households and small settlements were frequent in the region.

A Landsat-5 Thematic Mapper (TM) image acquired on 25 August 1987 (Figure 7.2; see also plate 1) and an ERS-1 SAR image acquired on 31 August 1994 were used. Colour aerial images at a 1:26 000 scale, acquired on 23 August 1995, were used to guide the training and accuracy assessment stages of the classification. Due to a lack of data in the same time-frame, the available data were used in spite of the large time gap between the image acquisitions. However, this approach is believed to be justifiable for two main reasons. Firstly, due to the low agricultural productivity and relative economic stagnation of the region, no major changes in land use and land ownership had occurred in the area. The main differences that occurred in the time between image acquisitions were differences in the crop mosaic which were unimportant as it was not intended to identify individual crops, but simply arable lands. Secondly, since all images were acquired at the end of August, after harvest, the arable lands consisted of exposed bare soils with a varied tillage pattern. The following land cover classes were therefore analysed: built-up areas, coniferous forests, deciduous forests, meadows and arable lands.

Pre-processing of the data was limited to geometric and radiometric correction of the data. Speckle noise removal was a necessary pre-processing step in the analysis of the microwave imagery to reduce the random variation present. To reduce the speckle present in the SAR image, a Gamma-Map (Maximum *a Posteriori*) filter (3×3 window) was applied to it (Frost *et al.*, 1982). The filter is based on a multiplicative noise model with non-stationary mean and variance and assumes Gamma distribution for the speckle noise. The ERS-1 SAR image was resampled to the pixel size of Landsat TM and compressed to 8-bit radiometric resolution (using a two standard deviations stretch). Similarly, ground data, acquired through photo-interpretation were converted to a raster format and resampled to the spatial resolution of the non-thermal Landsat TM data. All of the image data were georeferenced to a standard reference system using nearest neighbour resampling and a second-order transformation equation (RMSE = 0.34 for Landsat TM and RMSE = 0.47 for ERS-1 SAR data).

The pre-processing of the Landsat TM image did not influence the textural information content of the data (since nearest neighbour resampling was employed). On the other hand, the ERS-1 SAR data, were subjected to three pre-processing processes (filtering, resampling, rescaling), which, although necessary, might degrade the data's textural information content (via smoothing).

2.1 Spectral and textural characteristics of the studied land cover classes

The Landsat TM sensor was specifically designed for studies such as land cover mapping. However, in spite of a small number of classes analysed in this study, the land cover classes do not exhibit perfect separability (Figure 7.3). The within-class variance was typically very large, with considerable overlap of the ranges of the

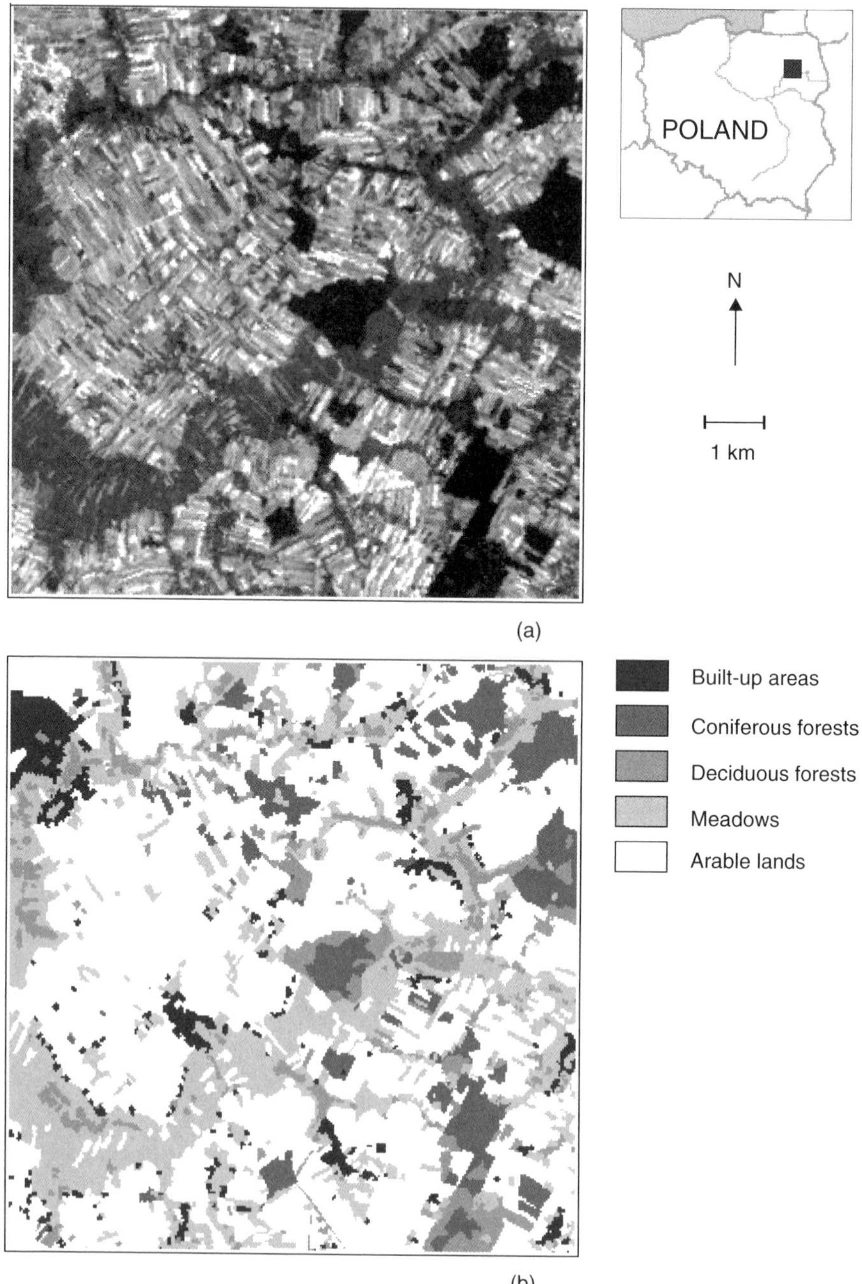

Figure 7.2 Test site. (a) Central part of Kolno Upland: red band of Landsat TM. (b) Ground data achieved through visual interpretation of aerial images.

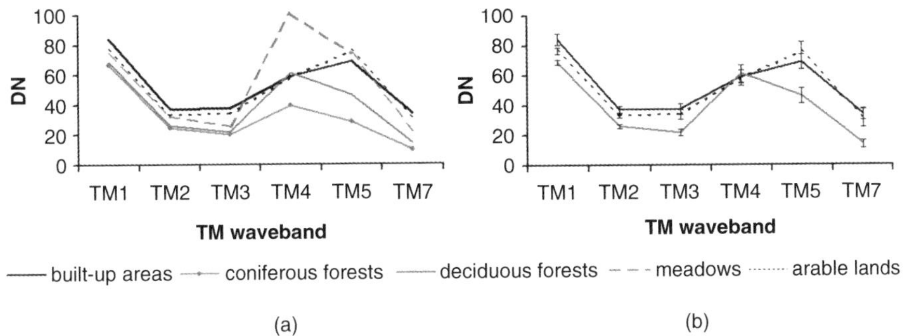

Figure 7.3 Spectral separability of the classes. (a) Spectral responses. (b) Within-class variation for three classes. Vertical bars mark standard deviation

observed spectral responses, particularly for the built-up areas and arable lands classes. The large within-class variance of the built-up class resulted from the varied land cover found within this class. Arable lands were highly varied due to the complex mosaic of small land parcels found at the site.

Spatial information is of particular benefit for mapping classes characterized by large within-class variance (even if the within-class variance is larger than the inter-class variance). For example, variograms derived from training samples show that built-up areas can be easily distinguished from arable lands using sills of variograms derived from the red waveband (Figure 7.4).

2.2 Spatial distribution of the studied land cover classes

The complexity of the boundaries in the study area was analysed from the land cover representation produced from the photointerpretation (ground data). Data layers with buffers (corridors) along boundaries were constructed using 3×3, 5×5 and

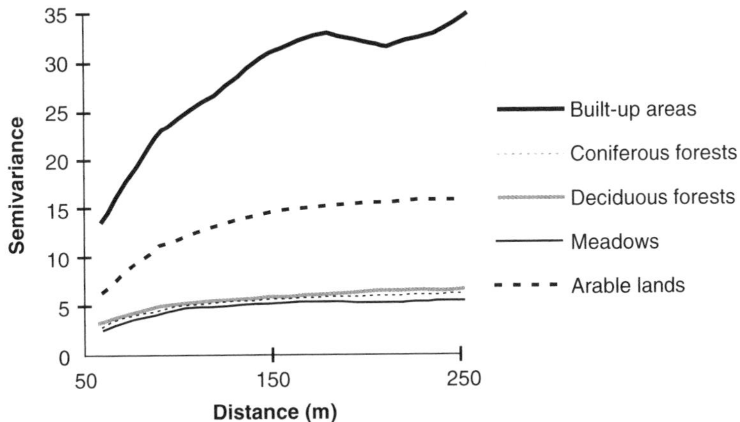

Figure 7.4 Spatial 'signatures' or responses. Variograms derived from red band (TM3)

7 × 7 windows. Nearly half of the pixels were located in the nearest proximity of borders (3 × 3 window) and as much as 77% of the data lay within a 7 × 7 boundary buffer (Figure 7.5). Table 7.1 presents the distribution of pixels in the proximity of boundaries for each class: deciduous forests form the smallest and most elongated patches, while patches of arable lands were generally the largest and most compact. The urban class was also characterized by a high degree of patchiness.

3 Methods

3.1 Computation of variogram-derived texture measures

To explore the capabilities of variogram-derived measures both univariate and multivariate estimators of the variogram were used (Cressie, 1993; Journel and Huijbregts, 1978; Wackernagel, 1995). Experimental univariate variograms were calculated for each pixel in each of the six non-thermal Landsat TM wavebands and the ERS-1 SAR image. Multivariate variograms (cross variograms and pseudo-cross variograms) were calculated between pairs of the first three principal components derived from the six non-thermal Landsat TM wavebands. A principal components analysis (PCA) was undertaken to reduce the number of possible wave-band combinations (from 15 to 3), while retaining as much as 98.7% of the original image variance. A bounded variogram model was assumed: if a range was not achieved during the calculation, it was set at the maximum allowable lag (lag = 34, corresponding to approximately 1 km on the ground). The maximum value of the range was, however, rarely reached, with a mean value below 200 m (5–7 pixels). Ranges calculated from data acquired in the infrared wavebands were slightly larger than those derived from those in the visible wavebands. The shortest ranges were noted for the ERS-1 SAR image (below 100 m). This suggested that the textural information content of the SAR image had not been lost during the pre-processing stage.

A large set of indices (a total of 89) was calculated for each central pixel within a moving window of a changeable size (Jakomulska and Clarke, 2001; Jakomulska and Stawiecka, 2002) from both univariate and multivariate variograms. These included sills and ranges, mean and sum of semivariances calculated up to a range, slope of the variogram, ratios of semivariances for consecutive pairs of lags, sum of absolute differences between variogram model (spherical, exponential and Gaussian) and experimental variogram model, ratios of sills for band pairs, and measures of variogram shape.

Table 7.1 Percentage of pixels within proximity of a border

Land cover class	3 × 3 window	5 × 5 window	7 × 7 window
Built-up areas	66	85	91
Coniferous forests	44	68	80
Deciduous forests	72	93	99
Meadows	59	83	91
Arable lands	30	50	65
Average	43.4	64.8	76.8

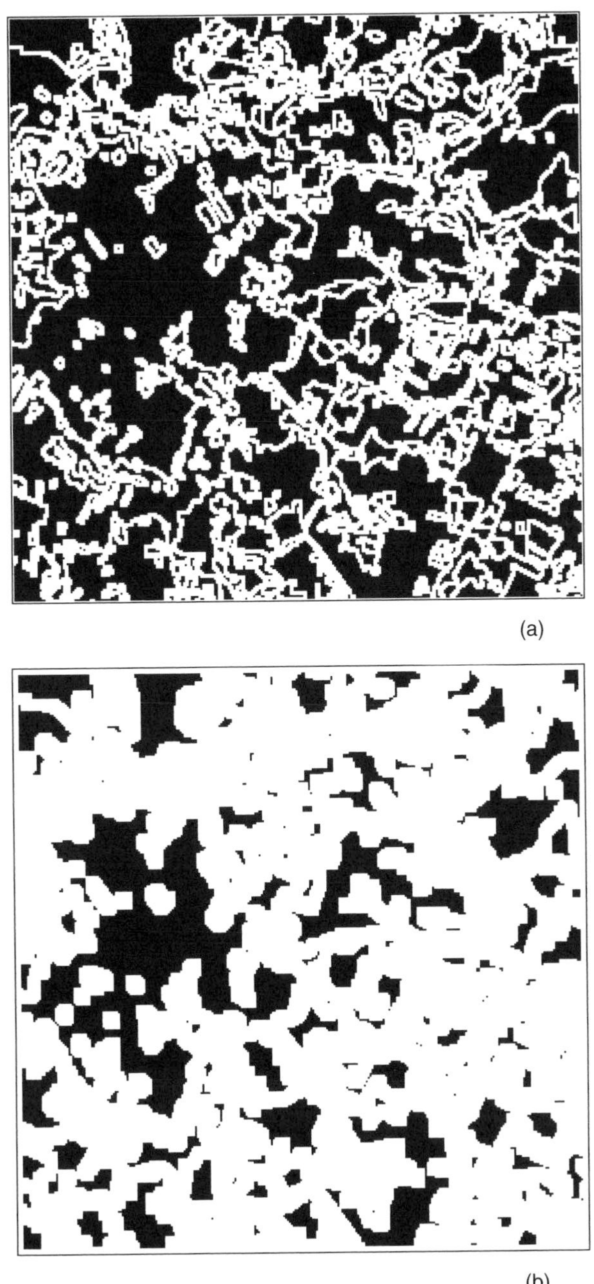

Figure 7.5 *Buffers along borders (in white) in (a) a 3 × 3 and (b) 7 × 7 window*

Due to the large number of variables, a subset of the most suitable for textural image classification was chosen for further analysis. Screening (through an assessment of the discriminating power of the variables) was achieved using decision binary trees (Jakomulska and Stawiecka, 2002), because their structure is explicit and, therefore, easily interpretable (Safavian and Landgrebe, 1991; Friedl and Brodley, 1997). Decision binary trees use a hierarchical framework to partition sequentially observations, where the criteria for partitioning the data space at each node can vary for each class. This is an important advantage in textural classification, since some classes may be classified using solely spectral information, some using only textural information, while other classes require both spectral and textural information in order to be classified accurately. Univariate decision trees were constructed using both spectral and textural wavebands. An example of a simplified, non-overlapping decision binary tree is shown in Figure 7.6.

Coniferous forests were distinguished from all the other classes by a large sill in the pseudo-cross variogram calculated between the first and the third principal component (Figure 7.6). A small sill in the pseudo-cross variogram calculated between the second and the third principal component distinguished vegetated areas (deciduous forest and meadows) from barren regions (arable lands and built-up areas). Arable lands were characterized by smaller mean semivariances (calculated up to a range) in the blue waveband relative to that observed for built-up areas. Finally, meadows were distinguished from deciduous forests by a spectral characteristic, namely a large reflectance in the green waveband (TM band 2).

Textural bands that did not participate in the construction of the decision trees were dropped from further analyses. All of the original spectral wavebands (TM1, TM2, TM3, TM4, TM5, TM7 and ERS SAR) and the 34 selected textural variables were, therefore, available for inclusion in further analyses (Table 7.2). Range, sill,

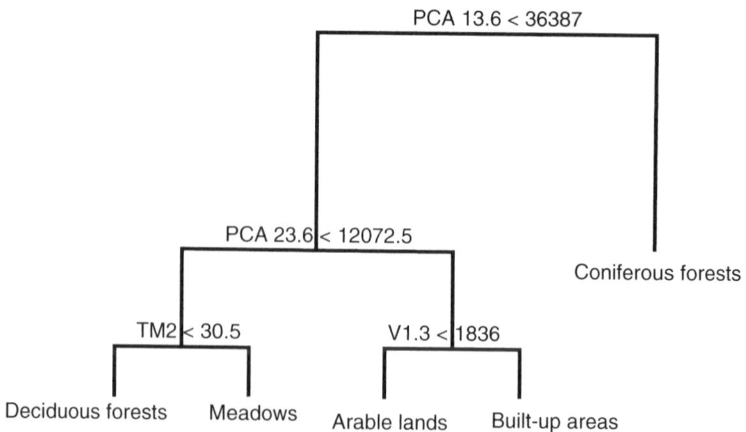

Figure 7.6 *A simplified non-overlapping decision binary tree constructed on textural and spectral bands. Variables used in decision rules: PCA13.6 – sill of a pseudo-cross variogram calculated between the first and the third PC, PCA23.6 – sill of a pseudo-cross variogram calculated between the second and the third PC, TM2 – original green band of TM, V1.3 – mean of semivariances calculated up to a range on blue band (TM1)*

Table 7.2 *Textural variables derived from the Landsat TM and ERS-1 SAR images*

Variables derived from:	Variogram	Variogram	Cross variogram	Pseudo-cross variogram
Calculated on:	6 Landsat TM bands	ERS-1 SAR image	3 PCs derived from Landsat TM bands	
Sill	6		3	3
Semivariance at lag = 0				3
Sum of semivariances			3	
Mean semivariances	6	1	3	
Number of variables	**12**	**1**	**9**	**6**
Total number of variables without ranges		**28**		
Range	6			
Total number of variables with ranges		**34**		

mean semivariances and sum of semivariances (up to a range) from univariate variograms (calculated on Landsat TM bands) and from cross variograms (calculated on the first three principal components (PCs) derived from the Landsat TM data) were used. For pseudo-cross variograms (on PCs) sill (semivariance at a range) and semivariance at lag = 0 were retained. Only one textural waveband (mean semivariance) calculated from the ERS-1 SAR imagery proved to have useful discriminating power and was used in the classification. With the exception of pseudo-cross variances at lag = 0, all texture indices were derived at a range or involve computation based on consecutive variances up to a range. Mean semivariances and cross-variances calculated up to a range incorporate both sill and range information and approximate shape of the variogram.

In the further analyses reported below three data sets were used:

1. spectral data only (7 variables: 6 Landsat TM wavebands and 1 ERS-1 SAR data set);
2. spectral data combined with textural measures derived from within the moving window: sills, mean semivariances and pseudo-cross variances at lag = 0 (7 spectral plus 28 textural variables, giving a total of 35 variables);
3. spectral data combined with all textural wavebands, including six ranges calculated for each Landsat TM spectral band, corresponding to the kernel sizes (7 spectral plus 34 textural variables, giving a total of 41 variables).

3.2 Artificial neural networks

A standard three-layered feed-forward artificial neural network (ANN) using a Levenberg–Marquardt second-order training method was employed for supervised classification (Haykin, 1994; Hagan and Manhaj, 1995; Svozil *et al.*, 1997). Since training data size and composition is of major importance to a classifier's ability to discriminate classes (Campbell, 1981; Foody *et al.*, 1995; Foody and Arora, 1997; Foody, 1999), a training set in which each class had equal representation was formed. This contained 200 pixels for each class, and was derived from training areas identified visually through on-screen digitizing. Pixels for training were selected both from within the core of patches as well as in the vicinity of patch boundaries. This process was based on expert decision, and no particular rule was adopted (since that would require precise definition for 'core' and boundary areas).

In the first experiment, only the spectral wavebands of the Landsat TM, with no textural information of any kind, were used. Several network architectures, represented by the number of neurons (or units) in the input (I) hidden (H) and output (O) layer (I: H: O) were tried, by varying the size of the hidden layer. Here, the number of hidden neurons was systematically varied from 4 to 40. However, in no case did convergence occur when utilizing solely the Landsat TM spectral data. This was in a striking contrast to the second and the third set of experiments outlined below, where summed square error per pattern during ANN training could have been reduced to a very small value (e.g. 5×10^{-2} or less).

In the second set of experiments, 35 selected spectral and textural wavebands (data set without range) sufficed to build small networks with what was perceived to be very good convergence properties. Very rapid convergence was observed for every training session with only four neurons in the hidden layer (a network architecture of 35:4:5).

In the third experiment, in addition to the 35 variables used in the second experiment, the data corresponding to the sliding window size (variogram range) were also used. The same number of hidden neurons as before, that is the network of the 41:4:5 architecture, sufficed for a similar, very rapid, convergence during every training session.

3.3 Accuracy assessment

Classification accuracy was evaluated with the aid of a confusion matrix that shows a cross-tabulation of the class labels in the output of a classification against those in the ground data. To assess classification accuracy, a testing set is typically chosen using some sampling design. In this study, the whole data set (excluding the training data) was used in assessing classification accuracy. This approach was chosen for two reasons. Firstly, this approach avoided complexities arising through the influence of the sampling design used to acquire the reference data and ensured that the composition of testing samples reflected the true class proportions (Congalton, 1991; Richards, 1996). Secondly, the approach aided the study of the spatial distribution of misclassified pixels. Due to the high density of boundaries in the studied data, the 'border effect' of textural classification was examined through the analysis of the spatial distribution of misclassified pixels with respect to buffers created along class boundaries (derived from the ground data).

Following Gopal and Woodcock (1994), two criteria, based on the membership (the network activation level) and difference in class memberships, were used to discriminate between pixels classified with (i) a high degree of confidence, (ii) a degree of uncertainty and (iii) those for which there was so little confidence in any allocation that they were left unclassified. The magnitude of the output neuron activation level was taken as a measure of the strength of membership to the class associated with that neuron. A threshold of 0.9 in the activation level of output neurons was applied to discriminate between pixels classified with a high level of confidence to the allocated class and those with a low maximum activation level. Pixels with the highest neuron activation above 0.9 were further divided into two subgroups. The first contained confidently classified pixels, where the difference between the maximum and the second largest output neuron activation levels was larger than 0.25. The second comprised pixels classified with some degree of uncertainty, with the difference between the highest and second highest output neuron activation levels <0.25.

4 Results

Since training on spectral wavebands only did not bring convergence, further analysis was limited to comparison of the second and third experiment of neural network classification. Both experiments were based on data sets containing a combination of spectral and textural measures, the former without and the latter with the range from the variogram used as a variable in the analysis. The results are summarized in Table 7.3.

4.1 Analysis of pixels classified with a high degree of confidence

Classification using 35 variables (without ranges) resulted in greater confidence in the classification (3% more pixels with highest neuron activation above 0.9 threshold) than classification using 41 variables (with ranges). However, the overall classification accuracy increased from 64.3%, to 71.0% (kappa coefficient of agreements of 0.48 and 0.53 respectively) after the inclusion of ranges. The addition of data on the variogram range to the textural set of bands increased either the user's or producer's accuracy of most of the classes (where user's accuracy is a probability that a pixel classified on the image actually represents that class on the ground and is a measure of the commission error while the producer's accuracy indicates the probability of a reference pixel being correctly classified and is a measure of the omission error (Congalton, 1991)). On average, the accuracy of each class except arable lands increased by several per cent (Table 7.4). The largest difference was noted for meadows

Table 7.3 *Neural network classification results: percentage of confident, uncertain and undecided (unclassified) data*

Threshold	Percentage of classified data	Without ranges	With ranges
Above 0.9	Single high activation (confident)	80	77
	Multiple high activations (uncertain)	4	12
Below 0.9	Undecided	16	11

Table 7.4 *Confusion matrices: training on 35 (without ranges; light columns) and 41 variables (including ranges; shaded columns). Columns indicate label in the ground data while rows show the label in the classified image*

	Built-up		Coniferous		Deciduous		Meadows		Arable lands		Total		User's accuracy (%)	
Built-up	**2011**	**1881**	170	866	1339	613	2451	816	6280	3796	12251	7972	**16.4**	**23.6**
Coniferous	51	2	**3714**	**2614**	343	171	59	49	74	28	4241	2864	**87.6**	**91.3**
Deciduous	162	238	1926	1392	**2694**	**2670**	1144	902	2485	1931	8411	7133	**32.0**	**37.4**
Meadows	399	128	19	9	1104	522	**9800**	**8412**	2054	707	13376	9778	**73.3**	**86.0**
Arable lands	649	1068	101	363	306	894	3322	4671	**25875**	**31271**	30253	38267	**85.5**	**81.7**
Total	3272	3317	5930	5244	5786	4870	16776	14850	36768	37733	**68532**	**66014**		
Producer's accuracy (%)	**61.5**	**56.7**	**62.6**	**49.8**	**46.6**	**54.8**	**58.4**	**56.6**	**70.4**	**82.9**		**100.0**	**64.3**	**71.0**

(increase by 13%) and built–up areas (increase by 7%), the kappa coefficient of agreement increased by 0.17 and 0.08, respectively.

In the data set without ranges, the major confusion between classes occurred for the following pairs of classes: built-up areas and deciduous forests, built-up areas and meadows, built-up areas and arable lands, coniferous and deciduous forests, deciduous forests and meadows, deciduous forests and arable lands and meadows and arable lands.

Most of the confusion present arose from the similarity of both spectral and textural response (e.g. between two types of forests or between built-up areas and arable lands). Confusion between meadows and arable lands appeared to arise from the generalization of the ground data (due to dispersed character of cultivated parcels, many small meadows were merged with larger arable lands classes in the generalization process). However, confusion between classes characterized by a 'good' separability of both spectral and textural responses (e.g. built-up areas and deciduous forests; built-up and meadows) was a direct result of the classification algorithm used. It is believed that the problem lies in the buffering side effect of the textural analysis.

The smallest patches in the study area were generally built-up areas, meadows and deciduous forests. The confusion between the three classes decreased after the addition of the six variogram ranges. Figure 7.7 (see plate 2) shows a large buffering effect in the results of classification without ranges, while the addition of variogram range reduced this effect. The classification without ranges over-estimated built-up areas and deciduous forests, which appeared erroneously at many borders. The major decrease was noted in the commission error of meadows and deciduous forests; and in the omission error of arable lands and deciduous forests.

To investigate border effects the spatial distribution of incorrectly allocated pixels was analysed. Proximity to boundaries was assessed through an analysis of the distribution of pixels with respect to buffers along class borders (Figure 7.8). It was noted, that in cases where commission error of a class decreased after ranges were added (built-up, deciduous and meadows), the number of pixels within the proximity of boundaries had decreased. On the other hand, with arable lands (for which there was a decrease of the omission error), the number of pixels within the buffer increased (Figure 7.8). Hence, the differences were not randomly distributed in space, but concentrated at the boundaries.

4.2 Analysis of pixels classified with uncertainty

The addition of variogram ranges increased the uncertainty of the classification, in terms of the percentage of pixels with more than one output neuron providing a high activation level (4% and 12% for data set without and with ranges, respectively). Pixels with multiple high neuron activations (highest activation > 0.9 and difference between the highest and second highest activation less than 0.25) were grouped in the 'uncertain' category. Accuracy quantified using only the highest neuron activation was low, 51% and 61% (without and with ranges, respectively). However, if either the highest or the second highest response was accepted, the accuracy increased to 85% and 94%.

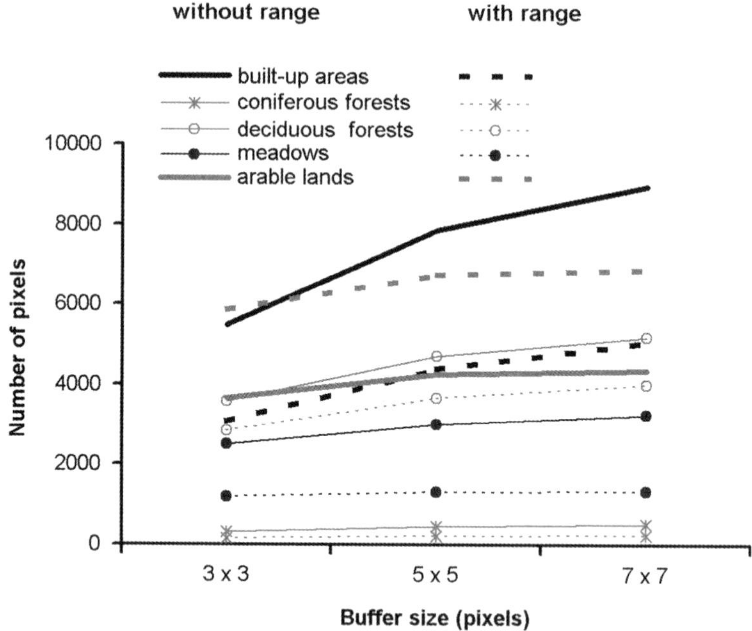

Figure 7.8 Amount of pixels within buffers along boundaries. Results for the textural classification without and with ranges are shown (marked in solid and dashed lines, respectively)

The spatial distribution of uncertain pixels for the data set without ranges was only slightly correlated with boundaries, and it is hypothesized that this subset comprised mixed pixels, being a result of the relatively large Landsat TM pixel size in relation to the size of the small land cover patches. For the data set with ranges, spatial correlation with borders was important, particularly for deciduous forests and meadows.

In spite of the present uncertainty, even within this dubiously classified group of pixels, it was possible to classify the data correctly. Assuming the condition that either of the two highest responses was correct, in combination with confidently classified pixels, the overall classification accuracy increased to 65.3% and 74.2% (without and with variogram ranges, respectively).

4.3 Analysis of unclassified pixels

Greater confidence in the classification was observed with than without variogram ranges, in terms of the total number of classified data (11% and 16% of data were left unclassified, respectively). Furthermore, cases were distributed fairly regularly over the whole data set (although not totally at random), while in the classification without ranges, the unclassified pixels were concentrated along boundaries. Again, this result indicated that inclusion of ranges decreased the uncertainty and error for pixels located along class boundaries.

Finally, a small group of pixels was observed with very low output neuron activation levels (the activation level of all output neurons was below 0.6). Most of these pixels belonged to the arable land class. Visual inspection of a sample of these pixels revealed that they were of a relatively distinct sub-class, wet arable lands, characterized by low reflectance in all spectral wavebands. Hence, in spite of the same land use and land cover, the wet arable lands differed with respect to spectral response of the more common agricultural lands. Note also that due to the small extent of the wet arable lands, sites of this class had not been included in training the classification.

5 Summary and Conclusions

The potential of textural classification has been demonstrated. An insight into the heart of the classification procedure demonstrated that the majority of the discriminating variables were textural bands extracted from the variogram. It has been further shown that the addition of the variables describing window size increased the accuracy of neural network classification both qualitatively and quantitatively: both the accuracy and confidence of the classification (in terms of the number of correctly classified pixels) increased by several per cent. It could be argued that the addition of any other information might increase the accuracy, since the dimensionality of the data used increased (unless the Hughes' phenomenon occurred). However, it has been shown that major differences between the two classifications were spatially correlated and occur along the boundaries of classes in the study area. It appears that the use of variogram ranges partially corrects for the problem of mixed pixels distributed along boundaries.

Despite the above, the overall classification accuracy remains low, with a large percentage of unclassified (11%) and uncertain (12%) data, and many misclassified pixels (28.9%). It is realized that further refinement of both textural quantification and classification techniques should be pursued. Foremost, the neural network analyses indicated that the within-class variance of the training sample was not fully representative, resulting in a low accuracy of the classification of the testing data. Derivation of very pure training samples reduces the generalization ability of the classifier, and in spite of the high training accuracy, reduces the accuracy of the prediction for unseen data (Richards, 1996; Foody, 1999). However, repeating analyses reported in this study but using randomly selected training data did not result in network convergence. Therefore, more attention should be paid to the choice of the training data set, since in the case of textural classification, both spectral and textural representativeness is important. Furthermore, considerable noise and some unavoidable discrepancies between satellite and aerial sensor images (ground data) introduced factors reducing the ability to assess precisely both the feasibility and accuracy of the methods tested.

Many studies have shown that textural information derived from within a window of an adapting size may be more useful than that derived from a kernel of a fixed size. However, as the window size varies the number of pixels and the number of lags used in the calculation of the variogram will change and this may influence the estimation of both the sill and range of the variogram (for small windows none of these may be achieved). This effect is partially accounted for when the size of the window is itself used as a variable in the classification. However, further improvement of the adapting window algorithm would be helpful. Ideally, pixels at the borders (for which a small window size is used) should be classified solely on the spectral or contextual information or directional variograms could be used to adapt the window size differently in different directions (St-Onge and Cavayas, 1995; Herzfeld and Higginson, 1996; Atkinson and Lewis, 2000). Although these techniques are computationally challenging, they show that there is potential to further extend the capabilities of textural analysis.

Acknowledgements

The authors acknowledge the European Space Agency for granting ERS-1 SAR image within the Third Announcement of Opportunity offered to conduct research and application development projects using data from the first and second European Remote Sensing satellites, ERS-1 and ERS-2. JPR was partially supported by the KBN-115/E343/2001/ICM grant. AMJ acknowledges The Foundation for Polish Science for granting a stipend for the young scientists.

References

Atkinson, P. M. and Lewis, P., 2000, Geostatistical classification for remote sensing: an introduction, *Computers and Geosciences*, **26**, 361–71.

Atkinson, P. M. and Tatnall, A. R. L, 1997, Introduction: neural networks in remote sensing, *International Journal of Remote Sensing*, **18**, 699–709.

Benediktsson, J. A., Swain, P. H. and Ersoy, O. K., 1990, Neural network approaches versus statistical methods in classification of multisource remote sensing data. *IEEE Transactions on Geoscience and Remote Sensing*, **28**, 540–52.

Berberoglu, S., Lloyd, C. D., Atkinson, P. M. and Curran, P. J., 2000, The integration of spectral and textural information using neural networks for land cover mapping in the Mediterranean, *Computers and Geosciences*, **26**, 385–96.

Campbell, J. B., 1981, Spatial correlation-effects upon accuracy of supervised classification of land cover, *Photogrammetric Engineering and Remote Sensing*, **47**, 355–63.

Congalton, R. G., 1991, A review of assessing the accuracy of classifications of remotely sensed data, *Remote Sensing of Environment*, **37**, 35–46.

Cressie, N. A. C, 1993, *Statistics for Spatial Data* (New York: Wiley).

Foody, G. M., 1999, The significance of order training patterns in classification by a feedforward neural network using back propagation learning, *International Journal of Remote Sensing*, **20**, 3549–62.

Foody, G. M. and Arora, M. K., 1997, An evaluation of some factors affecting the accuracy of classification by an artificial neural network, *International Journal of Remote Sensing*, **18**, 799–810.

Foody, G. M., McCulloch, M. B. and Yates, W. B., 1995, The effect of training set size and composition on artificial neural network classification, *International Journal of Remote Sensing*, **16**, 1707–23.

Franklin, S. E. and McDermid, G. J., 1993, Empirical relations between digital SPOT HRV and CASI spectral response and lodgepole pine (Pinus contorta) forest stand parameters, *International Journal of Remote Sensing*, **14**, 2331–48.

Franklin, S. E., Wulder, M. A. and Lavigne, M. B., 1996, Automated derivation of geographic window sizes for use in remote sensing digital image texture analysis, *Computers and Geosciences*, **22**, 665–73.

Friedl, M. A. and Brodley, M. A., 1997, Decision tree classification of land cover from remotely sensed data, *Remote Sensing of Environment*, **6**, 1399–409.

Frost, V. S., Stiles, J. A., Shanmugan, K. S., and Holtzman, J. C., 1982, A model for radar images and its application to adaptive digital filtering of multiplicative noise, *IEEE Transactions on Pattern Analysis and Machine Intelligence*, **4**, 157–66.

Gopal, S. and Woodcock, C., 1994, Theory and methods for accuracy assessment of thematic maps using fuzzy sets, *Photogrammetric Engineering and Remote Sensing*, **60**, 181–8.

Hagan, M. T. and Manhaj, M. B., 1995, Training feed-forward networks with the Marquardt algorithm, *IEEE Transactions Neural Networks*, **5**, 989–93.

Haralick, R. M., Shanmugam, K. and Dinstein, I., 1973, Textural features for image classification, *IEEE Transactions on Systems, Man, and Cybernetics*, **3**, 610–21.

Haykin, S., 1994, *Neural Networks – A Comprehensive Foundation* (Montreal: Maxwell Macmillan International).

Herzfeld, U. C. and Higginson, C. A., 1996, Automated geostatistical seafloor classification – principles, parameters, feature vectors, and discrimination criteria, *Computers and Geosciences*, **22**, 35–52.

Jakomulska, A. and Clarke, K. C., 2001, Variogram-derived measures of textural image classification. Application to large-scale vegetation mapping, in P. Monestiez, D. Allard and R. Froidevaux (eds), *geoENVIII – Geostatistics for Environmental Applications. Geostatistics for Quantitative Geology* (Dordrecht: Kluwer Academic), pp. 345–55.

Jakomulska, A. and Stawiecka, M., 2002, Integrating spectral and textural information for land cover mapping, in G. Begni (ed.) *Observing our Environment from Space. New Solutions for a New Millennium* (Lisse: A. A. Balkema), pp. 347–54.

Journel, A. G. and Huijbregts, C. J., 1978, *Mining Geostatistics* (London: Academic Press).

Miranda, F. P., Macdonald, J. A. and Carr, J. R., 1992, Application of the semivariogram textural classifier (STC) for vegetation discrimination using SIR-B data of Borneo, *International Journal of Remote Sensing*, **13**, 2349–54.

Miranda, F. P., Fonseca, L. E. N., Carr, J. R. and Raranik, J. V., 1996, Analysis of JERS-1 (Fuyo-1) SAR data for vegetation discrimination in northwestern Brazil using the semivariogram textural classifier (STC), *International Journal of Remote Sensing*, **17**, 3523–9.

Ramstein, G. and Raffy, M., 1989, Analysis of the structure of radiometric remotely-sensed images, *International Journal of Remote Sensing*, **10**, 1049–73.

Richards, J. A., 1996, Classifier performance and map accuracy, *Remote Sensing of Environment*, **57**, 161–6.

Safavian, S. R. and Landgrebe, D., 1991, A survey of decision tree classifier methodology, *IEEE Transactions on Systems, Man and Cybernetics*, **21**, 660–74.

St-Onge, B. A. and Cavayas, F., 1995, Estimating forest stand structure from high resolution imagery using the directional variogram, *International Journal of Remote Sensing*, **16**, 1999–2021.

Svozil, D., Kvasnicka, V. and Pospichal, J., 1997, Introduction to multi-layer feed-forward neural networks, *Chemometrics and Intelligent Laboratory Systems*, **39**, 43–62.

Wackernagel, H., 1995, *Multivariate Geostatistics* (Berlin: Springer-Verlag).

Wallace, C. S. A., Watts, J. M. and Yool, S. R., 2000, Characterizing the spatial structure of vegetation communities in the Mojave Desert using geostatistical techniques, *Computers and Geosciences*, **26**, 397–410.

Wharton, S. W., 1982, A context-based land-use classification algorithm for high-resolution remotely sensed data, *Journal of Applied Photographic Engineering*, **8**, 46–50.

Woodcock, C. E., Strahler, A. H. and Jupp, D. L. B., 1988a, The use of variograms in remote sensing: I real digital images, *Remote Sensing of Environment*, **25**, 323–48.

Woodcock, C. E., Strahler, A. H. and Jupp, D. L. B., 1988b, The use of variograms in remote sensing: II real digital images, *Remote Sensing of Environment*, **25**, 349–79.

Wulder, M. A., Franklin, S. E. and Lavigne, M. B., 1996, High spatial resolution optical image texture for improved estimation of forest stand Leaf Area Index, *Canadian Journal of Remote Sensing of Environment*, **22**, 441–9.

Wulder, M. A., Lavigne, M. B., LeDrew, E. F. and Franklin, S. E., 1997, Comparison of texture algorithms in the statistical estimation of LAI: first-order, second-order, and semi-variance moment texture (SMT), *Canadian Remote Sensing Society Annual Conference, GER'97, Geomatics in the Era of Radarsat*. May 24–30, 1997, Ottawa, Canada.

Wulder, M. A., LeDrew, E. F., Franklin, S. E. and Lavigne, M. B., 1998, Aerial image texture information in the estimation of northern deciduous and mixed wood forest leaf area index (LAI), *Remote Sensing of Environment*, **64**, 64–76.

8

Remote Monitoring of the Impact of ENSO-related Drought on Sabah Rainforest using NOAA AVHRR Middle Infrared Reflectance: Exploring Emissivity Uncertainty

Doreen S. Boyd, Peter C. Phipps, William J. Duane and Giles M. Foody

1 Introduction

It is known that the terrestrial biosphere exhibits variation in its properties over a range of both spatial and temporal scales (Van Gardingen *et al.*, 1997; Slaymaker and Spencer, 1998; Trudgill, 2001) and this variability is both naturally and anthropogenically driven. This variability has direct implications for human society, as well as the Earth system, since vegetation, the principal biospheric component, is tightly coupled to the radiative, meteorological, hydrological and biogeochemical processes and functions that operate within that system (Bonan, 1997; NASA, 2001). Thus, the quest to fully understand the causes of biosphere variability and to measure and predict terrestrial biospheric responses to natural and anthropogenic influences has assumed great scientific importance (Walker and Steffen, 1997; Rotenberg *et al.*, 1998). However, despite making huge strides in this regard, uncertainties in our knowledge of how the terrestrial biosphere is changing and our ability to estimate future changes are readily apparent and need to be overcome (Harvey, 2000; IGBP, 2001; IPCC, 2001).

Uncertainty in Remote Sensing and GIS. Edited by G.M. Foody and P.M. Atkinson.
 ISBN: 0–470–84408–6

In short, there is considerable uncertainty attached to our current understanding of terrestrial biospheric change. Three levels of uncertainty are evident: uncertainty as a result of data shortage, uncertainty as a result of model deficiencies and uncertainty as a result of environmental processes that are indefinable and indeterminate (O'Riordan, 2000). Realistically, only the first two of these uncertainty levels can be addressed and one approach to achieving this involves the routine collection of information on the terrestrial biosphere, particularly the spatial distribution, extent and temporal dynamics (intra-annual and inter-annual) of ecosystems. It has been suggested that much of this information may be provided by remote sensing (Wickland, 1989; Stoms and Estes, 1993; NASA, 2001) and the arguments for this suggestion are compelling (Curran and Foody, 1994), particularly with the launch of new Earth remote sensing instruments and sensors designed for the task (e.g. EOS Terra). Unfortunately, despite the potential offered by remote sensing, there are several uncertainties involved in using remotely sensed data. Thus, the uncertainties associated with our knowledge of terrestrial biospheric change are compounded by uncertainties in the use of the remote sensing approach. It is imperative, therefore, that all known uncertainties in the use of remotely sensed data are characterized and accounted for, if at all possible.

This chapter focuses on one example. It presents a case study in which an identified uncertainty in the use of remotely sensed data for the provision of information about the terrestrial biosphere is evaluated. Specifically, the uncertainty associated with the use of National Oceanographic and Atmospheric Administration (NOAA) Advanced Very High Resolution Radiometer (AVHRR) middle infrared (MIR) reflectance data to monitor the impact of El Niño/Southern Oscillation (ENSO)-induced drought on tropical rainforests is explored. The derivation of useful MIR reflectance data, from total radiant energy measured by the AVHRR sensor in channel 3, requires inversion of a basic physical model. The components of this model are specified in established laws of physics but include a parameter, emissivity, which is rarely known with complete certainty. The impact of using different values for the emissivity within the derivation of MIR reflectance is investigated. Here, therefore, the focus is on uncertainty that arises as a result of measurement error (i.e. not knowing for certain what the emissivity value should be) and ignorance (i.e. using a constant emissivity despite spatial and temporal variability). This chapter aims to ascertain whether uncertainty in the magnitude of the emissivity parameter in the derivation of MIR reflectance compromises its use for the monitoring of ENSO-related drought stress on a tropical rainforest ecosystem. Since MIR reflectance offers greater potential for this application than conventional approaches (e.g. based on Normalized Difference Vegetation Index (NDVI)), these results have significant implications for the furthering of our knowledge on change in tropical rainforests caused by drought events and our ability to use this knowledge to aid forest management.

2 The ENSO

The ENSO phenomenon is a coupled atmospheric and oceanic mechanism responsible for worldwide climatic anomalies (Diaz and Margraf, 1992). These climatic anomalies arise via teleconnections to the anomalous warming of oceanic waters in the central and eastern Pacific and changes in the Walker cell circulation system. The ENSO phenomenon occurs with a periodicity of typically 2 to 7 years, having profound impacts on the terrestrial biosphere (Rasmusson, 1985; Glantz, 1996), and is now widely acknowledged as the largest known climate variability signal on inter-annual timescales (Houghton *et al.*, 1990; Li and Kafatos, 2001). The period from June 1997 to May 1998 saw the global climate system perturbed by the largest ENSO event of the twentieth century (McPhadden, 1999). The resultant deranged weather patterns around the world killed an estimated 2100 people and caused at least 33 billion US dollars in property damage (Suplee, 1999). These sorts of impacts may be more commonplace with the news from some of the latest climatic model predictions that suggest ENSO events may intensify further in their severity and frequency (Houghton *et al.*, 1996; Hulme and Viner, 1998). The need to understand the nature of impacts of ENSO events on the terrestrial biosphere, and in turn how these impacts affect the Earth system, and subsequently to forecast future impacts is obvious. However, the precise nature of ENSO phenomenon impacts on the terrestrial biosphere is highly differentiated in space (Plisnier *et al.*, 2000). There is a need, therefore, for detailed investigations on the ENSO phenomenon that bring together climatic and ecosystem datasets.

One ecosystem particularly under threat from ENSO events is the ever-wet tropics (Whitmore, 1988; Walsh and Newbery, 1999). Such ecosystems face continuing anthropogenic pressures as well as a predicted increase in the frequency and severity of ENSO-related climatic events (e.g. drought, floods) (Harwell, 2000; Harrison, 2001; Couturier *et al.*, 2001). This, coupled with the significant role that the ever-wet tropics play within the Earth system (Philips, 1998; Artaxo, 1998; Houghton *et al.*, 2000) and the fact that these processes may change under ENSO conditions (Potter *et al.*, 2001), mean that they deserve specific attention.

3 The Role for Remote Sensing

Over the past decade remote sensing has emerged as the only realistic approach to measure much of the necessary data for studies of the terrestrial biosphere (Townshend *et al.*, 1994; Nemani and Running, 1997). It affords the accurate and systematic measurement of long-term instrumental and proxy observations of key environmental indicators of terrestrial biospheric change (Houghton *et al.*, 1996; Nemani and Running, 1997; Franklin and Wulder, 2002). Remote sensing provides the only means to obtain a view of the ecosystems of the terrestrial biosphere at appropriate spatial and temporal resolutions (Curran and Foody, 1994). Remotely sensed data can be linked to measurements acquired *in situ*, allowing detailed analysis of key processes and controlling factors of terrestrial biospheric change. Furthermore, since remotely sensed data are acquired under fully traceable conditions they can be used to establish a baseline from which any divergence from the global environmental norm can be quantified.

Since the study of impacts of ENSO events on the terrestrial biosphere requires information on vegetation properties at fine temporal resolutions across large areas, there is an obvious contribution from remote sensing. The general approach has been to couple remotely sensed data with indicators of ENSO strength and phase. ENSO indicators used include those measured in the Pacific Ocean: Pacific sea surface temperature (SST), the southern oscillation index (SOI) and outgoing long-wave radiation (OLR) (Philander, 1990; Trenberth and Tepaniak, 2001). Other indicators include those that are more localized in scale, for example, rainfall totals (Nicholson and Entekhabi, 1986; Nicholson, 1996) or air temperature and pressure statistics (Ropelewski and Halpert, 1986; Trenberth and Shea, 1987). The remotely sensed data used have been mainly those acquired in channels 1 (visible reflected radiation; 0.58–0.68 μm) and 2 (near infrared reflected radiation (NIR); 0.75–1.10 μm) of the NOAA AVHRR sensor, combined in the NDVI. The AVHRR NDVI is believed to provide an effective measure of photosynthetically active biomass (Goward *et al.*, 1985; Myneni and Williams, 1994) and has been highly correlated with vegetation biophysical properties and hence can be of considerable value for the study of ecosystem properties. As well as using the AVHRR NDVI over space (e.g. Foody and Curran, 1994; Fazakas and Nilsson, 1996; Goetz *et al.*, 1999), it has been realized that intra-annual time-series of AVHRR NDVI data can be used to resolve phenological canopy properties (e.g. Spanner *et al.*, 1990; Achard and Estreguil, 1995; Duchemin *et al.*, 1999; Moody and Johnson, 2001). Moreover, inter-annual time-series NDVI data have been correlated with several climatic variables in a wide range of environments (Liu *et al.*, 1994; Shultz and Halpert, 1995; Wellens *et al.*, 1997; Yang *et al.*, 1998), as well as to SST, SOI and OLR (Myneni *et al.*, 1996; Anyamba and Eastman, 1996; Plisnier *et al.*, 2000; Anyamba *et al.*, 2001). As a result of its successful use, it has been tentatively postulated that NDVI data be used as a supplementary index of ENSO activity, whereby the variability of time-series NDVI may serve as a proxy for climate variations, provide insights into understanding regional climate impacts of ENSO teleconnections on ecosystems and forecast their occurrence (Hess *et al.*, 1996; Verdin *et al.*, 1999; Plisnier *et al.*, 2000; Li and Kaftos, 2001).

The successful use of the NDVI for studying terrestrial biospheric responses to climatic anomalies lies partly in the inherent characteristics of the NOAA AVHRR sensor (Cracknell, 1997), the ease with which the NDVI index can be computed and the physical principles determining the reflectivity of vegetation in visible and NIR wavelengths. The anomalous climatic characteristics associated with an ENSO event can produce phenological and canopy biophysical property changes that determine the NDVI values that deviate from the norm. Consequently, the NOAA AVHRR NDVI has been touted as offering a long-term data set, starting in July 1981, that is unequalled for monitoring terrestrial land cover and condition (Los *et al.*, 1994; Batista *et al.*, 1997).

There are several uncertainties associated with the use of NOAA AVHRR NDVI data for the above purpose. Problems with aerosol contamination, instrument degradation and orbital drift are common (Eastman and Fulk, 1993; Gutman, 1995; Batista *et al.*, 1997) and thus the NDVI used may be a function of these as well as the ecosystem properties from which they have been acquired. Furthermore, low NDVI–precipitation correlations have been noted in some ecosystems, for example, tropical forest (Schultz and Halpert, 1995; Eklundh, 1998; Richard and Poccard, 1998). These uncertainties may, however, be overcome by modelling their influences, particularly those of a systematic nature (e.g. instrument degradation; Los, 1993), or combining NDVI data with other remotely sensed data to minimize their influence or using an alternative remotely sensed data set which is more suited for use in a particular ecosystem. Indeed, the NDVI is not the only remotely sensed product that can be generated from NOAA AVHRR data. Other products include surface temperature (Ts), generated from radiation acquired in AVHRR channels 4 and 5, and middle infrared (MIR) reflectance, derived from the total MIR radiation acquired in AVHRR channel 3. Ts measurements have been used to study drought impacts on ecosystems, both, alone (e.g. Plisnier *et al.*, 2000), and within vegetation indices such as NDVI/Ts (e.g. McVicor and Bierwirth, 2001) and the Vegetation and Temperature Condition Index (e.g. Unganai and Kogan, 1998). MIR reflectance data have also been shown to have potential for assessing landscape dryness (Boyd and Duane, 2001). The use of Ts and MIR reflectance data for the study of ENSO impacts is in its infancy, even though the data used to generate these products have been measured simultaneously to those used to derive the NDVI. Thus, we still do not wholly understand the properties of all these remotely sensed data being acquired in such great quantities and have, therefore, yet to realize their full potential for the study of ENSO impacts on the terrestrial biosphere.

4 MIR Reflectance at 3.75 μm

The total MIR radiation signal acquired in AVHRR channel 3 (between 3.55 and 3.93 μm) comprises both the radiation reflected from the Earth's surface and shorter wavelength radiation that has been absorbed at the surface and emitted at MIR wavelengths. This hybrid signal, part reflected and part emitted radiation, introduces a degree of uncertainty into its use and thus the use of MIR radiation for terrestrial biospheric research has been limited (Boyd and Curran, 1998). Nonetheless, the MIR radiation has several attributes that should encourage its wider use in global environmental research. It has been observed that the use of MIR radiation offers favourable atmospheric penetration capabilities over that of visible and NIR radiation and is relatively insensitive to scan angle effects (Kaufman and Remer, 1994; França and Setzer, 1998). Moreover, MIR radiation, used in conjunction with and in place of the NDVI or Ts, has been used successfully in the mapping, monitoring and prediction of forest types and their properties (e.g. Malingreau *et al.*, 1989; Laporte *et al.*, 1995; Foody *et al.*, 1996; DeFries *et al.*, 1997; Lucas *et al.*, 2000). Furthermore, it has been demonstrated that the removal of the emitted radiation portion of the MIR radiation to leave the reflected portion (known as MIR reflectance) increases the accuracy of remotely sensed prediction of ecosystem properties (Boyd *et al.*, 1999; Boyd *et al.*, 2000).

To derive MIR reflectance from MIR radiation there is the assumption that

$$L_{\mathrm{ch3}} = \rho_{\mathrm{ch3}} + E_{\mathrm{ch3}} \tag{1}$$

where L_{ch3} is the total radiation signal measured in AVHRR channel 3, ρ_{ch3} is the reflected component of the channel and E_{ch3} is the emitted component of the channel. Based on this assumption, Kaufman and Remer (1994) proposed a model developed from earlier work by Ruff and Gruber (1983) and Gesell (1984) that can be used to derive MIR reflectance from the total radiant energy measured in AVHRR channel 3. The model assumes that calculating and removing the emitted component from the total radiation signal can be used to derive MIR reflectance. It is assumed that the radiation measured in channel 4 (10.50–11.30 μm) of the AVHRR sensor, which is almost wholly composed of energy emitted by the Earth, is related to the emitted component of the total radiant energy measured in AVHRR channel 3 through Planck's function. The correction for surface temperature emission is accomplished by calculating the brightness temperature (T) in channel 4 by inverting Planck's function (equation 2)

$$T = \frac{C_2 \nu}{\ln\left(1 + \frac{C_1 \nu^3}{E}\right)} \tag{2}$$

where T is brightness temperature calculated from channel 4 (K), ν is the central wave number of channel 4 (cm^{-1}), C_1 is constant 1.1910659×10^{-5}, C_2 is constant 1.438833 and E is a radiance value from channel 4 (mW m^2 sr cm^{-1})

The brightness temperature calculated from channel 4 is then used to calculate the emitted component of the total radiant energy measured in AVHRR channel 3. The total radiant energy measured in AVHRR channel 3 (L_{ch3}) can be described (equation 3)

$$L_{ch3} = t'_{ch3} \rho_{ch3} \frac{F_0 \mu_0}{\pi} + t_{ch3} \epsilon_{ch3} B_{ch3}(T) \tag{3}$$

where t'_{ch3} is the two-way transmission function in MIR channel of AVHRR sensor in channel 3, ρ_{ch3} is the reflectance in MIR channel of AVHRR sensor in channel 3, F_0 is the incident solar radiation at top of atmosphere in channel 3, μ_0 is the cosine of solar zenith angle, t_{ch3} is a one-way transmission function in channel 3, ϵ_{ch3} is the emissivity of the surface in MIR channel of AVHRR sensor in channel 3, B_{ch3} is the channel 3 Planck function and T is the brightness temperature calculated from channel 4 (equation 2).

In equation (3) the second term now represents the emitted component of the signal. To obtain MIR reflectance the equation can be rearranged, thus (equation 4)

$$\rho_{ch3} = \frac{L_{ch3} - t_{ch3} \epsilon_{ch3} B_{ch3}(T)}{t'_{ch3} \frac{F_0 \mu_0}{\pi} - t_{ch3} \epsilon_{ch3} B_{ch3}(T)} \tag{4}$$

where L_{ch3} is the full radiation signal, t_{ch3} is the channel 3 one-way transmission function, t'_{ch3} is the channel 3 two-way transmission function, ϵ_{ch3} is the emissivity of the surface in MIR channel of AVHRR sensor in channel 3, B_{ch3} is the channel 3 Planck function, T is the brightness temperature, F_0 is the incident solar radiation at top of atmosphere and μ_0 is the cosine of solar zenith angle.

MIR reflectance from a vegetation canopy is thought to be principally a function of the liquid water content of the canopy. Thus, any change in the leaf biomass of the canopy, such as that induced by an ENSO event, will be accompanied by a change in the amount of canopy water, and thereby the ability of the canopy to absorb or reflect MIR radiation (Kaufman and Remer, 1994; Boyd and Curran, 1998). In addition to change in leaf biomass, an ENSO event that leads to drought conditions in an ecosystem could promote a change in the canopy condition promoting canopy senescence, which would increase the MIR reflectance from a canopy (Elvidge, 1988; Salisbury and D'Aria, 1994). Indeed a study of intra-annual temporal variability of

MIR reflectance from a forest ecosystem has been observed and attributed to both leaf area variability and landscape dryness, believed to be driven by climatic characteristics (Boyd and Duane, 2001).

MIR reflectance from an ecosystem has yet to be related to ground collected climatic measurements and so its potential for the task of monitoring ENSO impacts on the terrestrial biosphere is unrealized. This chapter presents a study in which the value of NOAA AVHRR MIR reflectance for the monitoring of ENSO-related drought stress of a tropical forest ecosystem is investigated. The possible correlation between precipitation values and remotely sensed response is explored. It is hypothesized that the MIR reflectance would be more sensitive than the NDVI to changes in the forest canopy properties and condition occurring as a result of anomalous rainfall, a product of an ENSO event. However, there are uncertainties with using NOAA AVHRR MIR reflectance; the effect of one, the accuracy of emissivity predictions, will be investigated here.

5 Study Area

The rainforests of Danum Valley, Sabah, Malaysia (117 °E 4 °N) during the 1997–98 ENSO event were the focus of study (Figure 8.1). Rainfall statistics show that during the 1997–98 ENSO event there was a marked reduction in rainfall, with the year of

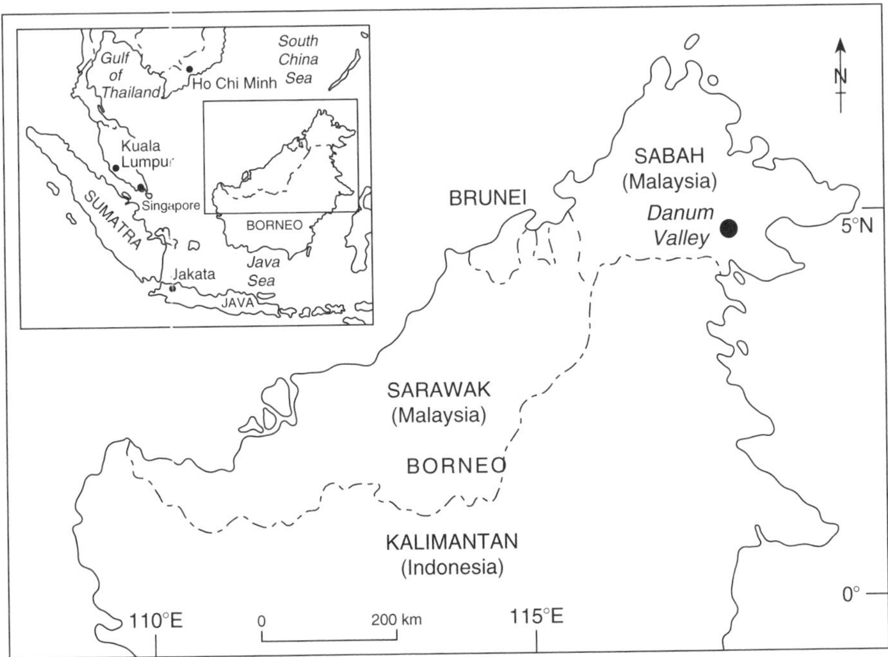

Figure 8.1 Location of the Danum Valley Conservation Area

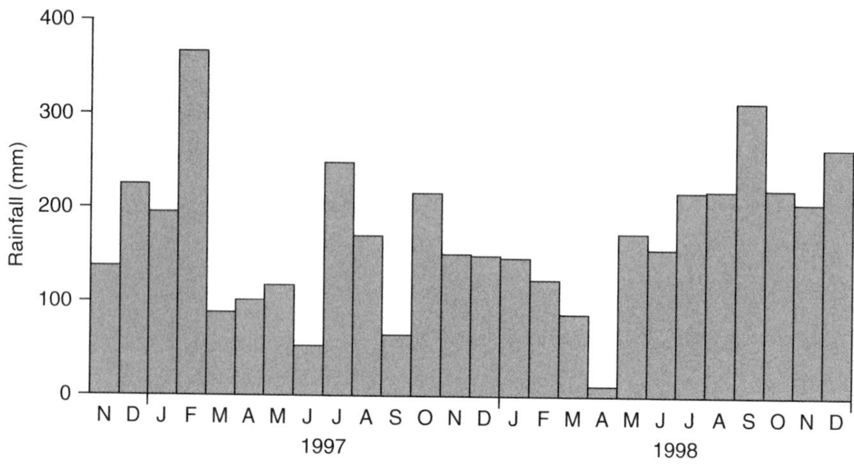

Figure 8.2 Monthly rainfall (mm) at Danum Valley for the period under study

1997 having the lowest rainfall total on record (Walsh and Newbery, 1999) (Figure 8.2). At this site is a conserved area of rainforest comprising 43 800 ha of primary forest, known as the Danum Valley Conservation Area (DVCA). The DVCA was established in May 1995 by an enactment in the Sabah State Legislature. It is a Class I Forest reserve to remain unlogged and otherwise undisturbed for posterity. The DVCA lies in a key position close to the drought-prone eastern coastlands of Borneo. During ENSO years, high atmospheric pressure persists over the Sabah region leading to more stable atmospheric conditions and a marked reduction in rainfall. Rainfall records of Sabah show that drought intensity and frequency have increased in the last few decades and this tallies with a recent increase in the frequency and magnitude of ENSO events. Furthermore, evidence acquired via climatic modelling suggests a further increase in the intensity of ENSO events (Walsh and Newbery, 1999). This, coupled with human interference of the forests of Sabah, mean that ENSO-related drought impacts on this ecosystem are likely to increase significantly. A monitoring system to analyse and describe forest response to drought stress and subsequently to forecast future impacts of ENSO events would be advantageous for the effective management of this threatened ecosystem.

6 Uncertainty Associated with the Use of MIR Reflectance

As with the use of the NOAA AVHRR NDVI there are several uncertainties with using NOAA AVHRR MIR reflectance. In particular, there are uncertainties associated with the method of deriving MIR reflectance (Nerry *et al.*, 1998; Roger and Vermote, 1998) from the total radiant energy measured in AVHRR channel 3, specifically, with the emissivity information that is required to determine the emitted portion of that signal. The term ϵ_{ch3} is introduced to equations (3) and (4) since any radiation emitted from a target, at a particular wavelength, is a combination of its spectral emissivity and brightness temperature, which can be expressed (without atmospheric effects) as

$$E\lambda = \epsilon\lambda T\lambda \tag{5}$$

where $E\lambda$ is the measured spectral emitted radiance at a particular wavelength (mW m^2 sr cm^{-1}), $\epsilon\lambda$ is the spectral emissivity at a particular wavelength and $T\lambda$ is the brightness temperature at a particular wavelength (K) (Price, 1989).

The emissivity of a target is a dimensionless number that is the ratio of the actual radiant emittance from a target and that from a blackbody at the same temperature (Norman *et al.*, 1995). Most natural targets are not blackbody emitters, but rather are greybody emitters (i.e. $\epsilon < 1.0$), thus only emit a portion of the energy they receive. It is the characteristics of the target itself that principally determine this portion (for example, surface roughness, architecture, LAI and water content) (Becker *et al.*, 1986; Elvidge, 1988; Norman *et al.*, 1990; Schmugge *et al.*, 2002) and so the Earth's surface has a highly variable emissivity, both temporally and spatially (Cracknell and Xue, 1996). If MIR reflectance is to be derived accurately from the total radiant energy measured in AVHRR channel 3, knowledge of the emissivity of the forest canopy over space and time is required. To acquire the emissivity value of a target remains a difficult task, particularly within MIR wavelengths (Givri, 1995). Furthermore, emissivity values in the MIR spectral region of vegetation have been found to range between 0.7 and 1.0 (Sutherland, 1986; Salisbury and D'Aria, 1994; Goita and Royer, 1997). This therefore, introduces a known uncertainty parameter to our proposal to use MIR reflectance to monitor ENSO-related drought stress of the rainforests of the DVCA.

One approach to overcoming uncertain emissivity is to assume an emissivity value and use this when deriving MIR reflectance over space and time or to ignore its influence (e.g. use $\epsilon = 1.0$). However, if the actual emissivity deviates from this constant value then the predicted MIR reflectance will be in error and ultimately any information on the ENSO impact of the forests of DVCA may be compromised. The problem of uncertainty in the magnitude of the emissivity has been evident in other studies (Becker, 1987; Coll *et al.*, 1994, Franca and Cracknell, 1994) and has been demonstrated to present more significant problems in using remotely sensed data than those induced by atmospheric effects (Wan and Dozier, 1989). It was important, therefore, to ascertain whether uncertainty in the magnitude of the emissivity in the derivation of MIR reflectance from the total radiant energy measured in AVHRR channel 3 over the DVCA compromised the use of MIR reflectance for the monitoring of ENSO-related drought stress.

7 Data and Image Pre-processing

The remotely sensed data used in this study were selected from a series of 41 NOAA-14 AVHRR images acquired during the period between November 1996 and December 1998, covering the 1997–98 ENSO event. Application of a cloudmask algorithm (Franca and Cracknell, 1995) reduced the number of useable images to 15. Data extracted from these images were compared to daily rainfall totals measured at the field site.

Data acquired in all five AVHRR channels were converted to radiances (Kidwell, 1991). Conversion of data in channels 1 and 2 took into account the orbit degradation effects associated with using pre-launch calibration coefficients (Rao *et al.*, 1996). The remotely sensed radiances in visible, NIR and MIR wavelengths were then processed to derive the desired variables for subsequent analyses. Here the ratio of exiting to incident surface radiance for each AVHRR pixel in the visible, NIR and MIR channels was obtained. The incident radiance was determined by applying an atmospheric correction model, based on work by Iqbal (1983), to the solar radiance within the limits of AVHRR channel 1 (visible), channel 2 (NIR) and channel 3 (MIR). The atmospheric variables accounted for included molecular and aerosol attenuation and absorption by ozone, water vapour (calculated from radiation in channels 4 and 5 following the Dalu (1986) procedure) and the remainder of the atmospheric gases, which were treated together as a well-mixed gas. The exiting radiance was determined by inversely applying the atmospheric model to the satellite sensor radiance measured at the top of the atmosphere. Running the atmospheric correction model required inputs describing the solar and satellite geometrics, predictions of aerosols and ozone amounts, aerosol optical properties and sensor characteristics.

The resultant atmospherically corrected signals from this image processing procedure constituted surface reflectance in visible and NIR wavelengths and a full MIR radiation signal (mW m^2 sr cm^{-1}) (i.e. inclusive of reflected and emitted components). The reflectances in visible and NIR wavelengths were combined in the NDVI.

To obtain MIR reflectance the model described in equations (2–4) was applied. In this case, the land surface temperature was determined using a split window algorithm described by Price (1989). By applying Planck's function to the surface temperature over the bandwidth of AVHRR channel 3 and the relevant channel response coefficients supplied by Planet (1988), the blackbody radiance was obtained. Emissivity was then applied to the blackbody radiance to derive the emitted surface radiance. The emitted surface radiance value was subtracted from the total exiting radiance to leave the purely reflected component at MIR wavelengths, which is used in conjunction with the incident solar radiance to determine the bi-directional reflectance value for the MIR channel (i.e. MIR reflectance).

8 Exploring Uncertainty in Emissivity

Pre-processing of the 15 images was undertaken twice using different estimates of the emissivity parameter to establish the impact of uncertainty on the derivation of MIR reflectance. One pre-processing run used the emissivity value of 1.0. The other pre-processing run used emissivity values calculated from an image-based approach. The selection of the image-based approach was based on a comparative study undertaken by Sobrino *et al.* (2001) of four of the most recently proposed algorithms for the independent calculation of emissivity from data acquired by the NOAA AVHRR sensor. For the purposes of this study, each of these algorithms was evaluated against a specified set of requirements (Table 8.1). These requirements are that the method

Table 8.1 *Evaluation of the four algorithms for the independent calculation of emissivity in MIR wavelengths from NOAA AVHRR imagery*

Method	Approach	Comment
1. The Thermal Infrared Radiance Ratio Model (TS-RAM) method proposed by Goita and Royer (1997).	Empirical version of TISI.	Empirically derived from existing studies.
2. The Temperature Independent Spectral Indices (TISI) method proposed by Li and Becker (1993).	Paired day- and night-time acquired imagery. Emitted radiation acquired at night used to evaluate the daytime emissivity.	Theoretically rigorous and produces satisfactory estimates of emissivity *but* the requirement for matched and accurately georectified day and night images is unlikely to be obtained for tropical areas which are frequently cloudy.
3. The Δday method proposed by Goita and Royer (1997).	Based upon the TISI method (Becker and Li, 1990) in that it uses images taken at different times.	Has two assumptions that are unlikely to be satisfied. One is that emissivity values in channel 3 and the two thermal infrared (TIR) channels (4 and 5) remain constant between images. Another is that there is a reasonable temperature gradient between images. This would be difficult to achieve in tropical areas.
4. The Normalized Difference Vegetation Index (NDVI) Threshold method proposed by Sobrino *et al.* (2001).	Uses the NDVI vegetation index to provide information about the state of the land surface. Assumes that pixels with a low value for NDVI are a mixture of soil and vegetation and high values represent a closed canopy and are assigned an emissivity value accordingly.	Easy to implement as NDVI straightforward to calculate and emissivity values could be produced efficiently on a pixel-by-pixel basis. However, NDVI has limited sensitivity in tropical forests and highly attenuated in tropical atmospheres.

adopted should be adapted easily to an operational situation, require minimal ancillary data and that emissivity values are produced on a pixel-by-pixel basis. From this, it was concluded that the TS-RAM model method (Goita and Royer, 1997) was the only approach that could be used in this case study.

The TS-RAM method allows for the calculation of emissivity values based on data acquired for each pixel, requires no ancillary data (we used what was available in the published literature) and could easily be implemented on an operational basis. The method is based on the theoretically rigorous Temperature Independent Spectral Index (TISI) method (Li and Becker, 1993) that uses the emitted radiation sensed at night, which consequently comprises no reflectivity information, to evaluate the daytime emitted radiation. Instead of using a matched pair of day and night images the TS-RAM model uses an empirical linear relationship derived between radiances in channel 3 and the thermal channels (4 and 5) to approximate the TISI parameter from which the emissivity can be calculated (equation 6)

$$\epsilon_{ch3} = 1 - \frac{B_{ch3}(T_{ch3}) - TISI_{ch3i} B_{ch3}(T_i)}{\frac{1}{\pi} f_{ch3}(\theta, \psi, \theta_s, \psi_s) E_{ch3} \tau_{ch3}(\theta_s, \psi_s) \cos(\theta_s) + R_{atch3}} \tag{6}$$

where $B_{ch3}(T_{ch3})$ is the radiance in channel 3, $B_{ch3}(T_i)$ is the surface emitted radiance calculated from the thermal channel, i is channel 4 or 5 of the AVHRR sensor, $f_{ch3}(\theta, \psi, \theta_s, \psi_s)$ is the angular form factor (assumed as 1 at view angles less than 30 °), E_{ch3} is the solar irradiance at the top of the atmosphere in channel 3, $\tau_{ch3}(\theta_s, \psi_s)$ is the atmospheric transmittance in solar direction in channel 3, $\cos(\theta_s)$ is the cosine solar zenith angle and R_{atch3} is the diffuse downwelling solar irradiation.

The TISI parameter ($TISI_{ch3i}$) is approximated from

$$TISI_{ch3i} = P(\theta_s) \frac{B_{ch3}(T_{ch3})}{B_i(T_i)} + I(\theta_s) \tag{7}$$

where P and I are slope and intercept values respectively derived from simulation modelling studies carried out by Sobrino *et al.* (2001). The simulation conditions were reproduced using the MODTRAN 2.0 code and an emissivity database containing the emissivities for 124 different materials. For this case study, these published P and I values reported for a standard tropical atmosphere at a solar zenith angle of 45.7 ° were used.

9 Data Analysis and Results

The 15 images were spatially co-registered using a nearest neighbour approach (with an RMSE of <1 pixel). The pixels representing the DVCA were identified, and their mean NDVI and MIR reflectance, derived using the assumed ($\epsilon = 1.0$) and calculated ($\epsilon =$ via TS-RAM method) emissivity values, extracted.

The two data sets of mean MIR reflectance derived for each image were compared, revealing that using an assumed emissivity value led to a lower MIR reflectance compared with the MIR reflectance derived using a calculated emissivity value (Figure 8.3). The magnitude of the difference between the two mean MIR reflectances

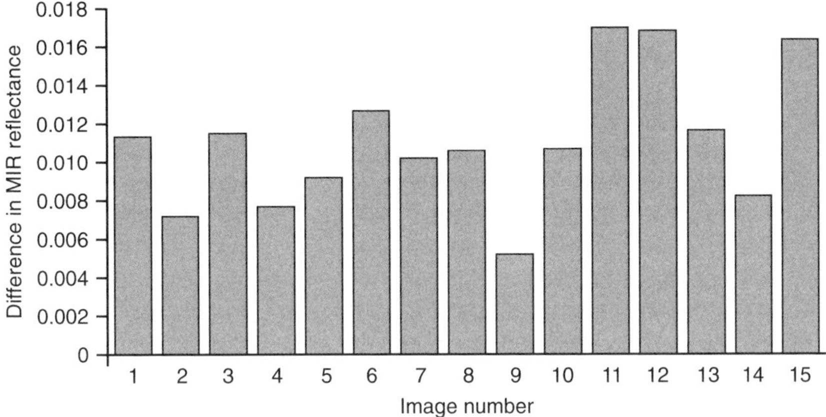

Figure 8.3 *Difference in MIR reflectance derived using the assumed and calculated emissivities*

for each image varies temporally. Further, the two data sets are different statistically at the 0.001 significance level.

To determine if the variability in mean MIR reflectance across the 15 images was a function of ENSO-induced drought conditions, correlation analyses between mean MIR reflectance and the amount of rainfall preceding each image were undertaken. Rainfall was summed for a range of periods from 1 to 20 days that were lagged 1, 3, 5, 10 and 20 days before each image (Figure 8.4) and the correlation between the mean MIR reflectance responses from each image was plotted against the derived rainfall totals. This was conducted for the mean MIR reflectances derived using assumed and calculated emissivity values. The mean MIR reflectances derived using an assumed emissivity value were weakly related to rainfall totals, with a maximum correlation achieved using a three-day rain period at a 13 day lag ($r_s = -0.42$, insignificant at 0.05 level) (Figure 8.5). The mean MIR reflectances derived using an emissivity value calculated using the TS-RAM method were more strongly related to rainfall, with a maximum correlation achieved using a five-day rain period at a 14 day lag ($r_s = -0.78$, significant at 0.01 level) (Figure 8.6).

The above results indicate that the emissivity parameter used when deriving MIR reflectance is important. Uncertainty in emissivity has implications for the derived relationships between remotely sensed MIR reflectance and environmental variables, which are ultimately used for monitoring purposes. In this case study, using the emissivity values calculated using the TS-RAM method significantly increased the correlation between MIR reflectance and rainfall. This is of particular relevance since MIR reflectance appears to offer more useful information about the impact of ENSO-drought on the rainforest of Danum Valley than the more commonly used NDVI (Figure 8.7). Relationships between NDVI and rainfall were weak and insignificant. These results indicate that MIR reflectance is more sensitive to change in canopy properties, such as leaf biomass, foliar chlorophyll activity and water concentration, associated with ENSO-induced drought at this site than the NDVI. There is a good case, therefore, for reducing the uncertainty in the emissivity parameter when

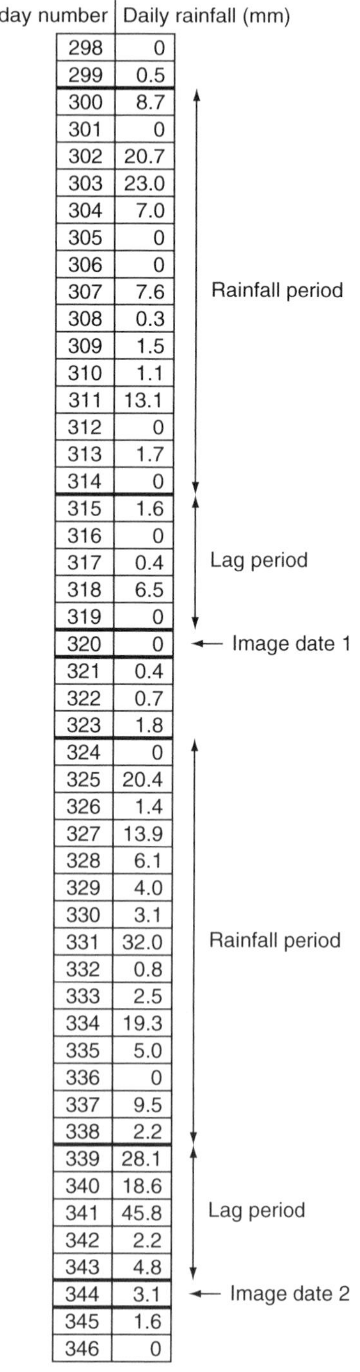

Julian day number	Daily rainfall (mm)
298	0
299	0.5
300	8.7
301	0
302	20.7
303	23.0
304	7.0
305	0
306	0
307	7.6
308	0.3
309	1.5
310	1.1
311	13.1
312	0
313	1.7
314	0
315	1.6
316	0
317	0.4
318	6.5
319	0
320	0
321	0.4
322	0.7
323	1.8
324	0
325	20.4
326	1.4
327	13.9
328	6.1
329	4.0
330	3.1
331	32.0
332	0.8
333	2.5
334	19.3
335	5.0
336	0
337	9.5
338	2.2
339	28.1
340	18.6
341	45.8
342	2.2
343	4.8
344	3.1
345	1.6
346	0

Figure 8.4 *Example of 15 Julian day rainfall period taken five days (lag period) before image date. The total rainfall used in analyses for image date 1 would be 84.7 mm. This process is repeated for each of the 15 images and the Spearman's rank correlation (r_s) between total rainfall and remotely sensed data derived for increasing lag periods ($n = 15$)*

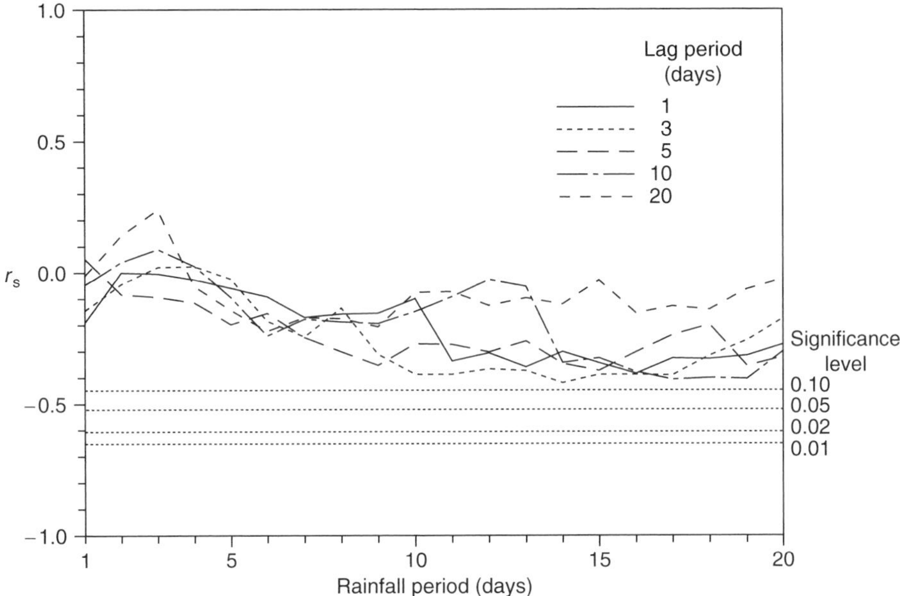

Figure 8.5 *Relationship between rainfall of different periods and at different lags to imagery and MIR reflectance derived using a constant emissivity value*

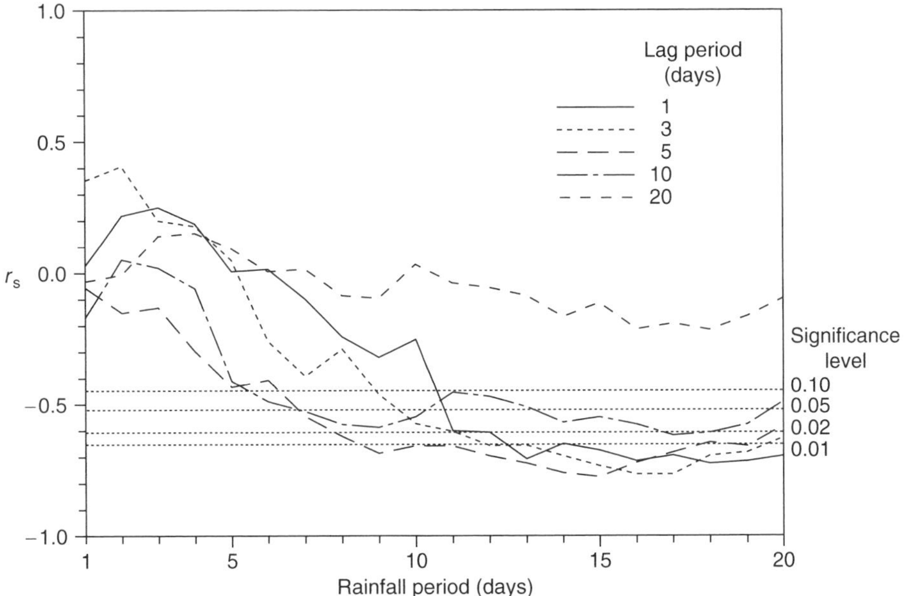

Figure 8.6 *Relationship between rainfall of different periods and at different lags to imagery and MIR reflectance derived using emissivity values calculated from the imagery using the TS-RAM algorithm*

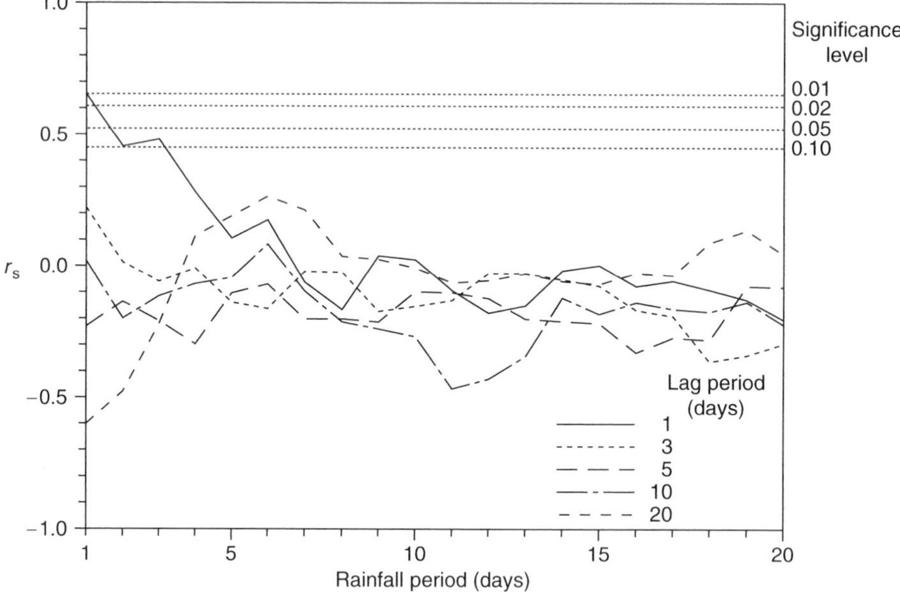

Figure 8.7 Relationship between rainfall of different periods and at different lags to imagery and NDVI

deriving MIR reflectance from the total radiant energy measured in AVHRR channel 3, and being more certain of the emissivity parameter used.

10 How Certain is Certain?

The above results illustrate that to use NOAA AVHRR data optimally to monitor the effects of ENSO on tropical rainforests such as those at the DVCA, MIR reflectance should be considered instead of the NDVI, but only if emissivity is known or can be predicted accurately. The mean emissivities calculated for the DVCA for each image vary temporally and spatially as would be expected. However, some values appear to be lower than would be expected, though they fall into the range reported in the literature (e.g. Nerry *et al.*, 1990; Wan and Dozier, 1989). It is very tempting to use these values in the derivation of MIR reflectance as they produce MIR reflectances that are highly correlated with rainfall and these relationships can be used in the development of a monitoring system. Thus, the scenario presented is one of further uncertainty arising as a result of measurement error (i.e. the emissivities produced by the TS-RAM method may be inaccurate). The question is whether or not these MIR reflectances should be used. To help answer that question, the accuracy of the calculated emissivities needs to be evaluated.

It was not possible to validate the emissivities obtained over the DVCA for each image in the time-series based on the data available. As an alternative, therefore, the emissivities produced by the TS-RAM method of one image were compared with those derived using a more rigorous approach, the Becker and Li (1990) TISI method. The TISI method requires a day–night pairing of images. However, only one of the 15 day images used in the time sequence had a corresponding night pairing data set and, due to cloud occurrence, this was of the area around the DVCA, rather than the DVCA itself, and therefore not ideal. The land cover of the Sabah state of Malaysia has a 60 % coverage of dipterocarp forests (Kasperon *et al.*, 1990), with the remainder of the land cover including logged rainforests, oil palm and cocoa plantations, degraded rainforest, grasslands and urban areas, thus providing a range of emissivities to be compared. The emissivities produced using the TS-RAM and TISI approaches for 31 283 pixels extracted from the image were compared (Figure 8.8). A significant and large correlation ($r = 0.86$; at 0.01 level) was obtained between the predicted emissivities. Overall, the TS-RAM approach underestimated the emissivity of a pixel by 15.177 % in relation to emissivities obtained using the TISI method. This error adjustment could be applied to the calculated emissivities using the TISI method and MIR reflectances derived. However, without the ability to test the emissivities of each pixel of the DVCA in the time-series there would still be a known uncertainty in our remotely sensed data set. It is apparent, therefore, that in an attempt to reduce a known uncertainty further uncertainties have been uncovered. At what point, therefore, do we accept that we have uncertainty in our remotely sensed data but continue to use them?

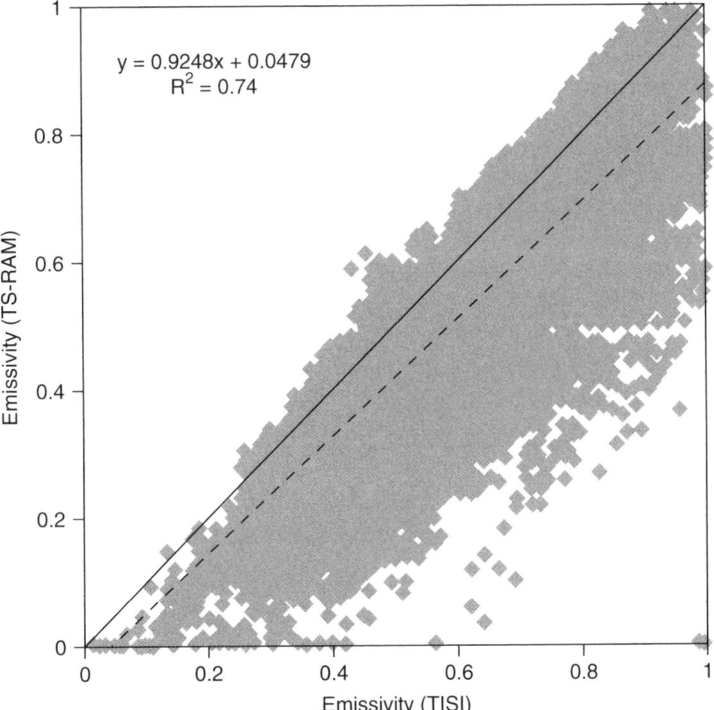

Figure 8.8 *Plot of the emissivities produced using the TS-RAM and TISI approaches applied to one image for 31 283 NOAA AVHRR pixels over Sabah, Malaysia*

11 Conclusions

This chapter explored one uncertainty associated with the use of NOAA AVHRR MIR reflectance data to monitor the impact of ENSO-induced drought on a tropical forest ecosystem of Sabah, Malaysia. It demonstrated that the use of MIR reflectance for this application shows considerable potential. However, this was compromised by the uncertainty associated with the emissivity parameter used when deriving MIR reflectance from the total radiant energy measured by the AVHRR sensor in channel 3. The processing of NOAA AVHRR imagery to derive MIR reflectance is a computationally demanding process and is still evolving (e.g. Nerry *et al.*, 1998; Roger and Vermote, 1998). This, coupled with the realization that there is an important uncertainty associated with it, may discourage its use. However, the use of an emissivity parameter increased the correlations between MIR reflectance and rainfall above those obtained with the widely advocated NDVI.

Whilst there is no doubt that remotely sensed data can contribute greatly to our knowledge of terrestrial biospheric change, and has the potential to go further, with more useful employment of the data (e.g. to drive ecosystem simulation models) and launch of new satellite sensors designed for the purpose in mind (e.g. EOS Terra), there are also many uncertainties involved with its use. This study has illustrated the impact of just one area of uncertainty on a specific application There are, however, uncertainties at every stage of the use of remote sensing data in environmental research, for example, in data acquisition, calibration, processing and in its application and these are compounded further when using a multi-temporal data set such as that used in this case study. Moreover, we have still to wholly understand the nature of the data that are acquired in such great quantities and have, therefore, yet to realize the full potential of remote sensing. By identifying the impacts of uncertainty of remotely sensed data and ultimately overcoming them there is a vast and growing archive of data from which important information about the terrestrial biosphere can be extracted.

From this case study it is recommended that the use of MIR reflectance become more commonplace. This will be facilitated by other sensors that acquire MIR radiation (Envisat Advanced Along Track Scanning Radiometer (AATSR) and Terra Moderate Resolution Imaging Spectroradiometer (MODIS)), methods to use the data (e.g. Petitcolin and Vermote, in press) and the generation of global data sets of MIR surface reflectance (e.g. AVHRR Pathfinder II; El Saleous *et al.*, 2000). However, before these data sets can be used on an operational basis the user must examine how the MIR reflectance values have been derived from full MIR radiation, in particular, how emissivity has been calculated and, thus, use the data with greater certainty.

Acknowledgements

This chapter relates in part to research undertaken in association with the Royal Society's South-East Asia RainForest Research Programme and is programme publication A/361. We are grateful to all involved, in particular Professor R. Walsh of University of Wales, Swansea.

References

Achard, F. and Estreguil, C., 1995, Forest classification of southeast Asia using NOAA AVHRR data, *Remote Sensing of Environment*, **54**, 198–208.

Anyamba, A. and Eastman, J. R., 1996, Inter-annual variability of NDVI over Africa and its relation to El Nino/Southern Oscillation, *International Journal of Remote Sensing*, **17**, 2533–48.

Anyamba, A., Tucker, C. J. and Eastman, J. R., 2001, NDVI anomaly patterns over Africa during the 1997/98 ENSO warm event, *International Journal of Remote Sensing*, **22**, 1847–59.

Artaxo, P., Ferandez, E. T., Martins, J. V., Yamasoe, M. A., Hobbs, P. V., Maenhaut, W., Longo, K. M. and Castanho, A., 1998, Large-scale aerosol source apportionment in Amazonia, *Journal of Geophysical Research*, **103**, 31837–47.

Batista, G. T., Shimabukuro, Y. E. and Lawrence, W. T., 1997, The long-term monitoring of vegetation cover in the Amazonian region of northern Brazil using NOAA–AVHRR data, *International Journal of Remote Sensing*, **18**, 3195–210.

Becker, F., 1987, The impact of spectral emissivity on the measurement of land surface temperature from a satellite, *International Journal of Remote Sensing*, **8**, 1509–22.

Becker, F. and Li, Z.-L., 1990, Temperature-independent spectral indices in thermal infrared bands, *Remote Sensing of Environment*, **32**, 17–33.

Becker, F., Nerry, F., Ramanantsizehena, P. and Stoll, P. M., 1986, Mesures d'emissivite angulaire par reflexion dans l'infrarouge thermique – implications pour la télédetection, *International Journal of Remote Sensing*, **7**, 1751–62.

Bonan, G. B., 1997, Effects of land use on the climate of the US, *Climate Change*, **37**, 449–86.

Boyd, D. S. and Curran, P. J., 1998, Using remote sensing to reduce uncertainties in the global carbon budget: the potential of radiation acquired in middle infrared wavelengths, *Remote Sensing Reviews*, **16**, 293–327.

Boyd, D. S. and Duane, W. J., 2001, Exploring spatial and temporal variations in middle infrared reflectance at 3.75μm measured from the tropical forests of west Africa, *International Journal of Remote Sensing*, **22**, 1861–78.

Boyd, D. S., Foody, G. M. and Curran, P. J., 1999, Reflected middle infrared radiation for estimating forest biomass of Cameroonian rainforests, *International Journal of Remote Sensing*, **20**, 1017–24.

Boyd, D. S., Wicks, T. E. and Curran, P. J., 2000, Use of middle infrared radiation to estimate the leaf area index of a boreal forest, *Tree Physiology*, **20**, 755–60.

Coll, C., Caselles, V. and Schmugge, T. J., 1994, Estimation of land surface emissivity differences in the split-window channels of AVHRR, *Remote Sensing of Environment*, **48**, 127–34.

Couturier, S., Taylor, D., Siegert, F., Hoffman, A. and Bao, M. Q., 2001, ERS SAR: a potential real-time indicator of the proneness of modified rainforests to fire, *Remote Sensing of Environment*, **76**, 410–17.

Cracknell, A. P., 1997, *The Advanced Very High Resolution Radiometer* (London: Taylor and Francis).

Cracknell, A. P. and Xue, Y., 1996, Thermal inertia determination from space – a tutorial review, *International Journal of Remote Sensing*, **17**, 431–61.

Curran, P. J. and Foody, G. M., 1994, Environmental issues at regional to global scales, in G. M. Foody and P. J. Curran (eds), *Environmental Remote Sensing from Regional to Global Scales* (Chichester: Wiley), pp. 1–7.

Dalu, G., 1986, Satellite remote sensing of atmospheric water vapour, *International Journal of Remote Sensing*, **7**, 1089–97.

DeFries, R., Hansen, M., Steininger, M., Dubayah, R., Sohlberg, R. and Townshend, J., 1997, Subpixel forest cover in central Africa from multisensor, multitemporal data, *Remote Sensing of Environment*, **60**, 228–46.

Diaz, H. F. and Margraf, V., 1992, Introduction, in H. F. Diaz and V. Margraf (eds), *El Niño – Historical and Paleoclimatic Aspects of the Southern Oscillation* (Cambridge: Cambridge University Press), pp. 1–7.

Duchemin, B., Goubier, J. and Courrier, G., 1999, Monitoring phenological key stages and cycle duration of temperate deciduous forest ecosystems with NOAA/AVHRR data, *Remote Sensing of Environment*, **67**, 68–82.

Eastman, J. R. and Fulk, M. A., 1993, Long sequence time series evaluation using standardized principal component analysis, *Photogrammetric Engineering and Remote Sensing*, **53**, 1649–58.

Eklundh, L., 1998, Estimating relations between AVHRR NDVI and rainfall in East Africa at 10-day and monthly time scales, *International Journal of Remote Sensing*, **19**, 563–8.

Elvidge, C. D., 1988, Thermal infrared reflectance of dry plant materials: 2.5– 20.0 μm, *Remote Sensing of Environment*, **26**, 265–85.

El Saleous, N. Z., Vermote, E. F., Justice, C. O., Townshend, J. R. G., Tucker, C. J. and Goward, S. N., 2000, Improvements in the global biospheric record from the Advanced Very High Resolution Radiometer AVHRR, *International Journal of Remote Sensing*, **21**, 1251–78.

Fazakas, Z. and Nilsson, M., 1996, Volume and forest cover estimation over southern Sweeden using AVHRR data calibrated with TM data, *International Journal of Remote Sensing*, **17**, 1701–9.

Foody, G. M. and Curran, P. J., 1994, Estimation of tropical forest extent and regenerative stage using remotely sensed data, *Journal of Biogeography*, **21**, 223–44.

Foody, G. M., Boyd, D. S. and Curran, P. J., 1996, Relationships between tropical forest biophysical properties and data acquired in AVHRR channels 1–5, *International Journal of Remote Sensing*, **17**, 1341–55.

Franca, G. B. and Cracknell, A. P., 1994, Retrieval of land and sea temperature using NOAA-11 AVHRR data in north-eastern Brazil, *International Journal of Remote Sensing*, **15**, 1695–712.

Franca, G. B. and Cracknell, A. P., 1995, A simple cloud masking approach using NOAA AVHRR daytime data for tropical areas, *International Journal of Remote Sensing*, **16**, 1697–705.

França, H. and Setzer, A. W., 1998, AVHRR temporal analysis of a savanna site in Brazil, *International Journal of Remote Sensing*, **19**, 3127–40.

Franklin, S. E. and Wulder, M. A., 2002, Remote sensing methods in medium spatial resolution satellite data land cover classification of large areas, *Progress in Physical Geography*, **26**, 173–205.

Gesell, G., 1984, An algorithm for snow and ice detection using AVHRR data: an extension to the APOLLO software package, *International Journal of Remote Sensing*, **10**, 897–905.

Girvi, J. R., 1995, Assessing the relation between emissivity and vegetation with AVHRR, *International Journal of Remote Sensing*, **16**, 2971–88.

Glantz, M. H., 1996, *Current of Change: El Niño's Impact on Climate and Society* (Cambridge: Cambridge University Press).

Goetz, S. J., Prince, S. S., Goward, S. N., Tawley, M. M., Small, J. and Johnston, A., 1999, Mapping net primary production and related biophysical variables with remote sensing: applications to the BOREAS region, *Journal of Geophysical Research*, **104 (D22)**, 27719–34.

Goïta, K. and Royer, A., 1997, Surface temperature and emissivity separability over land surface from combined TIR and SWIR AVHRR data, *IEEE Transactions on Geosciences and Remote Sensing*, **35**, 718–33.

Goward, S. N., Tucker, C. J. and Dye, D. G., 1985, North American vegetation patterns observed with the NOAA-7 Advanced Very High Resolution Radiometer, *Vegetatio*, **64**, 3–14.

Gutman, G., Tarpley, D., Ignatov, A. and Olson, S., 1995, The enhanced NOAA global land dataset from the Advanced Very High Resolution Radiometer, *Bulletin of the American Meteorological Society*, **76**, 1141–56.

Harrison, R. D., 2001, Drought and the consequences of El Niño in Borneo: a case study of figs, *Population Ecology*, **43**, 63–75.

Harvey, L. D. D., 2000, *Climate and Global Environmental Change* (Harlow: Prentice-Hall).

Harwell, E. E., 2000, Remote sensibilities: discourses of technology and the making of Indonesia's natural disaster, in M. Doornbos, A. Saith and B. White (eds), *Forests: Nature, People, Power*, (Oxford: Blackwell), pp. 299–332.

Hess, T., Stephens, W. and Thomas, G., 1996, Modelling NDVI from decadal rainfall in the north east arid zone of Nigeria, *Journal of Environmental Management*, **48**, 249–61.

Houghton, J. T., Jenkins, G. J. and Ephraums, J. J., 1990, *Climate Change. The IPCC Scientific Assessment. Intergovernmental Panel on Climate Change*, (Cambridge: Cambridge University Press).

Houghton, J. T., Meira Filho, L. G., Callander, B. A., Harris, N., Kattenberg, A. and Maskell, K., 1996, *Climate Change 1995 – The Science of Climate Change* (Cambridge: Cambridge University Press).

Houghton, R. A., Skole, D. L., Nobre, C. A., Hackler, J. L., Lawrence, K. T. and Chomentowski, W. H., 2000, Annual fluxes of carbon from the deforestation and regrowth in the Brazilian Amazon, *Nature*, **403**, 301–4.

Hulme, M. and Viner, D., 1998, A climate change scenario for the tropics, *Climatic Change*, **39**, 145–76.

IGBP, 2001, *Global Change Newsletter: LBA Special Edition*, No. 45 (Stockholm: The Royal Swedish Academy of Science).

IPCC, 2001, *IPCC Third Assessment Report – Climate Change 2001* (Geneva: WMO).

Iqbal, M., 1983, *An Introduction to Solar Radiation* (Canada: Academic Press).

Kasperson, J. X., Kasperson, R. E. and Turner, B. L. II, 1995, *Regions at Risk: Comparisons of Threatened Environments* (Tokyo: UN University Press).

Kaufman, Y. J. and Remer, L. A., 1994, Detection of forests using MIR reflectance: an application for aerosol studies, *IEEE Transactions on Geosciences and Remote Sensing*, **32**, 672–81.

Kidwell, K. B., 1991, *NOAA Polar Orbiter Data User's Guide* (Washington, DC: NOAA).

Laporte, N., Justice, C. and Kendall, J., 1995, Mapping the dense humid forest of Cameroon and Zaire using AVHRR satellite data, *International Journal of Remote Sensing*, **16**, 1127–45.

Li, Z.-L. and Becker, F., 1993, Feasibility of land surface temperature and emissivity determination from AVHRR data, *Remote Sensing of Environment*, **43**, 67–85.

Li, Z. T. and Kafatos, M., 2000, Interannual variability of vegetation in the United States and its relation to El Niño Southern Oscillation, *Remote Sensing of Environment*, **71**, 239–47.

Liu, W. T. H., Massambani, O. and Nobre, C. S., 1994, Satellite recorded vegetation response to drought in Brazil, *International Journal of Climatology*, **14**, 343–54.

Los, S. O., 1993, Calibration adjustment of the NOAA AVHRR normalized difference vegetation index without recourse to channel 1 and 2 data, *International Journal of Remote Sensing*, **14**, 1907–17.

Los, S. O., Justice, C. O. and Tucker, C. J., 1994, A global 1 by 1 NDVI data set for climate studies derived from the GIMMS continental NDVI data, *International Journal of Remote Sensing*, **15**, 3519–46.

Lucas, R. M., Honzak, M., Curran P. J., Foody, G. M., Milne, R., Brown, T. and Amaral S., 2000, Mapping the regional extent of tropical forest regeneration stages in the Brazilian Legal Amazon using NOAA AVHRR data, *International Journal of Remote Sensing*, **21**, 2855–81.

Malingreau, J. P., Tucker, C. J. and Laporte, N., 1989, AVHRR for monitoring global tropical deforestation, *International Journal of Remote Sensing*, **10**, 855–67.

McPhadden, M. J., 1999, Genesis and evolution of the 1997–98 El Niño, *Science*, **283**, 950–4.

McVicar, T. R. and Bierwirth, P. N., 2001, Rapidly assessing the 1997 drought in Papua New Guinea using composite AVHRR imagery, *International Journal of Remote Sensing*, **22**, 2109–28.

Moody, A. and Johnson, D. M., 2001, Land-surface phenologies from AVHRR using the discrete Fourier transform, *Remote Sensing of Environment*, **75**, 305–23.

Myneni, R. B. and Williams, D. L. 1994, On the relationship between FAPAR and NDVI, *Remote Sensing of Environment*, **49**, 200–11.

Myneni, R. B., Los, S. O. and Tucker, C. J., 1996, Satellite-based identification of linked vegetation index and sea surface temperature anomaly areas from 1982 to 1990 for Africa, Australia, South America, *Geophysical Research Letters*, **23**, 729–32.

NASA., 2001, www.earth.nasa.gov/visions/researchstrat/index.htm.

Nemani, R. and Running, S., 1997, Land cover characterization using multitemporal red near IR and thermal IR data from NOAA/AVHRR, *Ecological Applications*, **71**, 79–90.

Nerry, F., Petitcolin, F. and Stoll, M. P., 1998, Bidirectional reflectivity in AVHRR channel 3: application to a region in northern Africa, *Remote Sensing of Environment*, **66**, 298–316.

Nicholson, S. E., 1996, A review of climate dynamics and climate variability in Eastern Africa, in T. C. Johnson and E. O. Odada (eds), *The Limnology, Climatology and Paleoclimatology of the East African Lakes* (Amsterdam: Gordon and Breach), pp. 25–56.

Nicholson, S. E. and Entekhabi, D., 1986, The quasi-periodic behaviour of rainfall variability in Africa and its relationship to the Southern Oscillation, *Archiv fur Meteorologie Geophysik und Bioklimatologie*, **A34**, 311–48.

Norman, J. M., Chen, J. L. and Goel, N. S., 1990, Thermal emissivity and infrared temperature dependence of plant canopy architecture and view angle, *Proceedings of 10th Annual International Geoscience and Remote Sensing Symposium* (New Jersey: IEEE), pp. 1747–50.

Norman, J. M., Divakarta, M. and Goel, N., 1995, Algorithms for extracting information from remote thermal-IR observations of the Earth's surface, *Remote Sensing of Environment*, **51**, 157–68.

O'Riordan, T., 2000, Environmental science on the move, in T. O'Riordan (ed.), *Environmental Science for Environmental Management* (Harlow: Prentice-Hall), pp. 1–27.

Petitcolin, F. and Vermote, E., in press, Land surface reflectance, emissivity and temperature from MODIS middle and thermal infrared data, *Remote Sensing of Environment*.

Philander, G. S., 1990, *El Niño, La Niña and the Southern Oscillation* (New York: Academic Press).

Phillips, O. L., Malhi, N., Higuchi, N., Lawrence, W. F., Nunez, P. V., Vasquez, R. M., Laurence, S. G., Ferreira, L. V., Stern, M., Brown, S. and Grace, J., 1998, Changes in the carbon balance of tropical forests: evidence from long-term plots, *Science*, **282**, 439–42.

Planet, W. G., 1988, *Data Extraction and Calibration of TIROS-N/NOAA Radiometers. NOAA Technical Memorandum NESS 107* (Washington, DC: US Department of Commerce, National Oceanic and Atmospheric Administration).

Plisnier, P. D., Serneels, S. and Lambin, E. F., 2000, Impact of ENSO on East African ecosystems: a multivariate analysis based on climate and remote sensing data, *Global Ecology and Biogeography*, **9**, 481–97.

Potter, C., Klooster, S., de Carvalho, C. R., Genovese, V. B., Torregrosa, A., Dungan, J., Bobo, M. and Coughlan, J., 2001, Modelling seasonal and interannual variability in ecosystem carbon cycling for the Brazilian Amazon region, *Journal of Geophysical Research*, **106 (D10)**, 10423–46.

Price, J. C., 1989, Quantitative aspects of remote sensing in the thermal infrared, in J. Asrar (ed.), *Theory and Applications of Optical Remote Sensing* (New York: Wiley).

Rao, C. R. N., Chen, J., Zhang, N., Sullivan, J. T., Walton, C. C. and Weinreb, M. P., 1996, Calibration of meteorological satellite sensors, *Advances in Space Research*, **17**, 11–20.

Rasmusson, E. M., 1985, El Niño and variation in climate, *American Scientist*, **73**, 168–77.

Richard, Y. and Poccard, I., 1998, A statistical study of NDVI sensitivity to seasonal and interannual rainfall variations in Southern Africa, *International Journal of Remote Sensing*, **19**, 2907–20.

Roger, J. C. and Vermote, E. F., 1998, A method to retrieve the reflectivity signature at 3.75 μm from AVHRR data, *Remote Sensing of Environment*, **64**, 103–14.

Ropelewski, C. F. and Halpert, M. S., 1987, Global and regional scale precipitation and temperature patterns associated with El Niño–Southern Oscillation, *Monthly Weather Reviews*, **15**, 1606–26.

Rotenberg, E., Mamane, Y. and Joseph, J. H., 1998, Longwave radiation regime in vegetation-parameterisations for climate research, *Environmental Modelling and Software*, **13**, 361–71.

Ruff, I. and Gruber, A., 1983, Multispectral identification of clouds and Earth surfaces using AVHRR radiometric data, *Proceedings of the 5th Conference on Atmospheric Radiation, Baltimore, MD, 31 October – 4 November* (Washington, DC: AMS), pp. 475–8.

Salisbury, J. W. and D'Aria, D. M., 1994, Emissivity of terrestrial materials in the 3–5 μm atmospheric window, *Remote Sensing of Environment*, **47**, 345–61.

Schmugge, T., French, A., Ritchie, J. C., Rango, A. and Pelgrum, H., 2002, Temperature and emissivity separation from multispectral thermal infrared observations, *Remote Sensing of Environment*, **79**, 189–98.

Shultz, P. A. and Halpert, M. S., 1995, Global analysis of the relationships among a vegetation index, precipitation and land surface temperature, *International Journal of Remote Sensing*, **16**, 2755–77.

Slaymaker, O. and Spencer, T., 1998, *Physical Geography and Global Environmental Change* (Harlow: Longman).

Sobrino, J. A., Raissouni, N. and Li, Z.-L., 2001, A comparative study of land surface emissivity retrieval from NOAA data, *Remote Sensing of Environment*, **75**, 256–66.

Spanner, M. A., Pierce, L., Running, S. W. and Peterson, D. L., 1990, The seasonality of AVHRR data of temperate coniferous forests: relationship with leaf area index, *Remote Sensing of Environment*, **33**, 97–112.

Stoms, D. M. and Estes, J. E., 1993, A remote sensing research agenda for mapping and monitoring biodiversity, *International Journal of Remote Sensing*, **14**, 1839–60.

Suplee, K., 1999, Nature's vicious cycle, *National Geographic*, **March**, 73–95.

Sutherland, R. A., 1986, Broadband and spectral emissivity at 2–18 μm of some natural soils and vegetation, *Journal of Atmospheric and Oceanic Technology*, **3**, 199–202.

Townshend, J. R. G., Justice, C. O., Skole, D., Malingreau, J. P., Cihlar, J., Teillet, P., Sadowski, F. and Ruttenberg, S., 1994, The 1km resolution global dataset – needs of the International Geosphere Biosphere Program, *International Journal of Remote Sensing*, **15**, 3417–41.

Trenberth, K. E. and Shea, D. J., 1987, On the evolution of the Southern Oscillation, *Monthly Weather Review*, **115**, 3078–96.

Trenberth, K. E. and Tepaniak, D. P., 2001, Indices of El Niño evolution, *Journal of Climate*, **14**, 1697–701.

Trudgill. S, 2001, *The Terrestrial Biosphere: Environmental Change, Ecosystem Science, Attitudes and Values* (Harlow: Prentice-Hall).

Unganai, L. S. and Kogan, F. N., 1998, Drought monitoring and corn yield estimation in Southern Africa from AVHRR data, *Remote Sensing of Environment*, **63**, 219–32.

Van Gardingen, P. R., Foody, G. M. and Curran, P. J., (eds), 1997, *Scaling Up: From Cell to Landscape*, Society for Experimental Biology, Seminar Series 63 (Cambridge: Cambridge University Press).

Verdin, J., Funk, C., Klaver, R. and Roberts, D., 1999, Exploring the correlation between Southern Africa NDVI and Pacific sea surface temperatures: results for the 1998 maize growing season, *International Journal of Remote Sensing*, **20**, 2117–24.

Walker, B. and Steffen, W., 1997, *The Terrestrial Biosphere and Global Change: Implications for Natural and Managed Ecosystems* (Stockholm, Sweden: The International Geosphere-Biosphere Programme, International Council of Scientific Unions).

Walsh, R. P. D. and Newbery, D. M., 1999, The ecoclimatology of Danum, Sabah, in the context of the world's rainforest regions, with particular reference to dry periods and their impacts, *Philosophical Transactions of the Royal Society London B*, **354**, 1869–83.

Wan, Z. and Dozier, I., 1989, Land surface temperature measurement from space, physical principles and inverse modelling, *IEEE Transactions on Geoscience and Remote Sensing*, **27**, 268–78.

Wellens, J., 1997, Rangeland vegetation dynamics and moisture availability in Tunisia: an investigation using satellite and meteorological data, *Journal of Biogeography*, **24**, 845–55.

Whitmore, T. C., 1998, Potential impact of climatic change on tropical rainforest seedlings and forest regeneration, *Climatic Change*, **39**, 429–38.

Wickland, D. E., 1989, Future directions for remote sensing in terrestrial ecological research, in G. Asrar (ed.), *Theory and Applications of Optical Remote Sensing* (New York: Wiley), pp. 691–724.

Yang, L., Wylie, B. K., Tieszen, L. L. and Reed, B. C., 1998, An analysis of relationships among climate forcing and time-integrated NDVI of grasslands over the US northern and central plains, *Remote Sensing of Environment*, **65**, 25–37.

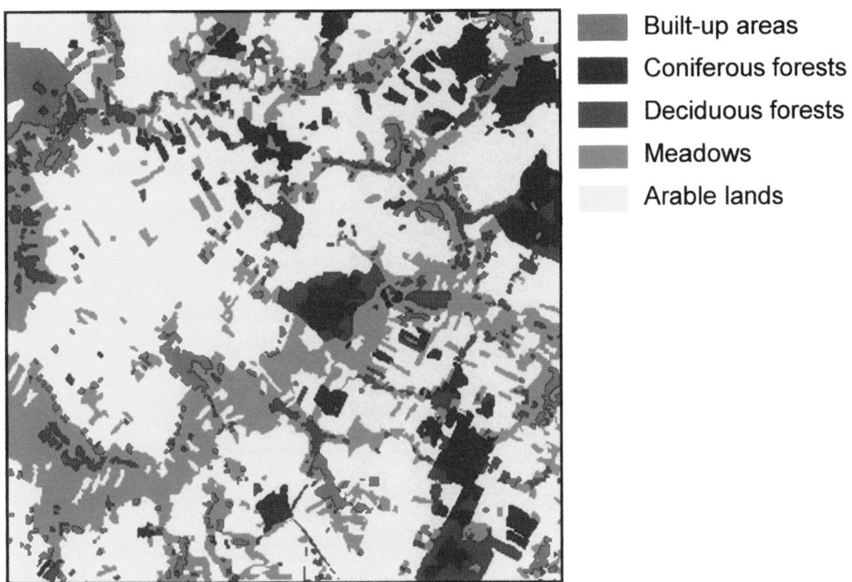

Plate 1 Central part of Kolno Upland. Ground data derived from visual interpretation of aerial images (see Figure 7.2)

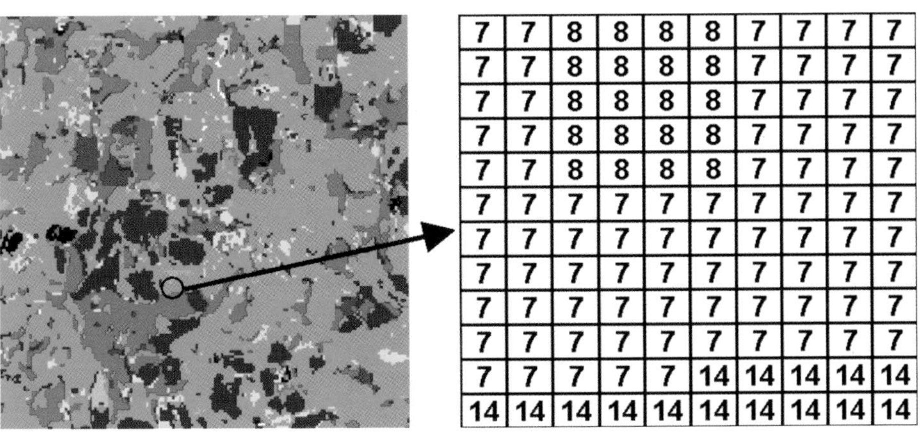

7	7	8	8	8	8	7	7	7	7
7	7	8	8	8	8	7	7	7	7
7	7	8	8	8	8	7	7	7	7
7	7	8	8	8	8	7	7	7	7
7	7	8	8	8	8	7	7	7	7
7	7	7	7	7	7	7	7	7	7
7	7	7	7	7	7	7	7	7	7
7	7	7	7	7	7	7	7	7	7
7	7	7	7	7	7	7	7	7	7
7	7	7	7	7	7	7	7	7	7
7	7	7	7	7	14	14	14	14	14
14	14	14	14	14	14	14	14	14	14

Plate 3 Pixel-based structure of the LCMGB with only a single land cover class code per pixel. In this example, codes 7, 8 and 14 represent rough grass, improved grass and deciduous woodland respectively (see Chapter 9)

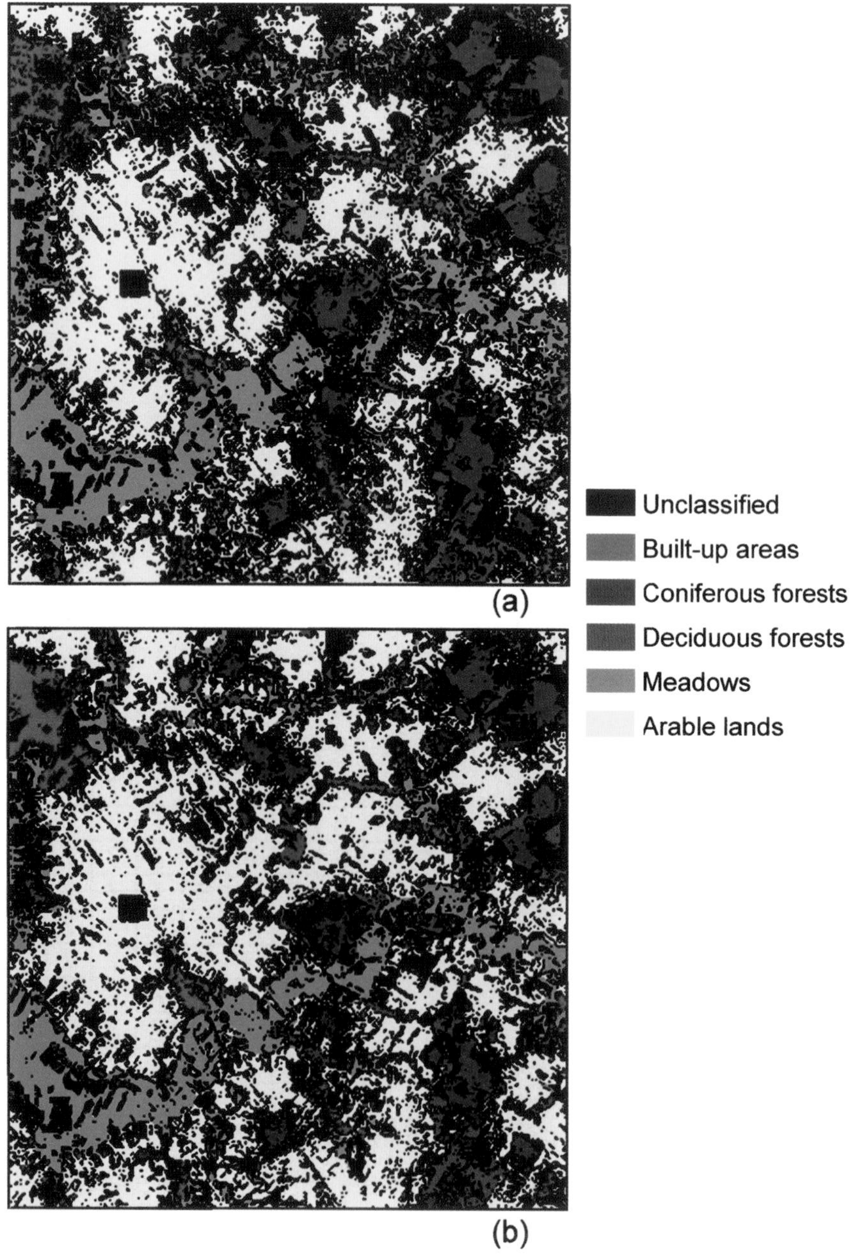

Plate 2 Results of ANN textural classification (a) without, and (b) with ranges (see Chapter 7)

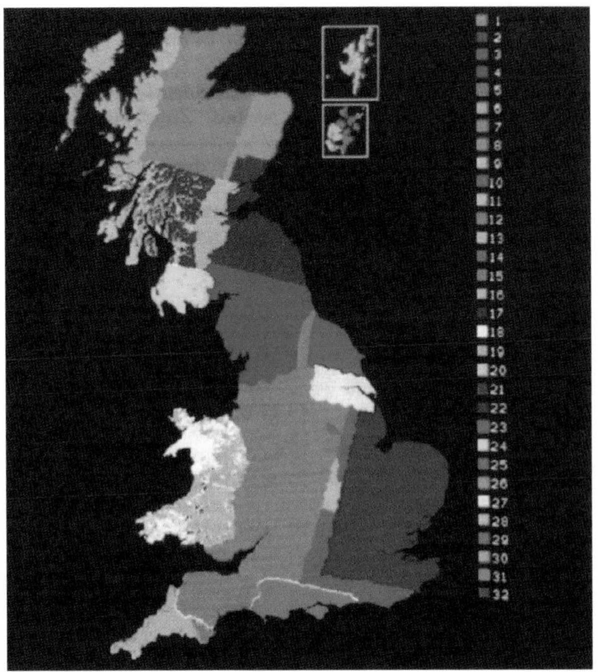

Plate 4 A crude map showing the areas classified from each image in the production of the LCMGB. Of the 32 areas, 16 represent summer–winter composite images and 16 single-date images (see Chapter 9)

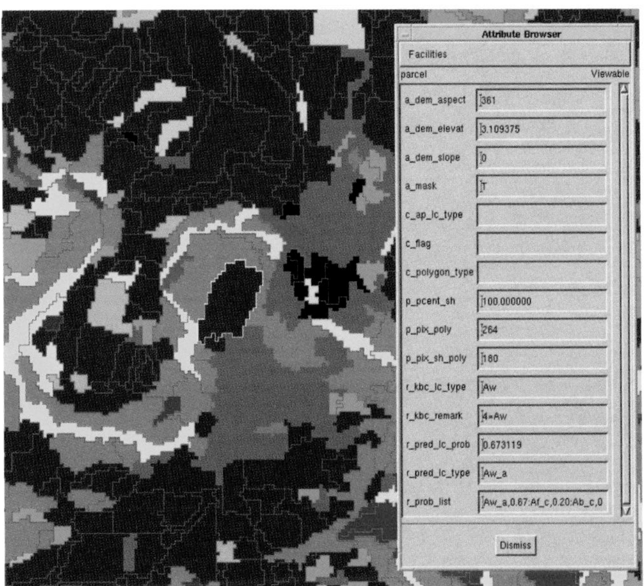

Plate 5 An example of the LCM2000 data, with an attribute browser showing a summary of the information held on a selected land parcel (see Chapter 9)

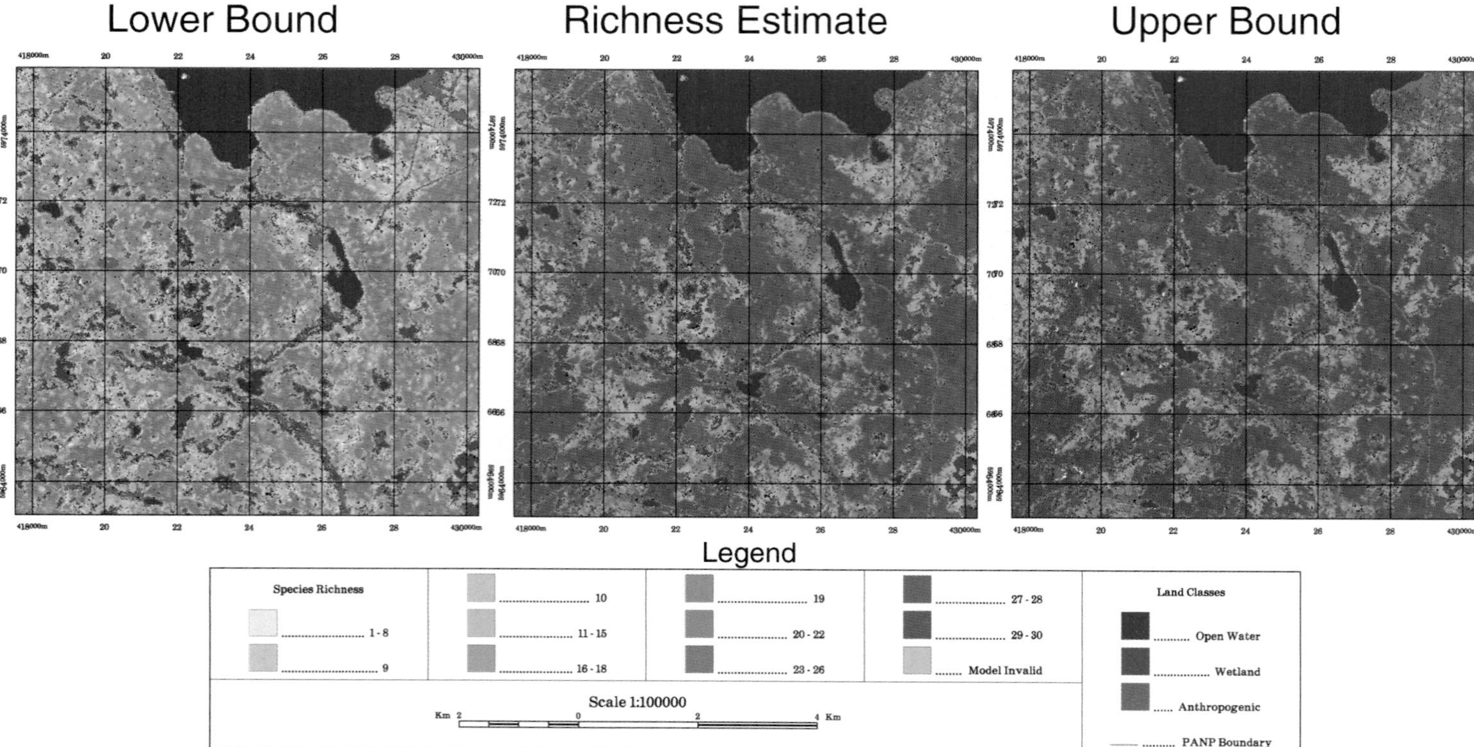

Plate 6 Species richness map together with upper and lower bounds derived from applying the estimated landscape variables in the multivariate regression model (see Chapter 11)

9

Land Cover Map 2000 and Meta-data at the Land Parcel Level

G. M. Smith and R. M. Fuller

1 Introduction

If available, the retention and provision of uncertainty information within large, especially national level, data sets is often limited to a single estimate for the whole data set. The uncertainty within the data set is not confined to the whole data set level, and is derived from uncertainty in the individual 'sampling units' or 'data items', such as fields or land parcels in a land cover map. The amount of data item uncertainty varies between data items due to the unique circumstances surrounding their generation. Parcel-based methodologies for land cover mapping provide a means of retaining uncertainty information and distributing it to the user at the data item level. An example of a parcel-based land cover mapping methodology is presented in this chapter to highlight the problems of deriving uncertainty or quality information within a complex production flowline.

Land Cover Map 2000 (LCM2000) was a survey of the United Kingdom (Fuller *et al.*, 2002) with the aim of mapping widespread Broad Habitats (Jackson, 2000). It updated and upgraded the pixel-based Land Cover Map of Great Britain (LCMGB) (Fuller *et al.*, 1994a). The LCM2000 was parcel-based, using 'land parcels' generated from image segmentation. The thematic classification of the land parcels was based primarily on composite images, derived mainly from Landsat Thematic Mapper (TM) data acquired in both the summer and winter, and the parcel-based classification used mean spectral values for each spectral band, based on the core pixels within a land parcel after shrinking to exclude edge pixels. The classification procedure

Uncertainty in Remote Sensing and GIS. Edited by G.M. Foody and P.M. Atkinson.
 ISBN: 0-470-84408-6

applied was a conventional maximum likelihood algorithm. Class membership probabilities, spatial context and external ancillary data were used in knowledge-based corrections to identify and correct land parcels classified with low confidence and/or those with classes out of their natural context.

The complex production flowline of LCM2000 required and produced a large amount of data, which provided a rich source of information compared with the single land cover class that was recorded previously by conventional pixel-based approaches. The philosophy behind the LCM2000 production was to retain as much information (e.g. intermediate results) as possible at the land parcel level. The 30 or so attributes attached to each land parcel can be grouped under the following headings: identification (e.g. unique identifier); context (non-remotely sensed ancillary information, e.g. elevation); control (information to guide the production, e.g. known (*a priori*) land cover type of a parcel); result (outputs of the production procedure, e.g. the predicted land cover type and a measure of its probability of membership to the predicted class), remark (additional information associated with the classification procedure and results) and image information (e.g. number of pixels extracted).

This chapter describes the information that can be stored on or made available from land parcel data structures. The use of this information as meta-data at the land parcel level is explored and the issues associated with determining uncertainty from meta-data at the land parcel level are considered.

2 Background

The LCMGB was produced in 1990–92 by semi-automated mapping, from Landsat TM images. The TM images, were geometrically registered to the British National Grid using cubic convolution resampling to give a spatial resolution of 25 m. A combination of summer and winter data was used to exploit seasonal differences in surface characteristics and thereby assist in distinguishing the required land cover types (Fuller *et al.*, 1994a, 1994b). For each summer–winter composite scene, examples of different land cover types were outlined manually to train a maximum likelihood classifier (Schowengerdt, 1997). Sub-classes were defined as a step towards the definition of 25 target land cover classes: for example, wheat and barley were sub-classes of 'tilled land' (Kershaw and Fuller, 1992). Typically, 80 sub-classes were required to encapsulate the range and variety of spectral classes in each scene. The reflectance statistics associated with the samples formed the basis of an extrapolation to classify the other pixels in the scene, using a maximum likelihood classifier. A number of simple knowledge-based correction (KBC) procedures helped to correct discrepancies (Groom and Fuller, 1996). The resulting classified data were a single layer raster data set with a single land cover class code for each pixel (Figure 9.1; see plate 3).

The classified images were smoothed to remove isolated single pixels, so as to map minimum units of two pixels or 0.125 ha, with land cover units greater than 1 ha generally mapped accurately. This allowed the major structural details of the landscape to be recorded, essentially at field-by-field scale. The 25 target land cover types were aggregated to 17 classes (Wyatt *et al.*, 1994) for accuracy assessment purposes. Despite evidence which refutes the concept of absolute accuracy (Fuller *et al.*, 1998), it was possible to apportion errors to the survey and to the reference data used in validation. Results suggested that the LCMGB as a whole was perhaps 79–84% 'accurate' with the field records about 90% 'correct' (Fuller *et al.*, 1998).

While the correspondence (confusion) matrices used in the accuracy assessment process identified the distribution of errors in thematic terms, there was no indication of the spatial distribution of error. Visual inspection can identify obvious errors, mostly 'noise' in the data, for example stray pixels of 'urban' cover in arable fields. There was no information which might help, for instance, to identify grass fields mis-classified as 'arable'; nor to suggest that one scene was classified more accurately than its neighbour or that land cover was mapped more accurately by two-date image composites rather than single-date infill. Indeed, there was no detailed record of the exact images used to classify each pixel. A crude map showing the areas classified from each image (Figure 9.2; see plate 4) was produced (Fuller *et al.*, 1994b), but this was not available in a geo-referenced form. There was no record of which pixels were corrected by KBC rules; nor of which pixels changed class through filtering out of isolated pixels. There was no option to record the magnitude of the probability for the most likely class of membership for a pixel, nor information about other classes that were rejected. The rejected classes may have had only marginally smaller probabilities of membership and, in real terms, been more likely.

Therefore, if it is considered that some form of correspondence analysis and a description of the source data can illuminate uncertainty, the LCMGB possessed only very generalized information related to uncertainty at the national level. This uncertainty information is of little use beyond giving a feel for the overall quality of the map. Based on the projects undertaken with the LCMGB and the papers published on its use, few, if any, users consulted this information or incorporated it within applications.

3 Land Cover Map 2000

The parcel-based LCM2000 project recorded the land cover of the United Kingdom in the form of vector land parcels using a methodology developed from the generation of the 1997 Land Cover Map of Jersey (Smith and Fuller, 2001). It updated but also upgraded the pixel-based LCMGB, with an altered classification scheme, an enhanced spatial structure and a refined methodology.

The LCM2000 project was again based on a combination of summer and winter satellite sensor images, taking the same spectral bands (red, near-infrared and mid-infrared) from each date. In the production of the LCM2000 16 target classes were identified (Table 9.1), these were subdivided into 27 sub-classes. The target classes and sub-classes were aggregated to give the widespread Broad Habitats demanded by users. Sub-classes were in turn divided giving 72 class variants; these were identified only where image dates and quality allowed it. The LCM2000 project aimed to map target classes with an accuracy of approximately 90% which was assessed by the correspondence with the results of the field survey component of the Countryside Survey 2000 (Haines-Young *et al.*, 2000).

To produce a parcel-based land cover map, image segmentation was used to identify 'uniform' areas which represented a single land cover type. The segmentation

Table 9.1 *A simplified version of the classes mapped in the LCM2000 (target and sub-classes) and their relationship to the widespread Broad Habitats (BH). The sub-class variants are excluded for clarity. Two widespread Broad Habitats were not mapped. These were (i) 3, boundary and linear features, which were generally too small relative to the spatial resolution of the imagery to be mapped and (ii) 14, rivers, which were merged with 13, standing water*

Widespread BH	LCM Target	LCM sub-classes
22. Inshore sublittoral	Sea / Estuary	Sea / Estuary
13. Standing water/canals	Water (inland)	Water (inland)
20. Littoral rock	Littoral rock and sediment	Littoral rock
21. Littoral sediment		Littoral sediment
		Saltmarsh
18. Supra-littoral rock	Supra-littoral rock / sediment	Supra-littoral rock
19. Supra-littoral sediment		Supra-littoral sediment
12. Bog	Bog	Bog
10. Dwarf shrub heath	Dwarf shrub heath	Dwarf shrub heath
		Open dwarf shrub heath
15. Montane habitats	Montane habitats	Montane habitats
1. Broad-leaved woodland	Broad-leaf wood	Broad-leaf wood
2. Coniferous woodland	Coniferous woodland	Coniferous woodland
4. Arable and horticulture	Arable and horticulture	Arable (cereal)
		Arable (non-cereal)
		Arable (other)
5. Improved grassland	Improved grassland	Improved grassland
	Abandoned and derelict grasslands	Setaside grass
		Rough neutral grass
6. Neutral	Semi-natural and natural grasslands and bracken	Managed neutral grass
7. Calcareous		Calcareous grass
8. Acid		Acid grass
9. Bracken		Bracken
11. Fen, marsh and swamp	Fen, marsh, swamp	Fen, marsh, swamp
17. Built up areas, gardens	Built up areas, gardens	Suburban/rural developed
		Continuous Urban
16. Inland rock	Inland Bare Ground	Inland Bare Ground

procedure consisted of two stages: (i) edge-detection to identify boundary features, and (ii) region growing from seed points. A Sobel edge-detection algorithm (Mather, 1999) was used to ensure that the seed points were selected away from boundaries. The region growing process accepted a new pixel into a segment where it fell within a threshold spectral range in the three selected spectral bands. Thresholds were adjusted to give field-by-field segmentations, though sub-segmenting some fields. Spatial generalizations were applied: non-segment edge pixels were dissolved into adjoining parcels; the pixels of small segments of fewer than 9 pixels (approximately 0.5 ha) were attached to the most spectrally similar adjoining segments using a spectral minimum distance rule; small, spectrally similar segments were merged. This resulted in a vector storage framework for any subsequent analysis.

Classification used sample ground reference ('training') data from which to determine the spectral characteristics of key cover types. Training for parcel-based analyses essentially operated in the same way as that used in conventional per-pixel classification (Lillesand and Kiefer, 2000; Mather, 1999; Schowengerdt, 1997). However, training areas were identified as whole segments and, therefore, their delineation was based objectively on the segmentation process, with no need to draw training area outlines. The parcel-based approach used a shrinking procedure when extracting reflectances for land parcels, to avoid edge pixels and to ensure the use of 'pure' core pixels in defining spectral characteristics. The shrinkage was a dynamic process whereby the required amount of shrinkage was applied and, if insufficient core pixels were acquired, the amount of shrinkage was reduced and the extraction repeated. This process iterated until a set number of pixels were extracted or the amount of shrinkage reached zero. The per-parcel classification used a maximum likelihood algorithm based on the spectral character of the training areas to determine class membership in the same way as per-pixel classification applied to the mean reflectance statistics.

A set of KBC procedures was used to identify and re-label those land parcels for which there was a high uncertainty in the class allocation, such as those which were classified with small membership probabilities and/or those which contained classes out of their natural context. Where any maximum likelihood class was allocated with a probability of less than 0.5, the second to fifth sub-class probabilities were examined to see if another class was more appropriate (i.e. if its spectral sub-classes cumulatively dominated the overall probability). If so, the next dominant target class (usually the second choice class) took precedence. As an example, if an 'urban' label was allocated with 0.35 probability but wheat, barley, oats and maize had respectively 0.3, 0.2, 0.1 and 0.05 probability, an arable class was deemed more likely, with wheat the most likely variant. Internal context was used by examining the classes of the land parcels surrounding the one being checked. For example, arable land, surrounded by dominant urban cover, was changed to a sub-urban sub-class which would appear more realistic. External context was used to accommodate classes 'out of normal context'. A digital elevation model was used variously to identify erroneous areas of 'marsh' above floodplain level or mis-classified 'urban' areas in high uplands. General manual recodes were also used where known inaccuracies could not be dealt with using external or internal context and with reference to maps and aerial photographs.

Construction of the full UK map required that all the individual classified areas were mosaicked together, with residual cloud-holes patched using single-date classifications. This was done, by first forming 100 km squares. Overlapping classified areas were loaded into a single dataset. A hierarchy of scenes was then formed, based on several factors including the dates of the input images and the amount of coverage. Each scene was made to 'erode' the ones below them in the hierarchy, deleting land parcels that were completely overlapped and cropping land parcels that were only partially overlapped. The erosion procedure operated through the hierarchy of scenes, until a single set of land parcels remained for each 100 km square. After completing the 100 km data set, further KBCs were applied: acid-sensitivity maps (Hornung *et al.*, 1995) were used to label semi-natural grasslands as 'acid', 'neutral' or 'calcareous'; peatland masks to distinguish heaths and bogs; CORINE data (Fuller and Brown, 1996) to identify orchards.

4 Meta-data at the Data Item Level

The complex production flowline of the LCM2000 project produced a large number of intermediate results and additional information. In a conventional per-pixel classification scheme this information would most likely be lost or at best stored separately from the final results. The vector land parcel structure of the LCM2000 allowed the easy retention of information on the land parcel to which it was related. The philosophy applied when developing the LCM2000 production flowline was to retain as much information as possible and store the majority at the land parcel level. An enquiry on the attributes of a land parcel gives equal access to all the stored information so the meta-data is as accessible as the class label.

The land parcels (Figure 9.3; see plate 5) had a set of intrinsic attributes as polygonal features in an object-oriented database (e.g. parcel name, area, perimeter, centre point etc.). Additional attributes were added to the land parcels which controlled the operation of the per-parcel classification procedure, reported the results of the procedure and summarized other spatially referenced data. These attributes were grouped under the following headings:

- *Identification:* Description of the land parcel and links to other data sets (e.g. unique parcel identification).
- *Context:* Ancillary information used to refine the per-parcel classification (e.g. elevation, drift or other land cover classifications).
- *Control:* Information to guide the training of the classifier and the validation of the results (e.g. where relevant, known land cover type of a parcel, type of parcel).
- *Result:* Outputs of the initial maximum likelihood classification, two phases of KBC and various representations of the hierarchical class structure.
- *Remark:* Outputs that summarize the process that has taken place.
- *Image information:* Image date combination, a record of the numbers of pixels used in the initial classification, the amount of shrinkage obtained and the mean vegetation index for each date used.

Many mapping projects generate data such as that discussed above. Key questions are how much is retained by the producer, how much of that is passed on to the user, in what form and how well is it documented? More importantly, can this information be used to give an indication of uncertainty for the products' main output, for instance land cover class label, at the data item level?

5 Assessing Uncertainty

As part of the development of the LCM2000, it was proposed to attach an index to each parcel to give an indication of the uncertainty of the land cover class label. To be of use within a national land cover product this uncertainty index must fulfil several criteria. It must be meaningful and scaled in such a way that ideally the value is representative of all the uncertainties that are present in the production. If some aspects of the production are ignored or unfairly biased then the uncertainty index will have less relevance to the results. The index must be consistent throughout the product so that data selected at different spatial locations, but with the same index value, will be likely to have the same uncertainty. To be incorporated within any end-user analysis the uncertainty index must be relatively easy to understand and require the minimum, if any, of additional processing.

An uncertainty index would be expected to be some function of the meta-data, such as input data (e.g. noise in the images) and the results (e.g. probabilities output from the spectral classification, application of KBCs). Problems arise when the formulation of the function which generates the index is considered. The attributes used may not have a straightforward interpretation and they will carry their own uncertainties. The experience gained in the production of the LCM2000 will be used to highlight some of the issues. As the LCM2000 was derived with the use of multiple images to obtain the full coverage of the UK, some of the issues may not apply to single image classifications, but they are still worth considering.

The image data were acquired by four similar but different instruments, each with their own uncertainties (for instance, optical and electronic performance) regarding their ability to record reflected radiance. A set of pre-processing stages were used to calibrate the images both radiometrically and spatially. These images were either used individually or combined into summer–winter composites. Each input image data set would, therefore, have its own uncertainty related to the source, pre-processing and the particular combination. Also depending on the geographical location of the image and the cover type to be mapped, the acquisition date could be more or less suitable. For instance, mapping deciduous woodland in Scotland may be difficult before June as the leaves are not fully developed, whereas in southern England it may be possible in early May. Therefore, understanding the uncertainties and fitness for purpose of the main input data was already very complex before any analysis of land cover types took place.

The use of class membership probabilities derived from the maximum likelihood classifier (MLC) has often been suggested (Foody *et al.*, 1992) as a means of determining the uncertainty of class label selection. The probabilities actually relate to spectral similarity and it is assumed that each land cover type has a unique spectral character. Unfortunately, this is not the case and, for instance, dwarf shrub heath in an upland situation can be classified as saltmarsh due to the spectral similarity of these two land cover types at a particular time. The image data uncertainties will propagate through to the training data and affect their fitness for purpose. This approach is also confounded by the uniqueness of the training data used for each image data set. The training data are often a subset of the full set of classes within the LCM2000, as most classes are location dependent; as there is no montane vegetation in southern England there is, therefore, no montane vegetation training data when mapping that region. The inclusion and exclusion of land cover types and individual training areas alters the means and variances which control the MLC and the subsequent probabilities used to select the most likely class of membership. To aid class separation it is often the practice to use spectral sub-classes which introduces a layer of abstraction from the land cover types and a sub-division of the probabilities. The whole training process is very subjective, relying on an operator to select training areas and allocate them to classes or sub-classes. Therefore, probabilities output from an MLC using different training sets cannot be compared to give a nationally consistent indication of the certainty with which a class label has been attached.

All the ancillary data used in support of the production flowline have their own uncertainties which may or, more likely, may not be provided by the data supplier. At best, estimates of uncertainty for the whole data set were provided for data used in the LCM2000. It was, therefore, difficult, if not impossible, to assess the effects of the uncertainty of the ancillary data on the uncertainty of the final LCM2000 class label allocations.

The KBC rules, although based on expert knowledge, also contained uncertainty. The identification of erroneous saltmarsh outside of the intertidal zone would appear a very robust rule, but the choice of replacement class is not so straightforward. This issue is linked to time of year, the specification of the MLC training data and unavoidable spectral ambiguity of certain land cover types at the spectral and spatial resolution of the instrument(s) used to record the image(s). Where continuous variables were used to control KBC rules, such as elevation, the thresholds that control the rules are hard. This does not allow for local variability in the rules when they are applied over extended areas. KBC rules, therefore, contain a combination of thematic and threshold uncertainty superimposed on the ancillary data uncertainty.

The production flowline is not perfect and errors occur at a number of points which may be trapped for later correction. The correction is normally by manual intervention, with the operator performing a visual interpretation of the image data and enquiry on contextual information before determining an appropriate class. These corrected land parcels, which amount to no more than 0.5% of the total, are subjective and possess few of the attributes that could be used for an uncertainty index.

The issues listed above are not exhaustive, but they already support the conclusion that the development of a meaningful and consistent uncertainty index from the land parcel meta-data is difficult if not impossible.

6 An Alternative Approach to Communicating Uncertainty

Because of the problems in indicating uncertainty discussed above, another means was, therefore, required to communicate information on the quality of the data within the LCM2000 to the end user. It was not possible to provide all of the meta-data as this would mean large data volumes, with some results difficult to interpret and with only limited amounts of uncertainty information. Also some of the data stored on the land parcels have licensing restrictions, preventing their being passed on to third parties with the product.

A compromise was found by forming an attribute which summarized the meta-data so that the user could attempt to make their own decision on the quality of the data set and its suitability for their application. The process history descriptor (PHD) was a string made up of six fields (Figure 9.4) containing information about the input data and the various stages of classification, knowledge-based correction and final data set compilation.

Field	1		2		3		4		5		6
Example	4	:	70	:	0	:	1	:	1	:	E

Figure 9.4 *An example of the process history descriptor*

The fields in the PHD represented in Figure 9.4 were as follows:

1. An integer number identifying the input image data and any scene-dependent processing. In this case, 4 represents a pair of Landsat TM scenes for East Anglia acquired in February and May 1998. It also provided a link to a calibration exercise where LCM2000 results were compared to independent field data.
2. An integer number giving the cumulative probability, expressed here as a percentage, of the top choice widespread Broad Habitat, calculated by summing probabilities for the Habitat's spectral sub-classes. This gives a relative measure of how similar the spectral information from the land parcel was to that of a widespread Broad Habitat.
3. An integer number to show whether or not a probability aggregation rule has been applied (0 or 1 respectively).
4. and 5. Integer numbers to show the number of KBC rules applied in the two phases of correction.
6. A single character flag identifying other characteristics, such as segment used to train the classifier. In this case, the letter E indicates a land parcel that has been eroded when data sets have been overlain.

The PHD may not be the ideal solution for communicating fully data quality information and uncertainty, but it at least goes some way to informing the user on these important issues. Users can then make their own decisions on data quality and fitness for purpose. The map is all too often the best information available and there is no real way of assessing its accuracy except at a generalized level. Indeed, if there were comprehensive data of higher quality, then the map would itself have no purpose.

7 Future Challenges

From this discussion a set of challenges can be identified that must be addressed by the interested parties in this debate. Data providers should always aim to provide some form of meaningful quality or uncertainty information with their data sets. This, along with any other relevant information, should preferably be provided as meta-data at the data item level with complete documentation. The research community should be developing methods for combining uncertainties between data sets and understanding how uncertainty in the production flowline manifests itself in the final product. Increasingly it will be necessary for software developers to provide visualization tools that not only display data, but also their associated uncertainty.

Finally, and most importantly, the user community should be educated and encouraged to incorporate uncertainty information into their analyses and the interpretation of the results.

8 Conclusions

An understanding of the issues of uncertainty and quality attributable to geographical data is crucial to the effective use of these data and the development of robust applications driven by them. This chapter has highlighted difficulties in both defining the uncertainty problem and in attempting to solve it for complex production flowlines. At present we are at the point of beginning to communicate product information at the level of the individual data item but we still face the challenge of making users aware of the information and teaching them how to use it. The move must also be made away from global estimates of uncertainty for the entire data set down to the level of the individual data item.

Uncertainty is especially important now that geographical information is being used to drive predictive models, where the outcome is influenced by uncertain inputs. Policy makers require hard and fast answers which are often not possible. Therefore, modellers must be able to give them a robust indicator of the uncertainties of the predictions.

The work on the LCM2000 project has not solved the problem of producing and communicating uncertainty information, but it has highlighted the complexity of the issues involved and suggested pragmatic solutions which go some way to addressing the main issues. Thus, parcel level meta-data were the most appropriate outputs that could be made at the time and their use represents an important first step on the way to producing meaningful and consistent quality information for national scale geographic data products.

Acknowledgements

Land Cover Map 2000 was funded by a consortium, comprising: the Countryside Council for Wales; the then Department of the Environment, Transport and the Regions; the Environment Agency; the then Ministry of Agriculture, Fisheries and Foods; the Natural Environment Research Council; Scottish Natural Heritage; Scottish Executive; the National Assembly for Wales; the Environment and Heritage Service (Department of the Environment, Northern Ireland); Department of Agriculture and Rural Development (Northern Ireland).

References

Foody, G. M., Campbell, N. A., Trodd, N. M. and Wood, T. F., 1992, Derivation and applications of probabilistic measures of class membership from the maximum-likelihood classification, *Photogrammetric Engineering and Remote Sensing*, **58**, 1335–41.

Fuller, R. M. and Brown, N., 1996, A CORINE map of Great Britain by automated means. Techniques for automatic generalization of the Land cover map of Great Britain, *International Journal of Geographical Information Systems*, **8**, 937–53.

Fuller, R. M., Groom, G. B. and Jones, A. R., 1994a, The Land Cover Map of Great Britain: an automated classification of Landsat Thematic Mapper data, *Photogrammetric Engineering and Remote Sensing*, **60**, 553–62.

Fuller, R. M., Groom, G. B. and Wallis, S. M., 1994b, The availability of Landsat TM images for Great Britain, *International Journal of Remote Sensing*, **15**, 1357–62.

Fuller, R. M., Smith, G. M., Sanderson, J. M., Hill, R. A. and Thomson, A. G., 2002, The UK Land Cover Map 2000: construction of a parcel-based vector map from satellite images, *Cartographic Journal*, **39**, 15–25.

Fuller, R. M., Wyatt, B. K. and Barr, C. J., 1998, Countryside Survey from ground and space: different perspectives, complementary results, *Journal of Environmental Management*, **54**, 101–26.

Groom, G. B. and Fuller, R. M., 1996, Contextual correction: techniques for improving land cover mapping from remotely sensed images, *International Journal of Remote Sensing*, **17**, 69–89.

Haines-Young, R. H., Barr, C. J., Black, H. I. J., Briggs, D. J., Bunce, R. G. H., Clarke, R. T., Cooper, A., Dawson, F. H., Firbank, L. G., Fuller, R. M., Furse, M. T., Gillespie, M. K., Hill, R., Hornung, M., Howard, D. C., McCann, T., Morecroft, M. D., Petit, S., Sier, A. R. J., Smart, S. M., Smith, G. M., Stott, A. P., Stuart, R. C. and Watkins, J. W., 2000, *Accounting for Nature: Assessing Habitats in the UK Countryside* (London: DETR).

Hornung, M., Bull, K. R., Cresser, M., Ullyett, J., Hall, J. R., Langan, S., Loveland, P. J. and Wilson, M. J., 1995, The sensitivity of surface waters of Great Britain to acidification predicted from catchment characteristics, *Environmental Pollution*, **87**, 207–14.

Jackson, D. L., 2000, *JNCC Report No. 307 Guidance on the Interpretation of the Biodiversity Broad Habitat Classification (Terrestrial and Freshwater Types): Definitions and the Relationships with Other Habitat Classifications* (Peterborough: Joint Nature Conservation Committee).

Kershaw, C. D. and Fuller, R. M., 1992, Statistical problems in the discrimination of land cover from satellite images: a case study in lowland Britain, *International Journal of Remote Sensing*, **13**, 3085–104.

Lillesand, T. M. and Keiffer, R. W., 2000, *Remote Sensing and Image Interpretation*, 4th edition (Chichester: Wiley).

Mather, P. M., 1999, *Computer Processing of Remotely Sensed Images – An Introduction*, 2nd edition (Chichester: Wiley).

Schowengerdt, R. A., 1997, *Models and Methods for Image Processing*, 2nd edition (London: Academic Press).

Smith, G. M. and Fuller, R. M., 2001, An integrated approach to land cover classification: an example in the Island of Jersey, *International Journal of Remote Sensing*, **22**, 3123–42

Wyatt, B. K., Greatorex-Davies, N. G., Bunce, R. G. H., Fuller, R. M. and Hill, M. O., 1994, Comparison of land cover definitions. *Countryside 1990 Series: Volume 3* (London: Department of the Environment).

10

Analysing Uncertainty Propagation in GIS: Why is it not that Simple?

Gerard B. M. Heuvelink

1 Introduction

To introduce the problem addressed in this chapter, let us look at the following example. Figure 10.1a shows a Digital Elevation Model (DEM) of a 2 × 2.5 km area in the Vorarlberg region in the Austrian Alps. Figure 10.1b shows the corresponding slope map, computed with a second-order finite difference method (Skidmore, 1989).

As we are well aware, the DEM is only an approximation of the real elevation in the area. It contains errors. Our interest now is in how these errors propagate to the slope map. Let us assume that the DEM error at each location is ±10 m. We now use the Monte Carlo method to analyse how the error propagates.

For those readers who are not familiar with the Monte Carlo method, let us briefly describe how it works. First, we generate a realization of the uncertain DEM, by adding, at each location, a random number drawn from the error probability distribution to the DEM elevation. Once this is done for all locations (i.e. all grid cells), we compute the slope map for this realization, and we store it. This procedure we repeat many times (typically between 50 and 200 times). From the collection of slope maps we can then determine how the uncertainty in the DEM has propagated to the slope, for instance by looking at the width of the histogram of the slope values at locations. This is illustrated in Figures 10.2 and 10.3.

The above simple example illustrates the ease with which an error propagation analysis may be carried out. The example is simple, but in fact the Monte Carlo method can as easily and as straightforwardly be applied to more complicated GIS

Uncertainty in Remote Sensing and GIS. Edited by G.M. Foody and P.M. Atkinson.
 ISBN: 0–470–84408–6

operations or spatial models (see Nackaerts *et al.*, 1999; De Genst *et al.*, 2001; Endreny and Wood, 2001; Canters *et al.*, 2002, for some recent examples). The only disadvantage is that it is computationally demanding, but nowadays this can hardly be considered a real problem, particularly as the method lends itself perfectly to a parallel computing approach.

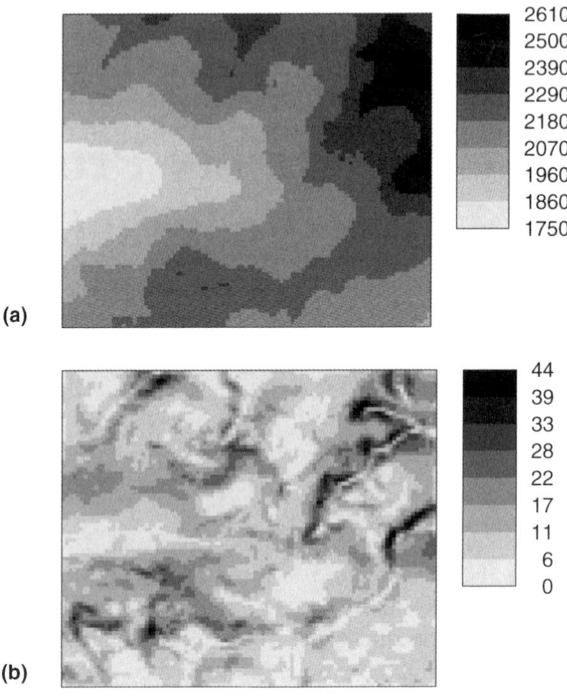

Figure 10.1 *Example data sets. (a) Digital Elevation Model (metres) of a 2 × 2.5 km area in the Vorarlberg region in the Austrian Alps; (b) slope map (per cent) of the same area*

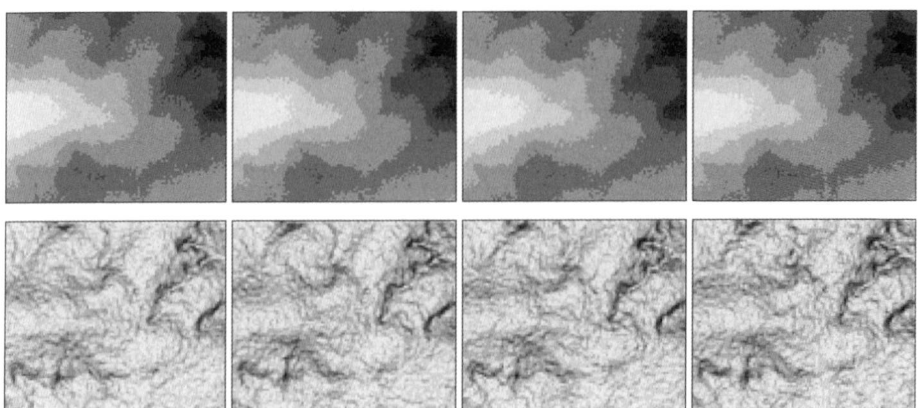

Figure 10.2 *Top: four simulated DEMs constructed by adding a simulated error map to the DEM of Figure 10.1. Bottom: associated slope maps*

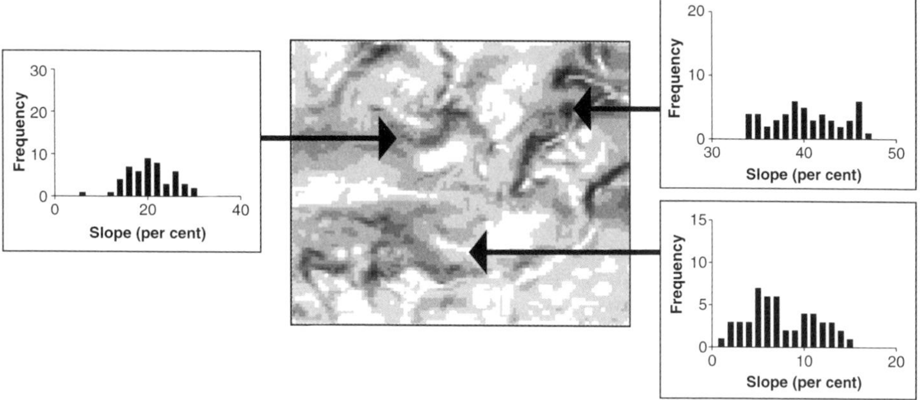

Figure 10.3 *Histograms of simulated slope values (for three arbitrary locations and computed from 50 Monte Carlo runs) characterize how uncertainty in the DEM propagates to slope*

From the above, some may be tempted to conclude that the problem of error propagation in GIS is solved: 'Monte Carlo simulation solves all our problems'. Is this really true? Unfortunately, it is not. There are still a large number of important problems to be resolved. Some of them fairly easy, many of them very difficult. In this chapter we will review several of them. We will explain what these problems are, and try to indicate how they may be tackled. Although the aim is to address a variety of problems, no attempt is made to be complete. Also, the discussion is biased towards problems that have a geostatistical angle to them.

2 Spatial Error Cannot be Reduced to a Single Number

In the Vorarlberg example we stated that the error in the DEM was ±10 m. We did not discuss how we came to this number, and neither did we discuss whether we had completely characterized the DEM error by just one number. To answer this last question, we definitely had not. To completely specify the DEM error at some location we need its full probability distribution (i.e. its shape and its parameters), otherwise we could not conduct the Monte Carlo analysis. To simulate the DEM error it was assumed that the error was normally distributed with zero mean and a standard deviation equal to 10 m. It was also assumed that this was the case for all locations, regardless whether it concerned a relatively flat or steep part of the area, or whether there were control points nearby or not. By sampling at each grid cell independently, it was also assumed implicitly that the error was spatially uncorrelated. This is a critical assumption. If we had assumed that there was spatial autocorrelation in the DEM error, say such that it was characterized by a semivariogram with zero nugget variance and a correlation length (i.e. 'range') of 1 km, then the results of the uncertainty analysis would have been entirely different (see Figure 10.4). This is because slope calculations are extremely sensitive to the degree of spatial autocorrelation in the DEM errors (Heuvelink, 1998, section 5.2). Clearly, if the results of the

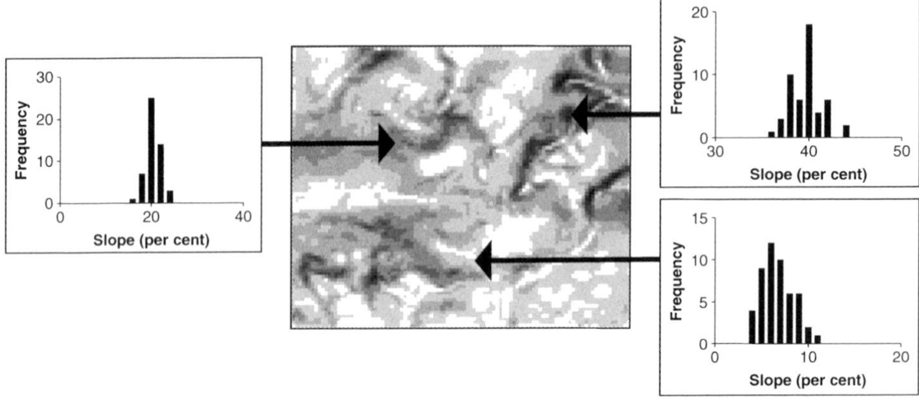

Figure 10.4 *Histograms of slope values (for the three example locations and computed from 50 Monte Carlo runs) characterize how the uncertainty in DEM propagates to slope predictions. In this case the DEM error is spatially autocorrelated with zero nugget variance and a spatial correlation length of 1 km. Comparison with Figure 10.3 shows that taking positive spatial autocorrelation in DEM error into account reduces the uncertainty in slope predictions*

uncertainty analysis are to have any value, it is crucially important to assess realistically the degree of spatial autocorrelation in the DEM error.

By failing to specify completely the DEM error, we in fact used 'default' values for the unspecified parameters. However, assuming by default that the error standard deviation is the same throughout the area and assuming by default that spatial autocorrelation is absent may not be very sensible. Instead, we should use realistic parameter values. How can we do this? How can we estimate these parameters? The most appropriate way to go about this is to rely on geostatistics. Geostatistics is a well-developed discipline that is dedicated to the statistical analysis of spatially varying properties (Kitanidis, 1997). Geostatistics provides the theory to define properly spatial errors, and it provides the instruments to estimate the parameters of the error model from point observations or other information sources. It even aids in carrying out Monte Carlo uncertainty analyses, by offering tools that can generate realizations of spatially correlated errors (Goovaerts, 2002). The body of theory and methods encompassed by geostatistics is large and continues to grow. It allows the handling of anisotropy, non-stationarity, non-normality and much more (Goovaerts, 1997).

A problem is that to use geostatistics responsibly requires a fair amount of background in statistics. Also, users must invest time and show a willingness to learn geostatistics. Can we reasonably expect this of the 'ordinary' GIS user who just wants to do an uncertainty propagation analysis? Often, this will be too much to ask of the ordinary user. The solution is to make sure that the error specification of the inputs to the uncertainty analysis are derived by experts and then stored in the GIS database for later use. Indeed, the GIS research community has been working already for quite some time on establishing a so-called 'error-aware' GIS, that also stores quality information about the data stored in the GIS database (Duckham, 2000; Qiu and Hunter, 2002). However, 'In spite of a growing demand from the user community for more appropriate documentation on spatial data quality, and the existence of established standards for doing so, in most cases meta-information on the accuracy of spatial data is lacking or is limited to simple, overall measures that do not describe spatial variation in error' (Canters *et al.*, 2002). We will return to this issue in the concluding section of this chapter.

3 The Case of Multiple Uncertain Inputs

Again, for illustrative purposes, let us now turn to another, more complicated error propagation example.

The soil acidification model SMART2 (Kros *et al.*, 1999) is a fairly simple, strictly vertical, one layer dynamic model that predicts the long-term response of aluminium (and nitrate) concentration below the root zone to changes in atmospheric deposition. The model was developed to analyse how atmospheric deposition causes soil acidification and how this, in turn, affects the groundwater quality.

At the European scale, the inputs to SMART2 have to be derived from general-purpose soil and vegetation type maps. These maps are not very accurate. In addition, the mapping procedure involves the use of transfer functions to transform map information to specific soil and vegetation-related parameters required by SMART2, and this further deteriorates the quality of the model input. In order to judge whether the application of SMART2 at the European scale yields results that are sufficiently accurate, a study was carried out to analyse how the uncertainty in the source maps and the transfer functions propagate to the model output. This was done using a Monte Carlo approach.

The uncertainty analysis that was carried out treated 11 soil-related variables and eight vegetation-related inputs as uncertain. To run the analysis, for all 19 inputs error probability distributions and spatial correlation functions had to be specified. This was not an easy job, because little 'hard' information (i.e. field observations) was available to estimate the accuracy of the input data and transfer functions. For many parameters of the error distribution, one simply had to rely on expert judgements. This is acceptable when there is nothing else to rely on, but it is far from ideal. Expert judgements are no substitute for objective measurements.

One important class of parameters that needed to be specified as well were the correlations between the uncertainties in the input variables. With 19 inputs, there are $1/2 \times 19 \times (19 - 1) = 171$ correlations to be assigned. Or even worse, there are 171 cross-correlation functions to be specified, because the uncertainty analysis requires the correlation between the error in input variable ν_i at location x_k and the error in input ν_j at location x_l. Again, geostatistics provides the means to assess these cross-correlations, but only when there are sufficient observations of all attributes involved at a fairly large number of locations. And even then, identifying all these semivariograms and cross-semivariograms is a laborious and tedious job, which cannot easily be automated.

In the SMART2 case, a large number of cross-correlation functions were either known to be or assumed to be zero. This is not strange, because when two spatial attributes, such as the nitrification and denitrification factors, are correlated, the errors associated with these attributes may well be uncorrelated. Nonetheless, increasing the number of uncertain inputs causes a much greater increase in the number of estimated parameters. Also, storing such information in a spatial database is tricky, because whenever an uncertain variable is added to the database, the spatial cross-correlation with all other uncertain variables already stored in the database must also be assessed. Moreover, to avoid ending up with negative error variances for combinations of inputs, the so-called non-negative definiteness condition must always be satisfied. For instance, when the correlation between variables X and Y equals 0.9 and that between X and Z equals 0.9 as well, then the correlation between Y and Z cannot be smaller than 0.62 or negative-definite correlation matrices may result. Fast checking for non-negative definiteness is not easy when the number of uncertain variables is large and when, in addition, correlation varies with distance. Should we expect the non-statistician to be able to deal with these complex issues? No, we should not.

4 Incorporating Uncertainty in Categorical Inputs

In the SMART2 case study there were also two uncertain categorical inputs. These were soil and vegetation type. Including uncertain categorical inputs into the uncertainty analysis increases dramatically the complexity of the analysis, as we shall now see.

The soil map had only seven classes, the vegetation map only four. But soil and vegetation are not independent and, therefore, it was necessary to consider the 28 combined soil–vegetation classes. The soil–vegetation map is uncertain, and this means that when the map states that at some location the soil–vegetation is in class sv_j, in reality the soil–vegetation at that location may be in class sv_i $(i \neq j)$. To quantify the uncertainty, we must provide the conditional probabilities $p_{ij} = P(\text{reality} = sv_i|\text{map} = sv_j)$. In the case of 28 classes, there are $28^2 = 784$ conditional probabilities to be predicted. These conditional probabilities can be stored in the so-called error or confusion matrix, as it is often termed in the remote sensing literature (Foody, 2000). How can these probabilities be obtained? One possibility is to derive them from a comparison of the map with ground-based observations. However, this requires quite a large number of observations. Another possibility is to use a much more accurate map as a surrogate for the 'truth' (Finke *et al.*, 1999). Yet another possibility is to embed the computation of the probabilities in the mapping procedure. For instance, discriminant analysis gives the probabilities as a byproduct in a supervised classification procedure. The latter also allows the error matrix to vary with location.

Even though there are methods to assess the many parameters that quantify the conditional probability distributions, it is important to recognize that when input variables are dependent (as is often the case) one quickly ends up with very large numbers of parameters. In the SMART2 case, we purposely reduced the number of soil classes to $m = 7$ and the number of vegetation classes to $n = 4$, yielding a manageable number of $m \cdot n = 28$ combined classes, but what if m and n had been much greater than the numbers used? Or what if a third categorical variable had been added to the uncertainty analysis? The number of parameters to be estimated quickly explodes, rendering their reliable estimation unfeasible, and so we must find ways to get around this problem.

An obvious solution is to make additional assumptions that reduce greatly the number of parameters to be estimated. To illustrate some approaches, let us continue the example where we seek the conditional probability that soil type at some location equals s_i and that vegetation equals ν_q, given that the soil map states that the soil type at that location equals s_j and that vegetation type equals ν_r $(i, j = 1 \ldots n;\ q, r = 1 \ldots m)$. There are $n^2 \cdot m^2$ parameters to be estimated. One way to reduce the number of parameters is to assume that:

$$\begin{aligned} P(S = s_i,\ V = \nu_q | SM = s_j,\ VM = \nu_r) \\ = P(S = s_i | SM = s_j) \cdot P(V = \nu_q | VM = \nu_r) \end{aligned} \tag{1}$$

where S stands for soil, V for vegetation, SM for soil map and VM for vegetation map. With equation (1), the number of parameters is reduced greatly from $n^2 \cdot m^2$ to $n^2 + m^2$, but only at the expense of the very strong assumption of statistical independence, which does not seem very realistic for the example of soil type and vegetation type.

An alternative approach is to assume that:

$$\frac{P(S = s_i,\ V = \nu_q | SM = s_j,\ VM = \nu_r)}{P(S = s_i | SM = s_j) \cdot P(V = \nu_q | VM = \nu_r)} \cong \frac{P(S = s_i,\ V = \nu_q)}{P(S = s_i) \cdot P(V = \nu_q)} \tag{2}$$

This reduces the number of parameters to be estimated to $n^2 + m^2 + n \cdot m$, while not denying the dependence between soil type and vegetation. Instead, it assumes that the dependence between the two is not affected by the conditioning. Whether this is a more realistic assumption remains to be seen. As far as I know this has not yet been investigated.

Although many map users would be very happy if an error matrix were available to summarize the accuracy of the categorical map, it is important to realize that in fact the error matrix does not provide all the information required to carry out an error propagation analysis. What is lacking is information about spatial dependence in the errors. After all, it is more than likely that the errors are spatially dependent, because similar errors are made at nearby locations. Recently, geostatistical procedures have been proposed to handle the spatial dependencies in errors in categorical maps (Finke *et al.*, 1999; Kyriakidis and Dungan, 2001). These procedures, that are largely based on indicator geostatistics, are promising but they suffer from the same problem mentioned in the previous sections: to use them requires a solid background in (geo)statistics. Again, can we expect this from the ordinary GIS user who just wants to carry out an uncertainty analysis?

5 Scale Problems

Let us continue the SMART2 example. Although SMART2 is intended to be used at the regional scale (De Vries *et al.*, 1998), it is still a model operating at the point support. This is because the model is based on differential equations that may be discretized but not to too large time and space steps and because the model parameters were calibrated on measurements taken on a point support (i.e. measurements the size of about 1 dm^3). In order to assess groundwater quality at the European scale (5 km × 5 km and larger blocks), the model was therefore applied to many point locations within each block (Heuvelink and Pebesma, 1999). After the model was run at these point locations, the model outputs were aggregated to yield a single block value. For a single block, a single Monte Carlo run gives the output at all points in the block. These can be represented in a single graph by a cumulative distribution function. However, due to input uncertainty there will be a population of graphs, since each Monte Carlo run produces a different graph. In Figure 10.5, the population of graphs is represented by the sides of a 90% prediction interval. From it, we can construct 90% prediction intervals for quantiles, such as the median (Figure 10.5a) or for the block areal fraction exceeding a fixed concentration (Figure 10.5b).

It now turns out that the results of the uncertainty analysis depend greatly on the support of the data used in the analysis (Heuvelink, 1999). The larger the block, the smaller the uncertainty. This is because more and more variability averages out when the block size is increased. Readers that are familiar with kriging will see here the similarity with block kriging: block kriging variances are smaller than point kriging variances and become smaller and smaller as the block size increases. In other words, the results of an uncertainty analysis are only meaningful when the support of the output is stated, and when it matches the support required by the user.

In practice, it may occur frequently that the support of the obtained output does not match the desired output support. This calls for a 'change of support', which is more often in the form of an aggregation than a disaggregation. Existing techniques can be used to perform the aggregation or disaggregation, but it is not always very easy (Bierkens *et al.*, 2000).

It is crucially important to recognize that the inputs and their associated errors must also be provided on the correct support. For instance, let precipitation and

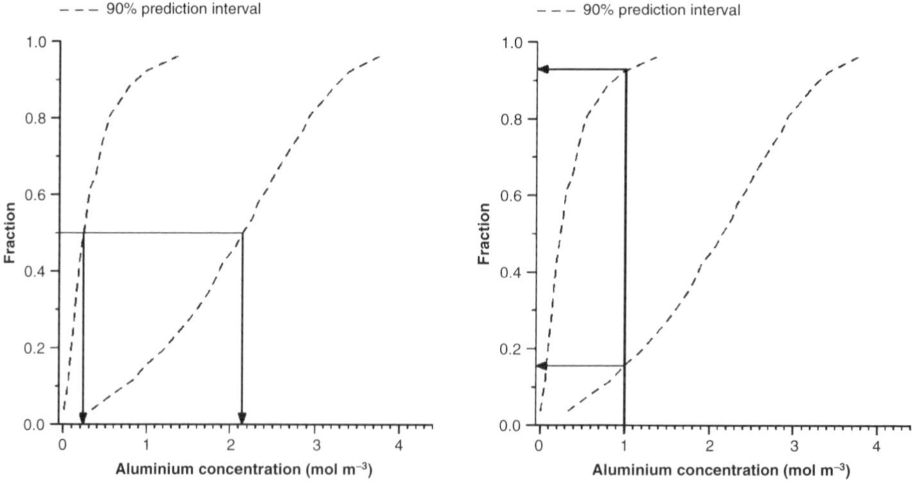

Figure 10.5 *Prediction intervals: (a) the 90% prediction interval for the median concentration in the block is [0.23, 2.15], (b) the 90% prediction interval for the areal fraction where point concentration values do not exceed* 1 mol m^{-3} *is [0.155, 0.930]*

infiltration capacity be two uncertain inputs to a hydrological model. Let infiltration capacity be measured on a support of 1 m^2, and let precipitation be measured using rain gauges with a surface area of 0.01 m^2. Both attributes are interpolated to create spatially exhaustive inputs to the hydrological model. Interpolation error is quantified and its propagation through the model is analysed. It should now be clear that either infiltration capacity and its associated uncertainty must first be disaggregated or precipitation and its associated uncertainty must first be aggregated before the uncertainty propagation analysis can commence. Otherwise the results of the uncertainty analysis are effectively meaningless.

The effect of using wrong (combinations of) supports on the validity of the results of uncertainty analyses can hardly be underestimated. Here we have only stipulated it, but it deserves much attention. It can create tremendous problems, even in seemingly very simple applications. A good example of this for lead pollution is given in Heuvelink (1999).

6 Conclusions

We started this chapter with the observation that to carry out an uncertainty analysis seemingly is not that difficult, because 'Monte Carlo simulation solves all our problems'. However, by looking a bit further and by addressing several aspects of spatial uncertainty analysis, we then saw that there are still a large number of serious problems to be resolved. In particular, we discussed difficulties in the assessment of input error, both for continuous and categorical spatial attributes, and problems concerning the scale or support of model entities. We did not pretend to have a complete list of problems. For instance, we did not mention that in many cases the results of a GIS analysis are used as input to an ensuing operation. If we want to be able to keep track of how uncertainties propagate in chains of operations, then we must store the uncertainty of the output of each and every operation. This means storage of the output error probability distribution, its support, its spatial autocorrelation and, most importantly, its cross-correlation with all other uncertain attributes already stored in the database. Do we have procedures and databases that can accomplish this?

A returning theme of all problems discussed is that we require a lot of the user who wants to carry out an uncertainty analysis. Too much perhaps. We more or less expect the user to be an expert in (geo)statistics, or at least to have an expert in his or her immediate surrounding. We also expect him or her to accept that the uncertainty analysis requires lots more time (both computing time and user time) than the operation itself. The calculations take much longer, there are so many more parameters to be determined and stored and there are so many more output parameters to be stored and analysed. Can we reasonably expect the 'ordinary' GIS user to spend so much extra time? And all this just to get to know how accurate the result of an analysis is?

It is now time to become realistic and admit that the majority of GIS users are not prepared or even allowed to go to so much trouble. Many of them need to account for every hour spent and are expected to run projects on a cost-effective basis. In many cases, running a complicated uncertainty analysis simply does not pay off. This seems to be the main reason why the large-scale application of error propagation tools in GIS is as yet unrealized, in spite of the high hopes expressed now more than a decade ago (Goodchild and Gopal, 1989; Openshaw *et al.*, 1991; Burrough, 1992).

What can be done about it? One solution that we may look into in the coming years is to seek ways that dramatically simplify the uncertainty propagation analysis, *without* seriously damaging the validity of the results of the analysis. Finding such ways is a true challenge. There is no guarantee for success. If successful, the result is not likely a set of simple rules such as 'ignore spatial cross-correlation', or 'pretend that there are no scale problems'. It will undoubtedly be more refined, with different rules in different situations.

The degree of simplification may also be user-dependent. We could imagine this as follows. At the start of a session, the system would ask the user to qualify himself as an error propagation expert, a GIS professional or an amateur. Depending on this, the system would decide whether to ask the user to specify a parameter or to use a default value instead (of course, the difficulty is finding the right default values). After all, there is no sense in asking an amateur 'warning: linear model of coregionalization not satisfied. Do you wish to continue with the current settings?'. Amateurs want to be asked as little as possible, the uncertainty analysis has to run almost by itself. Only then can we truly speak about having extended GIS with a general purpose 'error button'. In spite of the negative sides to a dramatic simplification of the error propagation analysis, this may still be preferred over a complete ignorance of how errors propagate in GIS analyses.

References

Bierkens, M. F. P., Finke, P. A. and De Willigen, P., 2000, *Upscaling and Downscaling Methods for Environmental Research* (Dordrecht: Kluwer).

Burrough, P. A., 1992, The development of intelligent geographical information systems, *International Journal of Geographical Information Systems*, **6**, 1–11.

Canters, F., De Genst, W. and Dufourmont, H., 2002, Assessing effects of input uncertainty in structural landscape classification, *International Journal of Geographical Information Science* **16**, 129–49.

De Genst, W., Canters, F. and Gulinck, H., 2001, Uncertainty modelling in buffer operations applied to connectivity analysis, *Transactions in Geographical Information Systems*, **5**, 305–26.

De Vries, W., Kros, J., Groenenberg, J. E., Reinds, G. J. and Van Der Salm, C., 1998, The use of upscaling procedures in the application of soil acidification models at different spatial scales, *Nutrient Cycling in Agroecosystems*, **50**, 225–38.

Duckham, M., 2000, Error-sensitive GIS development: technology and research themes, in G. B. M. Heuvelink and M. J. P. M. Lemmens (eds), *Proceedings of the 4th International Symposium on Spatial Accuracy Assessment in Natural Resources and Environmental Sciences* (Delft: Delft University Press), pp. 183–90.

Endreny, T. A. and Wood, E. F., 2001, Representing elevation uncertainty in runoff modelling and flowpath mapping, *Hydrological Processes*, **15**, 2223–36.

Finke, P. A., Wladis, D., Kros, J., Pebesma, E. J. and Reinds, G. J., 1999, Quantification and simulation of errors in categorical data for uncertainty analysis of soil acidification modelling, *Geoderma*, **93**, 177–94.

Foody, G. M., 2000, Accuracy of thematic maps derived from remote sensing, in G. B. M. Heuvelink and M. J. P. M. Lemmens (eds), *Proceedings of the 4th International Symposium on Spatial Accuracy Assessment in Natural Resources and Environmental Sciences* (Delft: Delft University Press), pp. 217–24.

Goodchild, M. F. and Gopal, G. (eds), 1989, *Accuracy of Spatial Databases* (London: Taylor & Francis).

Goovaerts, P., 1997, *Geostatistics for Natural Resources Evaluation* (New York: Oxford University Press).

Goovaerts, P., 2002, Geostatistical mapping of satellite data using P-field simulation with conditional probability fields, *International Journal of Geographical Information Science*, **16**, 167–78.

Heuvelink, G. B. M., 1998, *Error Propagation in Environmental Modelling with Geographical Information Systems* (London: Taylor & Francis).

Heuvelink, G. B. M., 1999, Aggregation and error propagation in Geographical Information Systems, in K. Lowell and A. Jaton (eds), *Spatial Accuracy Assessment. Land Information Uncertainty in Natural Resources* (Chelsea, MI: Ann Arbor), pp. 219–25.

Heuvelink, G. B. M. and Pebesma, E. J., 1999, Spatial aggregation and soil process modelling, *Geoderma*, **89**, 47–65.

Kitanidis, P. K., 1997, *Introduction to Geostatistics: Applications to Hydrogeology* (Cambridge: Cambridge University Press).

Kros, J., Pebesma, E. J., Reinds, G. J. and Finke, P. F., 1999, Uncertainty assessment in modelling soil acidification at the European scale: a case study, *Journal of Environmental Quality*, **28**, 366–77.

Kyriakidis, P. C. and Dungan, J. L., 2001, A geostatistical approach for mapping thematic classification accuracy and evaluating the impact of inaccurate spatial data on ecological model predictions, *Environmental and Ecological Statistics*, **8**, 311–30.

Nackaerts, K., Govers, G. and Van Orshoven, J., 1999, Accuracy assessment of probabilistic visibilities, *International Journal of Geographical Information Science*, **13**, 709–21.

Openshaw, S., Charlton, M. and Carver, S., 1991, Error propagation: a Monte Carlo simulation, in I. Masser and M. Blakemore (eds), *Handling Geographical Information* (Harlow: Longman), pp. 78–101.

Qiu, J. and Hunter, G. J., 2002, Managing data quality information, in W. Shi, P. F. Fisher and M. Goodchild, (eds), *Spatial Data Quality* (London: Taylor & Francis).

Skidmore, A. K., 1989, A comparison of techniques for calculating gradient and aspect from a gridded digital elevation model, *International Journal of Geographical Information Systems*, **3**, 323–34.

11

Managing Uncertainty in a Geospatial Model of Biodiversity

Anthony J. Warren, Michael J. Collins, Edward A. Johnson and Peter F. Ehlers

1 Introduction

The boreal forest spans the northern hemisphere through Canada, Russia and Alaska and plays an important role in the societies which depend on it. In addition to its vital role in Earth–atmosphere interactions, the boreal forest is of great economic importance as a renewable natural resource. It provides valuable and numerous pulp, paper and lumber products that are used by all facets of society. It is in our interest to sustain this resource and it is, therefore, the subject of much scientific research geared toward understanding the mechanisms and processes of which it is a part.

Fire has always played an important role in the proper functioning of the boreal forest. The serotinous pine cones of the tree species, for instance, will not open and release their seeds without exposure to extremely high temperatures and, therefore, rely on fire for regeneration (Cameron, 1953; Johnson, 1992). Although the boreal forest is a dynamic ecosystem with continual disturbance by fire, it has been capable of maintaining itself in a relatively stable state. With this in mind, harvesting practices seek to mimic this form of natural disturbance. However, only by understanding the processes that govern natural mortality and regeneration, can this goal be realized. Biological diversity (or biodiversity) is one important aspect of this research and it is the subject of this study. The biodiversity of the boreal forest is not a dynamic process itself, rather, it is the product of many processes. These include internal processes operating within the ecosystem as well as external processes operating on the ecosystem.

Currently, ecologists are able to measure and map biodiversity of flora in the boreal region by collecting various data on the vegetation in forested stands from

Uncertainty in Remote Sensing and GIS. Edited by G.M. Foody and P.M. Atkinson.
 ISBN: 0-470-84408-6

ground-based surveys. Although this method is by far the most accurate, it is a very costly, time-consuming and labour intensive process. From these ground observations, ecologists can also study and characterize the processes operating in the boreal forest. If the underlying processes which shape biodiversity can be determined and parametrized in a numerical model, they can potentially be used to predict biodiversity. However, whether biodiversity is estimated directly from the ground data or predicted with numerical models from ground-measured input variables, only stand-based estimates of diversity are derived. Our understanding of biodiversity over the landscape is, therefore, limited by these sparse data. To increase our understanding, there is a need to map diversity over an entire landscape.

1.1 Biodiversity

In the broadest sense, biological diversity is natural variation. This natural variation can occur at the level of the molecule, gene and species. At these levels, biological diversity can be described across many scales such as the forest stand or over landscape units such as a drainage basin (Huston, 1994).

Ecologists generally use two components to describe biological diversity or biodiversity. The first is species richness (*S*) which is simply a count of the number of different species present in a given area. The second is known as evenness which is also referred to as species relative abundance and describes the proportions of species in the area. A natural ecosystem is usually composed of a few species that are very abundant within an area and many species which are much less abundant. A measure of evenness accounts for this aspect of diversity (May, 1975; Magurran, 1988). Both components are accepted representations of diversity and we have chosen to use richness in this research. Having said that, we should mention that relying strictly on a measure of richness to describe diversity is potentially dangerous because it does not describe the distribution of species in the area.

In the context of forests, it is important to note that trees usually contribute a greater quantity of usable natural resources to a site than herbaceous plants. High diversity areas, therefore, are not necessarily beneficial. For example, shortly after an area has been harvested, herbaceous plants and shrubs that were once starved for light and competing for nutrients and moisture with the mature canopy above would be free to grow. In addition, invasive species that have migrated in from nearby areas will be able to grow. Within a short time span (which could be years or months), the number of species in the area would be very high but not necessarily beneficial to the organisms that use the area (Harper and Hawksworth, 1995; Bormann and Likens, 1979). For example, Andow (1991) reviewed the literature on arthropods in monocultures versus polycultures and found that increases in vegetation diversity adversely affect herbivorous arthropod populations.

1.2 Objectives

The overall objective of our ongoing research was to understand the processes governing the spatial pattern of herbaceous plants on forested landscapes. One possible dimension of this pattern is biodiversity and we are currently developing methods for mapping biodiversity. This work involves a multidisciplinary team of ecologists, mathematicians, geographers and engineers. The research aims to reveal the landscape characteristics that are controlling diversity and development of empirical models both for estimating these characteristics from geospatial data such as remotely sensed imagery and digital elevation models and to estimate the diversity itself using predictive vegetation mapping systems. An integral component of this research is the estimation of the uncertainty in both the landscape characteristics and the biodiversity.

The model used currently to map biodiversity is a simple multivariate linear regression that is largely intended to explore the importance of the four controlling landscape variables (Chipman and Johnson, 2002). This model has been implemented at the landscape level by estimating (mapping) the input variables using geospatial data (Warren, 1999). The objective of the work reported in this chapter is to present methods for estimating the uncertainty in landscape variables and for propagating this uncertainty through the biodiversity model and produce uncertainties in the final map. In particular the work aims to answer three questions:

1. How well can richness be estimated using estimated versus observed landscape variables? In other words, to what extent do the errors in the estimated landscape variables influence the richness predictions?
2. How accurate are richness predictions relative to field observations?
3. What is the level of uncertainty in richness predictions?

2 Study Area and Data

The study site is part of the boreal mixed-wood forest located in and around Prince Albert National Park (PANP), central Saskatchewan (53° 35′ N to 53° 20′ N and from 106° 47′ W). PANP is located on the southern fringe of the boreal forest which gives way to an expansive agricultural region to the south (Figure 11.1). The following physical description of the area has been summarized from Bridge (1997).

The geomorphology of the area has been defined primarily by the glacial events of the Pleistocene (~12 000 years ago). Although glacial tills dominate the area, there are also substantial organic glaciofluvial (of glacial river origin) and glaciolacustrine (of glacial lake origin) deposits. The topography is composed of rolling hills with an elevation range of 500 to 800 m above sea level. Long cold winters and short cool summers are characteristic of the regional climate. Between 400 and 500 mm of precipitation is received by the area with approximately 70 % falling as rain.

There are two major disturbance regimes at work in the area: forest fires and forest harvesting. The first includes both human induced and natural lightning caused fires although the latter make up the majority. Weyerhaeuser Canada operates a Forest Management License Area (FMLA) in the region and is responsible for the regeneration of harvested areas. However, within PANP there is no harvesting and the only major source of natural disturbance is fire.

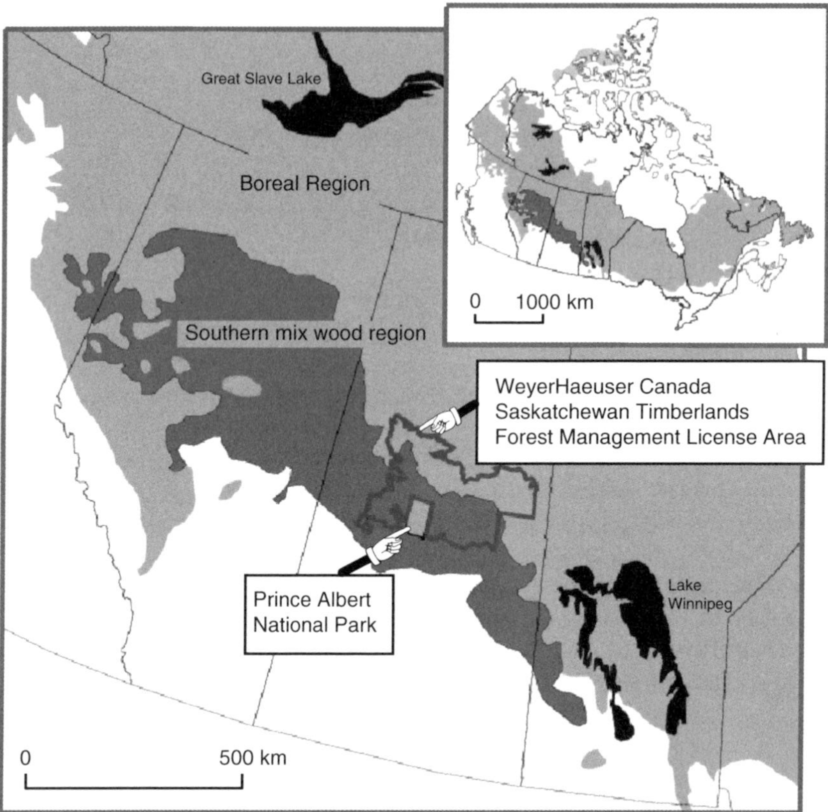

Figure 11.1 Location of the study area. The boreal region is shown by the grey shaded areas with the southern mixed-wood boreal forest delineated by the darker grey belt

Eight tree species dominate the area which comprise both coniferous and deciduous species. The coniferous species are *Picea glauca* (Moench) Voss, *Picea mariana* (Mill.) B.S.P., *Pinus banksiana* Lamb., *Abies balsamea* (L.) Mill., *Larix laricina* (DuRoi) K. Kock. *Populus tremuloides* Michx., *Populus balsamifera* L. and *Betula papyrifera* Marsh. make up the deciduous species. There are also many herbaceous shrubs and ground cover plants that contribute to the vegetation composition.

2.1 Stand data

More than 150 stands were sampled over a period of three years. Stand data collected in the field included species counts of canopy and sapling trees, shrubs and herbaceous plants as well as stand density. These data were used to calculate species richness for each stand. In addition, estimates of the age of each stand were made from tree cores. Details of the stand-based field sampling methods may be found in Warren (1999) and Chipman and Johnson (2002). For each stand the distance to the nearest ridgeline from topographic maps was estimated. Landscape properties which were measured in the field are referred to in this chapter as observed. These included richness, canopy stem density, canopy species type, distance to ridgeline and canopy age.

2.2 Geospatial data

Vector data representing elevation contours, rivers and lakes were made available by Parks Canada for PANP and surrounding adjacent lands. These were used to interpolate an elevation model (DEM) of the area using the ArcInfo software. The contour interval of the topographic vectors was 8 m. These vector GIS files were not provided with any metadata so that errors of unknown type and quantity existed within these files. More details on the interpolation and processing of the DEM may be found in Warren (1999).

Landsat Thematic Mapper (TM) imagery of the PANP acquired on 10 June 1996 was used. The image was virtually cloud free and included all seven spectral bands with a 30 m spatial resolution (120 m for channel 6). The imagery was referenced to the WGS-84 ellipsoid in a UTM projection system. This processing resulted in resampling the image pixels to 25 m ground sample spacing to match the DEM.

In addition to electro-optical data, SIR-C (Shuttle Imaging Radar) polarimetric synthetic aperture radar (SAR) image data were obtained from the Jet Propulsion Laboratory (JPL). These data were acquired on 4 and 6 October 1994. Software supplied by the JPL (NASA Jet Propulsion Laboratory, 1993, 1994; Chapman, 1995; Vuu *et al.*, 1995) was used to strip the raw data from the tape, synthesize the resulting Stokes matrix for each pixel into linear polarization images (HH, HV, VH and VV) and finally to perform multi-looking (the effective number of looks was seven). The ground resolution for both images was 12.5 m. Together, the two images cover a substantial portion of the PANP study site.

The original SAR imagery was filtered using a gamma-gamma filter to remove some of the speckle noise (Lopes *et al.*, 1993). Close inspection of the imagery prior to filtering revealed great local variation in backscatter amplitude. This was especially apparent in the L-band imagery. After filtering, this variation had been reduced. Using approximately 40 ground control points for each SAR image, the images were transformed into the WGS-84 ellipsoid with an RMS error of approximately one pixel (25 m).

A raster map of time-since-fire was obtained from the University of Calgary, Biological Sciences Department, Ecology Division. This map was derived from data records acquired and maintained by Weir (1996).

3 Methods

3.1 Biodiversity model

Chipman and Johnson (2002) developed an ecological model which is able to predict biological diversity of herbaceous plants in boreal forest ($R^2 = 0.52$). The variables of interest are (1) the distance of a stand from the drainage basin ridgeline, (2) the time since the last fire (i.e. forest age), (3) the canopy species type and (4) the density of canopy species. These four variables are surrogate variables which represent underlying ecological processes which relate to the use of plant resources. Specifically, the resources of interest are moisture, nutrients and light. The model was founded on theories which relate the availability and use of these resources to biodiversity. The ecological significance of these four variables is discussed in Warren (1999) and Chipman and Johnson (2002).

The general form of the species richness prediction equation developed by Chipman and Johnson (2002) is:

$$S = \beta_0 + \beta_1 TSF + \beta_2 DFR + \beta_3 CSD + \delta_1 JP + \delta_2 BS + \delta_3 TA \tag{1}$$

where β_0 is the intercept, $\beta_{1\ldots3}$ are parameter estimates for the variables time-since-fire (*TSF*), distance-from-ridgeline (*DFR*) and canopy-stem-density (*CSD*) respectively, and $\delta_{1\ldots3}$ are parameter estimates for the categorical dummy variables jack pine (*JP*), black spruce (*BS*) and trembling aspen (*TA*) respectively. The categorical forest type variables are binary. Only one of these variables may have a value of one at any one time while all others must have a zero value. The mixed class is not included in the equation but is represented when all other species have a value of 0. The effect of these binary variables is to change the intercept of the model. The values of the model parameters were estimated from field observations of the four landscape characteristics and richness.

To estimate richness at the landscape scale, the four landscape characteristics were estimated from geospatial data and the model run for each pixel. A schematic diagram describing the overall system for producing a richness map is given in Figure 11.2 and further details may be found in Warren (1999). In this section, the methods for estimating the uncertainty in the estimated landscape characteristics and for propagating this uncertainty through the model are discussed.

3.1.1 Time-since-fire

The time-since-fire variable (TSF) was used as a surrogate for the manipulation of resources as a result of a disturbance process. Weir (1996) produced a 1:50 000 scale map of 1249 (TSF) polygons for the PANP based on a combination of aerial photograph interpretation and field reconnaissance. He also estimated the uncertainty for polygon fire ages and these are summarized in Table 11.1. Where a range of uncertainty was specified, the upper limit of the range was used in order to be conservative.

In addition to the measurement uncertainty associated with the fire ages, there was also spatial uncertainty in the location of the polygon boundaries. This spatial

Table 11.1 *Expert opinion estimates of time-since-fire measurement uncertainty*

Fire age (years)	Estimated uncertainty (years)
0–149	$\pm$ 1
150–200	$\pm$ 1–5
$>$ 200	$\pm$ 15–20

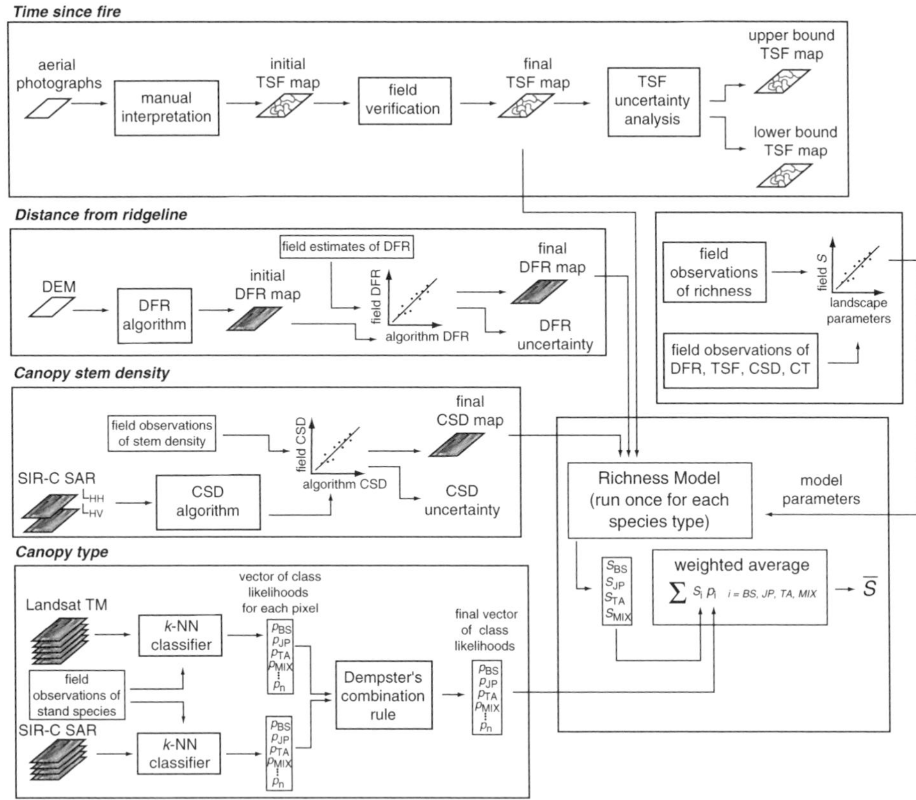

Figure 11.2 *A schematic diagram showing the overall system for deriving a richness map*

uncertainty manifested itself in the form of additional uncertainty in the ages of the forests in the polygons.

Although TSF is a continuous variable that can be any value greater than or equal to zero it must in effect be treated as ordinal data. That is, the fire ages within each polygon are essentially ranked, discrete attributes. To illustrate, consider a point on the exact boundary between two adjacent polygons of ages 10 and 205 years. This point must take on the fire age of one of the two polygons (either 10 or 205 years) and cannot take on an intermediate value. If there is any uncertainty in the location of the fire polygon boundaries, there must be uncertainty attached to the ages assigned to points which are close to or on these boundaries. The magnitude of this uncertainty will be expected to increase with an increase in the magnitude of the difference in adjacent polygon fire ages. For instance, the border point between a polygon representing a region with a large TSF that neighbours one with a small TSF will have a very high age uncertainty relative to a border point between two polygons representing areas of similar age.

A question that must still be answered is what constitutes a boundary point? As one follows a straight line transect from a shared polygon boundary to the centre of the polygon, the expected uncertainty will decrease such that the uncertainty estimate will be partly a function of distance. However, this uncertainty–distance relationship is not known so a conservative worst-case approach to uncertainty assessment was adopted. At some point along this transect there will be a very high probability that the TSF is correct. It is here that the spatial uncertainty in the polygon border was assumed to no longer have an effect on the estimated TSF. The distance from the boundary to this point along the transect will define the width of a buffer zone within the polygon as shown in Figure 11.3. The buffer zones represent areas in which the amount of uncertainty is in question. In the cross-section view, this is shown with the rectangle labelled as the zone of unknown uncertainty (Figure 11.3). Since the uncertainty at specific points within this zone is unknown, the uncertainty interval is taken as the lower and upper limits regardless of distance from the boundary.

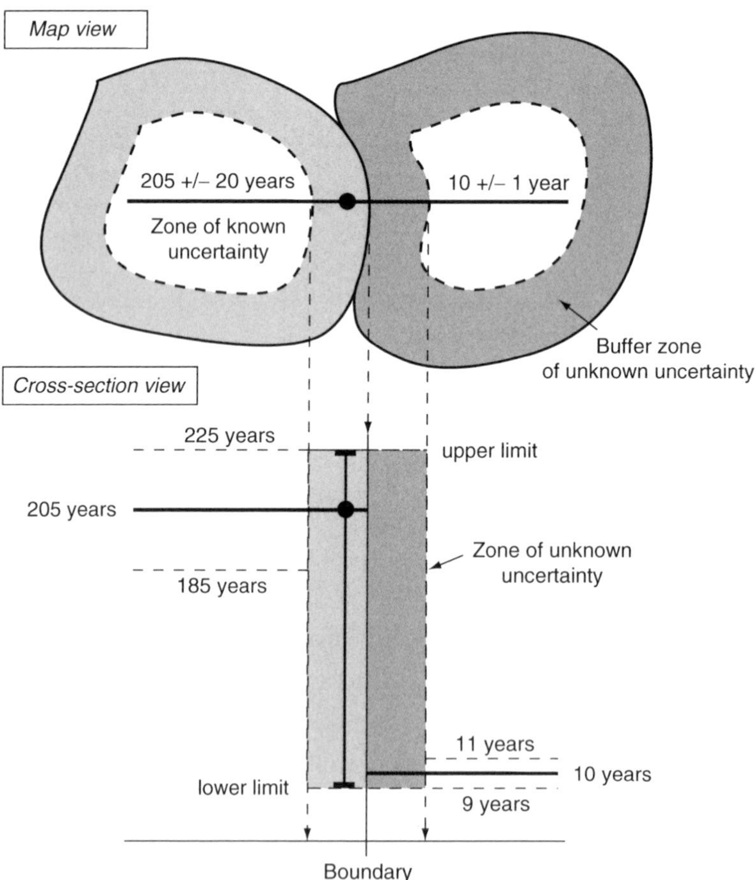

Figure 11.3 *Uncertainty in boundary regions. Illustrated here is the idea that the location of a boundary line between time-since-fire polygons is fuzzy. The inner polygons represent areas for which the uncertainty can be reasonably estimated. The shaded buffer zones within the perimeter of the polygons represent areas for which uncertainty cannot be accurately defined. There is some unknown probability that the point within the 205-year polygon should actually belong within the 10-year polygon which as a result, manifests itself in a wide range of uncertainty as shown by the upper and lower limits*

Considering the point near the shared polygon boundary in Figure 11.3, the age could be as old as 225 years or as young as 9 years. This gives rise to asymmetric uncertainty intervals such that a point labelled as representing a forest that was 205 years old could be +20 or −196 years from this estimate:

$$\begin{aligned} 205 - 196 \text{ years} &= 9 \text{ years} = \text{lower limit} \\ 205 + 20 \text{ years} &= 225 \text{ years} = \text{upper limit} \end{aligned}$$

Using this conservative approach we have the following two methods for estimating TSF uncertainty:

1. For areas inside a polygon in which we are reasonably sure that the TSF is correct (areas with attribute uncertainty but not spatial uncertainty), the uncertainty can be represented with a traditional interval based on the attribute uncertainties from Table 11.1. For example, a TSF of 170 years can be represented by 170 ± 5 years or alternatively by a lower and upper limit (165, 175) respectively.
2. For the buffer zones (areas with attribute and spatial uncertainty), the uncertainty can be represented by a lower and upper limit. These limits are defined by the youngest and oldest possible ages of any neighbouring polygons to which a point might belong. This was illustrated by the example above. Notice that these limits include the known attribute uncertainty from Table 11.1. In the event that more than two polygons intersect, only the youngest and oldest TSFs will be of interest in defining the uncertainty bounds.

These two methods are the basis for estimating and mapping the uncertainty in the TSF map.

3.1.2 Canopy type

The canopy type variable was used as a surrogate for the processes which control light transmission to the forest floor. As mentioned earlier, there were four forest classes used in the biodiversity model. These were jack pine, black spruce, trembling aspen and a mixed class (MIX) of white spruce, balsam fir and trembling aspen. In addition to the forest classes, water (WAT), anthropogenic (ANT), and wetland (WET) classes were also defined (each with a training and testing set of data). The anthropogenic class included towns, roads, recently harvested land as well as other artificial features.

The reference data were used to perform two *k*-nearest neighbour supervised classifications. The first used the seven channels of the Landsat TM imagery and the second used the six SAR image channels. The results for these two classifiers (Table 11.2) show that the two classifiers complemented each other. While the accuracy derived from the SAR data was fairly low, the accuracy for black spruce was higher than observed in the Landsat TM-based classification. One way to combine the information in the SAR and Landsat TM data sets was to combine the 13 channels in a single feature space but the results obtained were poor. An alternative was to combine the class-conditional likelihoods of the individual classifiers. A Dempster–Shafer evidential reasoning algorithm was used and the accuracies, shown in Table 11.3 as class-conditional kappa values as a measure of the uncertainty of the canopy type prediction, were increased.

3.1.3 Distance from ridgeline

The distance from ridgeline (DFR) variable was used as a surrogate for moisture and nutrient gradients which are a result of hillslope process. Estimates of the distance from each image pixel to the nearest ridgeline were required. The algorithm (path 2 in Figure 11.4) for estimating this variable was a compromise between the shortest

Table 11.2 *Summary of kNN results for testing sites using (a) 7 band Landsat TM data and (b) SIR-C SAR data. In each matrix the rows indicate the label in the reference data while the columns that in the classified image. Note that due to incomplete overlap of the SAR and TM data sets the sample size differs slightly*

Class	JP	BS	TA	MIX	WAT	ANT	WET	Kappa
JP	173	2	26	15	0	0	3	0.78
BS	14	43	0	75	0	0	0	0.31
TA	6	0	335	9	0	0	1	0.95
MIX	10	16	13	61	0	0	0	0.59
WAT	0	0	0	0	2018	0	0	1.00
ANT	1	0	32	0	0	397	50	0.80
WET	19	4	27	0	0	117	544	0.72

(a)

Class	JP	BS	TA	MIX	WAT	ANT	WET	Kappa
JP	67	25	74	4	0	49	0	0.25
BS	31	68	6	6	0	15	6	0.48
TA	55	25	204	17	0	11	4	0.60
MIX	18	37	6	30	0	9	0	0.28
WAT	0	0	0	0	788	0	0	1.00
ANT	35	41	39	11	0	336	17	0.63
WET	0	4	1	0	0	113	593	0.79

(b)

Table 11.3 *Summary of classification results after SAR and TM classifications were combined. The rows indicate the label in the reference data while the columns that in the classified image. The change in kappa coefficient, relative to that derived to the use of the TM data alone is also shown*

Class	JP	BS	TA	MIX	WAT	ANT	WET	kappa	Change in kappa
JP	164	6	36	9	0	1	3	0.73	−0.05
BS	17	61	2	40	0	6	6	0.44	+0.13
TA	1	0	312	1	0	0	2	0.99	+0.04
MIX	10	20	17	50	0	3	0	0.50	−0.09
WAT	0	0	0	0	788	0	0	1.00	0
ANT	1	0	39	0	0	408	32	0.85	+0.05
WET	0	1	12	0	0	141	557	0.78	+0.06

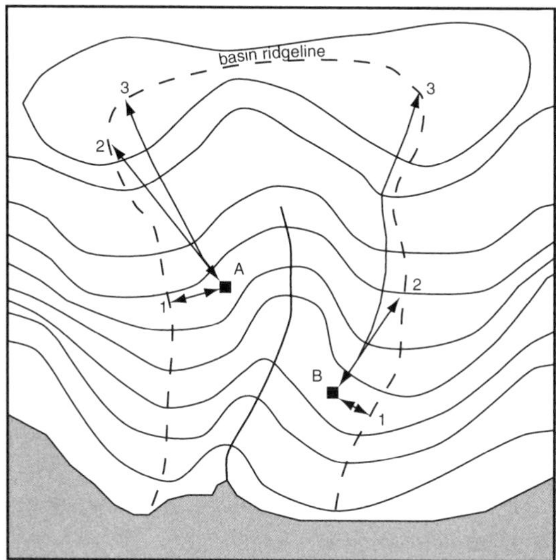

Figure 11.4 Three possible approaches for measuring distance from two pixels (A and B) to a basin ridgeline. The first (used by Bridge (1997)) finds the closest ridge. The second (from this work) finds the closest ridge based on a path direction calculated from local aspect. The third traces a path which cuts perpendicularly through contour lines and is considered to be the best estimate

Euclidean distance (path 1 in Figure 11.4) (Bridge, 1997) and the continuous path that crosses each elevation contour at right angles (path 3 in Figure 11.4).

Calculation of DFR was subject to many sources of uncertainty. For the DEM, these sources include uncertainty from the contour lines used for interpolation, the errors produced during interpolation and resampling the grid cell size. For the basin and ridge delineation, these sources include uncertainty from the flow direction and flow accumulation. Uncertainty was also introduced in calculating the distance from the ridgelines. In some situations, there were no river vectors to ensure that a basin was properly divided such that distances would only be measured from a pixel to its respective ridge.

The quantity and qualitative nature of the uncertainties present in the data and processes used to calculate DFR made it impossible to model the uncertainty in the DFR. The assessment of uncertainty for the DFR map, therefore, was of an empirical nature. A comparison data set consisting of manually measured DFR from 1:50 000 topographic maps was acquired for each of the sampled forest stands. Uncertainty in the DFR data was assessed by comparing the values estimated manually from the maps with those computed algorithmically. For this, forest stands were located on the topographic maps and from a visual assessment of the terrain around a forest stand, the nearest ridge was located on the map and the distance measured. The measured transect running from stand up to ridge always cut perpendicularly through the contour lines. This was repeated for 31 forest stands such that a full range of distances were sampled. Figure 11.5 shows the relationship between map-measured DFR and computed DFR. The dashed

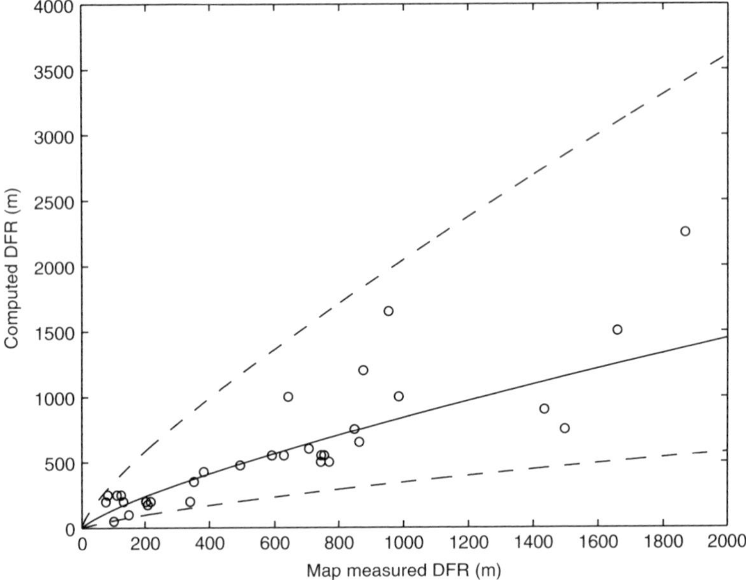

Figure 11.5 *The correlation between the two methods used to determine DFR for 31 ground sample forest stands. For the manually calculated method, 1:50 000 topographic maps were interpreted and basin ridgelines were traced out. The distance was measured from forest stand to ridge by cutting a path perpendicularly through contours. Included on the plot is the loglinear regression line and error bars (dotted lines) based on a 95 % confidence level*

lines are the 95% confidence intervals which were used as estimates of DFR uncertainty.

3.1.4 Canopy stem density

Estimating canopy stem density (CSD) from satellite sensor imagery was extremely difficult. Reports of canopy stem density estimates for boreal forest are sparse in the literature. Many studies have demonstrated the use of SAR to predict forest characteristics such as species composition, biomass, leaf area index (LAI) or diameter at breast height (DBH) (Ranson *et al.*, 1995; Ranson and Sun, 1994; Le Toan *et al.*, 1992; Wilson, 1996). However, within this literature there is little or no attention given to the estimation of stem density. It seems plausible that estimation of stem density has been largely unsuccessful and thus, reporting of results has not followed.

Kurvonen *et al.* (1999) estimated stem volume using L-band JERS-1 and ERS-1 SAR data. However, their prediction equations required the use of vegetation and ground moisture and were not practical for this work.

Ranson *et al.* (1995) and Ranson and Sun (1994) found correlations between SAR band ratios and forest biomass. They proposed that using a ratio of SAR bands (as opposed to unratioed bands) may increase signal dynamic range and thus increase the correlation with forest properties. Ranson *et al.* (1995) related the logarithm of forest biomass to SIR-C SAR LHV backscatter with a coefficient of determination (R^2) of 0.846 for the BOREAS study site which is adjacent to PANP, but such results were not reproduced at this site. Collins and Livingstone (1996) used SAR polarization ratios (same band) for mapping thin sea ice. These techniques were examined for this research: all possible combinations of SAR polarization and band ratios were regressed with the natural logarithm of canopy stem density. The L-band polarization ratio (LHH/LHV), had the largest, although still small, correlation coefficient with CSD ($R^2 = 0.13$). The basic form of the model relationship was:

$$\ln CSD = m \times (LHH/LHV) + b \quad (2)$$

and the relationship is depicted graphically in Figure 11.6. To determine the error interval, the ln(error) was added and subtracted from the ln(CSD) to gain the upper and lower error bounds on the prediction. The exponential of each was then calculated which produced the error in terms of CSD. These bounds are plotted in Figure 11.6 as the dashed lines. An important feature to notice is that the error bounds are not symmetric about the prediction value. This is a product of using logarithmically transformed data in the model development. As can be seen in the graph SAR input values below 0.8 and above 1.0 are prone to extremely high uncertainty.

3.2 Uncertainty propagation

The parameters in the linear model were estimated using field observations of richness, DFR, TSF, canopy species type and CSD. The model parameters are listed in Table 11.4, note that the model is in reality four models with the delta values (δ) changing the slope for each canopy type.

An outline of the overall richness estimation procedure is given in Figure 11.2. To produce a map of estimated richness, the model was run for each pixel using the estimates of the DFR, CSD and TSF. The canopy type classification algorithm generated a vector of class membership likelihoods for each pixel. We summed the likelihoods for the four forest classes and if this sum was greater than a threshold the pixel was determined to be a forest pixel otherwise it was labelled with one of the three non-forest classes.

Table 11.4 *Parameter estimates for the species richness equation (equation (1))*

Term	Estimate
Intercept (β_0)	25.995261
TSF (β_1)	−0.036611
DFR (β_2)	−0.000926
CSD (β_3)	−0.004021
JP (δ_1)	0.8042173
BS (δ_2)	−4.295932
TA (δ_3)	4.7861226

For a pixel labelled as forest, the richness model was run four times, once for each of the canopy types, and derived a weighted average richness in which the weights were the class likelihoods for the forest classes as follows,

$$S = \frac{P_{JP}S_{JP} + P_{BS}S_{BS} + P_{TA}S_{TA} + P_{MIX}S_{MIX}}{\Sigma_i p_i} \tag{3}$$

where p_i are the class likelihoods.

The uncertainties for DFR and CSD are non-symmetric confidence intervals derived from regressions of estimates against observations. The uncertainties for TSF are symmetric intervals in the interior of the TSF polygons and non-symmetric upper and lower bounds in the polygon boundary zones. There was no straightforward way to propagate these disparate errors through the model. We decided, therefore, to be conservative and use upper and lower bounds on the richness as our estimates of uncertainty for our richness predictions. These two bounds represent two worst-case scenarios when we have all the errors adding together rather than cancelling each other as normal error analysis assumes. Note that given the negative values for $\beta_{1...3}$, S increases in value with a decrease in CSD, DFR and TSF. Hence, the lower bound on S was generated by using upper bounds on the three landscape variables in equation (1). As before, a weighted average lower bound was calculated. The same procedure was followed for calculating the weighted average upper bound on richness.

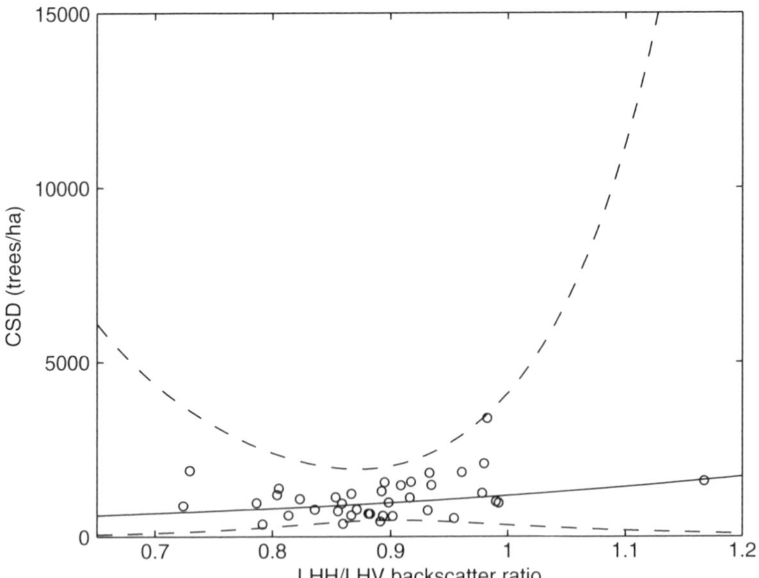

Figure 11.6 Measurement data and the linear regression model used to predict canopy stem density from the SAR backscatter ratio. The solid line is the model prediction line (of best fit) and the dashed lines represent error bars two standard deviations from the prediction line

4 Results and Discussion

The first result was intended to answer question 1: How well can richness be estimated using estimated versus observed landscape variables?

Figure 11.7 is a comparison between the richness predicted from the estimated landscape variables versus the richness predicted from the observed variables. The

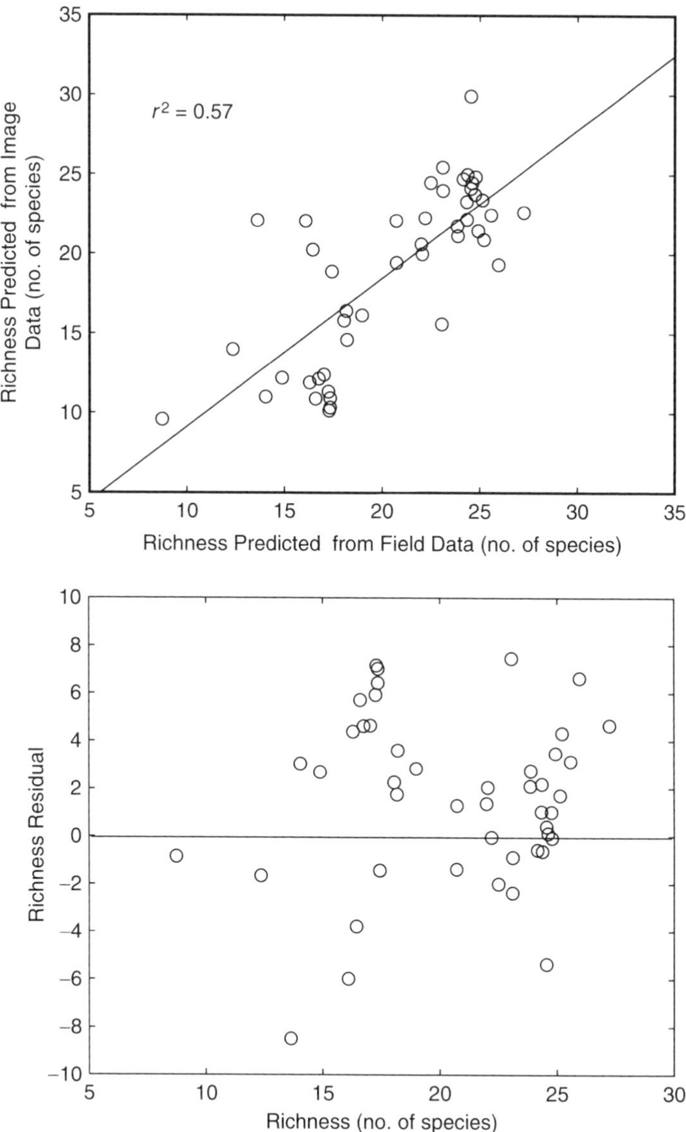

Figure 11.7 Comparison of richness estimates: (a) the relationship between richness predicted using observed variables versus richness predicted from estimated variables, (b) the residuals, these are estimated – observed

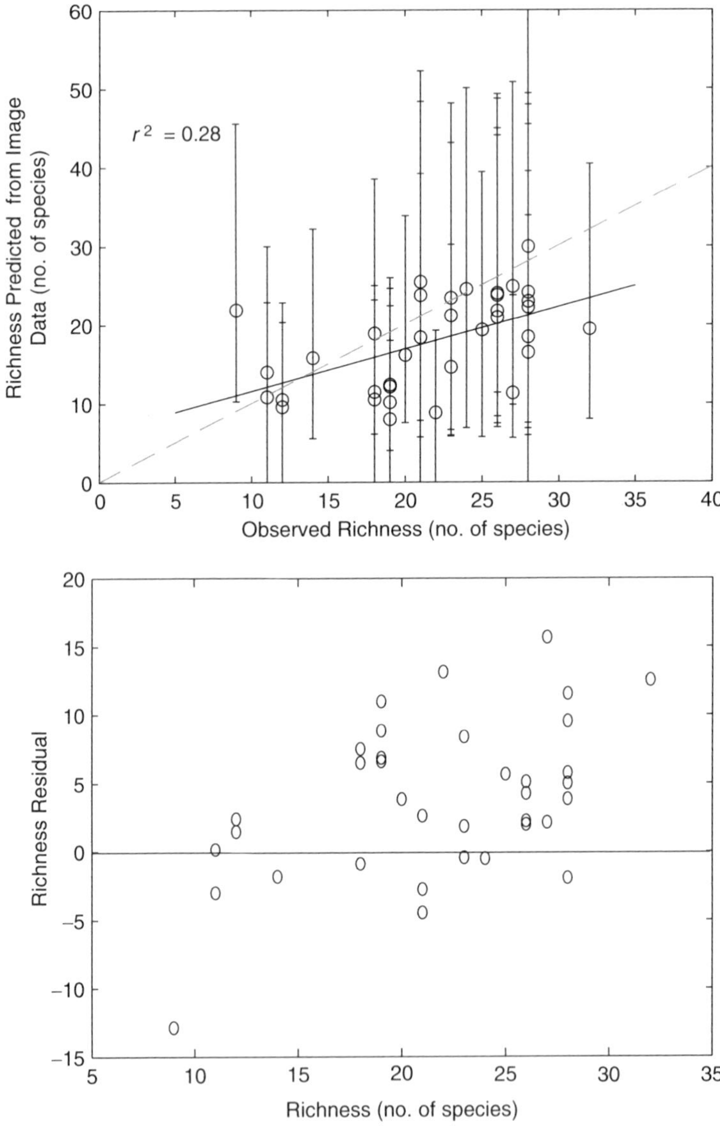

Figure 11.8 *Comparison of richness estimates: (a) the relationship between richness observed in the field with that predicted from the model, (b) the residuals from the relationship*

degree of correlation, while not particularly large ($R^2 = 0.57$), indicated that we should have some confidence in the predictive model when we apply it to the whole landscape. The residual plot in Figure 11.7b shows that the estimated-variable prediction tends to over-estimate the richness relative to the observed-variable prediction (which is itself higher than the observed richness); the residual RMS error was 3.8 species. Hence, the effect of using estimated versus observed landscape variables in the richness model was to over-estimate richness.

The second question addressed was: How accurate are richness predictions relative to field observations? Figure 11.8 is a comparison between the richness predicted from the model versus the richness observed in the field. Again, the degree of correlation was relatively small ($R^2 = 0.28$) but it was significant.

The residual plot in Figure 11.8 shows largely positive residuals (RMS error is 6.8 species) that indicated that the model tended to under-estimate the richness relative to the observed richness for the testing samples. Hence, the model was conservative in its estimation of richness.

Finally, the level of uncertainty in the richness predictions was evaluated. The error bars in Figure 11.8 are the upper and lower bounds on the estimated richness. There are two features to note. First, the size of the bars plotted. A consistently conservative approach to the estimation and propagation of error has been used and this led to very large levels of uncertainty in the richness estimates. This should not erode the value of the richness map. While the levels of uncertainty in the map are high they are at least quantified which is regarded as a step forward in biodiversity mapping. The second feature to note is that the errors are non-symmetric with the upper bound being larger than the lower bound. This is consistent with the finding that the estimated richness values were smaller than the observed values. The predicted richness was conservative which may be viewed as a positive result.

Figure 11.9 (see plate 6) shows a sample of the predicted species richness map for the PANP together with the upper and lower bounds. The wetter areas of the park tend to have lower diversity. This observation coincides with the reasoning for the inclusion of the distance from a ridgeline variable into the model. However, the actual patterns in the distance from a ridgeline occur at a finer scale than on the species richness map and, therefore, there are no apparent spatial correlations between the two maps. The same is true for canopy stem density. In general, areas which were classified as trembling aspen are highly correlated with relatively high biodiversity. Lower biodiversity areas tend to coincide with the spatial distribution of jack pine and the mixed class. The lowest biodiversity coincides with black spruce. There was no apparent spatial correlation between the patterns of species richness and the time-since-fire map. Since fire is suppressed within the park, there tends to be very little young forest area. This might account for the lack of apparent variation in species richness due to this variable.

5 Conclusions

This work has shown that a simple richness model is, to some extent, capable of estimating the correct levels of biodiversity in a variety of landscapes. The form of the uncertainties in the input landscape variables differ from one another making it impossible to implement standard methods of error propagation. In the absence of a standard solution a deliberately conservative approach has been used. A method for representing and propagating uncertainty through a predictive model for species richness was also developed and used. This has led to clear, if conservative levels of uncertainty. Nevertheless, by propagating and reporting uncertainty associated with a biodiversity map, a realistic understanding of the limits of both the resulting map and model itself have been obtained. This can prevent misuse of such a map and allows the refinement of the model inputs to reduce uncertainty.

References

Andow, D. A., 1991, Vegetational diversity and arthropod population response, *Annual Reviews of Entomology*, **36**, 561–86.

Bormann, F. and Likens, G., 1979, *Patterns and Process in a Forested Ecosystem* (New York: Springer-Verlag).

Bridge, S., 1997, *The Landscape Scale Spatial Distribution of Vegetation Gradients in a Mixed-wood Boreal Forest: Linking Ecological Patterns to Geomorphic Processes Across Scales*, Unpublished MSc thesis, Department of Biology (Calgary, AB: University of Calgary).

Cameron, H., 1953, *Melting Point of the Bonding Material in Lodgepole Pine and Jack Pine Cones* (Silviculture Leaflet No. 86). Canada Department of Resources and Development, Forestry Branch.

Chapman, B., 1995, *SIR-C Data Compression Software User Guide* (Pasadena, CA: Jet Propulsion Laboratory).

Chipman, S. and Johnson, E. A., 2002, Understory vascular plant species diversity in the mixedwood boreal forest of western Canada, *Ecological Applications* (accepted for publication).

Collins, M. and Livingstone, C. E., 1996, On the dimensionality of multiparameter microwave image data from thin sea ice in the Labrador Sea, *IEEE Transactions on Geoscience and Remote Sensing*, **34**, 114–36.

Harper, J. L. and Hawksworth, D. L., 1995, *Biodiversity Measurement and Estimation* (London: The Royal Society and Chapman and Hall).

Huston, M., 1994, *Biological Diversity: The Coexistence of Species on Changing Landscapes* (Cambridge: Cambridge University Press).

Johnson, E., 1992, *Fire and Vegetation Dynamics* (Cambridge: Cambridge University Press).

Kurvonen, L., Pullianinen, J. and Hallikainen, M., 1999, Retrieval of biomass in boreal forests from multitemporal ERS-1 and JERS-1 SAR images, *IEEE Transactions on Geoscience and Remote Sensing*, **37**, 198–205.

Le Toan, T., Beaudoin, J. and Guyon, D., 1992, Relating forest biomass to SAR data, *IEEE Transactions on Geoscience and Remote Sensing*, **30**, 403–11.

Lopes, A., Nezry, E., Touzi, R. and Laur, H., 1993, Structure detection and statistical adaptive speckle filtering in SAR images, *International Journal of Remote Sensing*, **14**, 1735–58.

Magurran, A., 1988, *Ecological Diversity and its Measurement* (Princeton, NJ: Princeton University Press).

May, R, M. 1975, Patterns of species abundance and diversity, In M. L. Cody and J. M. Diamond (eds), *Ecology and Evolution of Communities*, (Cambridge: Belknap Press).

NASA Jet Propulsion Laboratory, 1993, SIR-C CEOS *Tape Reader v2.3* (Pasadena, CA: Jet Propulsion Laboratory).

NASA Jet Propulsion Laboratory, 1994, SIR-C CEOS *Data Compression* (Pasadena, CA: Jet Propulsion Laboratory).

Ranson, K., Saatchi, S. and Sun, G., 1995, Boreal forest ecosystem characterization with SIR-C/XSAR, *IEEE Transactions on Geoscience and Remote Sensing*, **33**, 867–76.

Ranson, K. and Sun, G., 1994, Mapping biomass of a northern forest using multifrequency SAR data, *IEEE Transactions on Geoscience and Remote Sensing*, **32**, 388–95.

Vuu, C., Wong, C. and Barret, P., 1995, SIR-C CEOS *Tape Reader User's Guide* (Pasadena, CA: Jet Propulsion Laboratory).

Warren, A. J., 1999, *Integrating Remote Sensing and GIS Techniques with Ecological Models to Map Biological Diversity in Boreal Forest*, Unpublished MSc thesis (Calgary, AB: University of Calgary).

Weir, J., 1996, *The Fire Frequency and Age Mosaic of a Mixedwood Boreal Forest*, Unpublished MSc thesis (Calgary, AB: University of Calgary).

Wilson, B., 1996, Estimating coniferous forest structure using SAR texture and tone, *Canadian Journal of Remote Sensing*, **22**, 382–9.

12

The Effects of Uncertainty in Deposition Data on Predicting Exceedances of Acidity Critical Loads for Sensitive UK Ecosystems

E. Heywood, J. R. Hall and R. A. Wadsworth

1 Introduction

Acidification can lead to harmful environmental effects, including depletion of fish stocks from lakes and streams, soil degradation and deforestation due to phytotoxicity (Rodhe *et al.*, 1995). The main sources of sulphur emissions, that in turn lead to sulphur deposition, are the burning of fossil fuels, for example, from coal-burning power stations and other stationary sources (e.g. industry). Oxidized nitrogen emissions can also originate from the same sources as sulphur, but additionally come from vehicle exhaust fumes. Both of these pollutants can contribute to acidification and their impact is commonly assessed relative to a critical load.

The critical load concept is based on the maximum load of pollutant deposition that an element of the environment can tolerate without harmful effects occurring. The amount of excess deposition above the critical load is called the exceedance. Within Europe, protocols controlling the emissions of sulphur and nitrogen are being implemented and further reductions in these pollutants have an associated cost. Consequently, it is imperative that uncertainties in the calculation of exceedances are quantified before further reductions in pollutants can be justified fully. The further development of methods to assess the uncertainties in critical load exceedance

Uncertainty in Remote Sensing and GIS. Edited by G.M. Foody and P.M. Atkinson.
 ISBN: 0–470–84408–6

calculations is one of the priority tasks identified by the UN/ECE Convention on Long-range Transboundary Air Pollution Working Group for Effects. The Task Force on Integrated Assessment also requires this information for inclusion in their assessments and optimization studies on the European scale. Policy makers both within the UK and within Europe will use this uncertainty information.

Ecosystems differ in their sensitivity to acidification. The spatial distribution of ecosystems at risk from acidification is often derived using the critical loads approach. For the purposes of this chapter a critical load may be defined as 'a quantitative estimate of an exposure to one or more pollutants below which significant harmful effects on specified sensitive elements of the environment do not occur according to present knowledge' (Nilsson and Grennfelt, 1988, p. 8). Critical loads have formed the basis for national and international deliberations on the control of atmospheric emissions and depositions of sulphur and nitrogen. The excess deposition above the critical load, the critical load exceedance, is used to quantify the potential harmful effects.

It should be noted that the empirical and mass balance methods on which national critical loads are based, define long-term critical loads for systems at steady-state. Therefore, exceedance is an indication of the potential for harmful effects to systems at steady-state. This means that current exceedance of critical loads does not necessarily equate with damage to the ecosystem.

Previously, critical load exceedances have been calculated as deterministic estimates. However, each variable in the calculation of critical load and exceedance has some uncertainty associated with it. Data inaccuracy may arise from numerous sources and may generally be divided into two main categories depending on their characteristics: precision or bias. Until recently exceedance calculations in the UK have not accounted for these inherent uncertainties in input parameter values.

The uncertainty of exceedance values is a product of many factors. This chapter outlines the results of a preliminary study undertaken to quantify the variation in acidity critical load exceedance caused by the uncertainty in sulphur and nitrogen deposition values only. The additional effect of uncertainty in critical load values themselves has not been considered here. The uncertainty analysis involved a fixed value analysis where deposition was varied by a fixed amount and a Monte Carlo analysis where deposition was varied randomly. Both types of analysis used the same two baseline deposition scenarios. The response of the exceedance calculation to variations in sulphur alone, nitrogen alone and sulphur and nitrogen together was investigated (i.e. a sensitivity analysis). Nitrogen and sulphur were varied together to investigate any synergistic effects.

Exceedance predictions have been measured in two ways. Firstly, the total area of sensitive ecosystems for which the critical load was exceeded (i.e. exceeded ecosystem area) in each grid square. Sensitive ecosystems are defined as semi-natural and natural upland and lowland ecosystems that are sensitive to acidification and/or eutrophication. Secondly, the accumulated exceedance (AE). This takes account of both the exceeded ecosystem area and the magnitude of exceedance and is defined as:

$$\text{AE (keq year}^{-1}) = \text{exceedance (keq ha}^{-1}\text{ year}^{-1}) \times \text{ exceeded ecosystem area (ha)}$$

The aims of the study reported in this chapter were to quantify the effect of uncertainty in nitrogen and sulphur deposition on the exceeded area and AE of sensitive UK ecosystems. Additionally, the study sought to investigate whether there is a linear relationship, as suggested in a previous study, between deposition uncertainty and the area of critical load exceedance or accumulated exceedance. In doing this, consideration was given to the relative sensitivity of critical load exceedance estimation to sulphur or nitrogen deposition uncertainty. Two deposition scenarios were studied and compared; one for the period 1995–97 and a predicted one for 2010.

2 Methods

Currently in the UK, critical loads data are calculated for a specified set of ecosystems. These are acid grassland, calcareous grassland, heathland, coniferous woodland and deciduous woodland. In addition, acidity critical loads have been defined for 1445 freshwater sites throughout Great Britain. The deposition data are total nitrogen and sulphur values mapped for every 5×5 km grid square of the UK. The exceedance values are then calculated for every 1×1 km grid square of the UK. These data can be aggregated to a 5 km spatial resolution or summed across grid squares to give national statistics. It should be noted here that the data are at two spatial resolutions, which itself introduces uncertainties. The comparison of data obtained with different spatial resolutions can be a major problem when attempting to measure spatial variation. This should be explored thoroughly in future work. However, Smith *et al.* (1995) have already simulated 1 km deposition from 20 km data (the scale of the deposition data current at the time of their work) and showed that the uncertainty introduced by using 20 km scale estimates of deposition is small. Smith *et al.* (1995) found that the major problem was the uncertainty in estimates of deposition, the issue addressed in this chapter.

2.1 Input data

2.1.1 Critical loads data

To examine the acidifying effects of both sulphur and nitrogen deposition simultaneously, the critical loads function (CLF) was developed in Europe (Posch *et al.*, 1999; Posch and Hettelingh, 1997; Posch *et al.*, 1995; Hettelingh *et al.*, 1995). The CLF defines separate acidity critical loads in terms of sulphur and nitrogen and compares them with sulphur and nitrogen deposition.

The intercepts of the CLF on the sulphur and nitrogen axes define the 'maximum' critical loads of sulphur and nitrogen (Figure 12.1). The maximum critical load of sulphur (CLmax(S)) is the critical load for acidity expressed in terms of sulphur only (i.e. when nitrogen deposition is zero). Similarly, the maximum critical load of nitrogen (CLmax(N)) is the critical load of acidity expressed in terms of nitrogen only (when sulphur deposition is zero). CLmin(N) is the deposition independent critical load of acidity solely due to nitrogen removal processes in the soil (nitrogen uptake and immobilization, Figure 12.1). These 'minimum' and 'maximum' critical

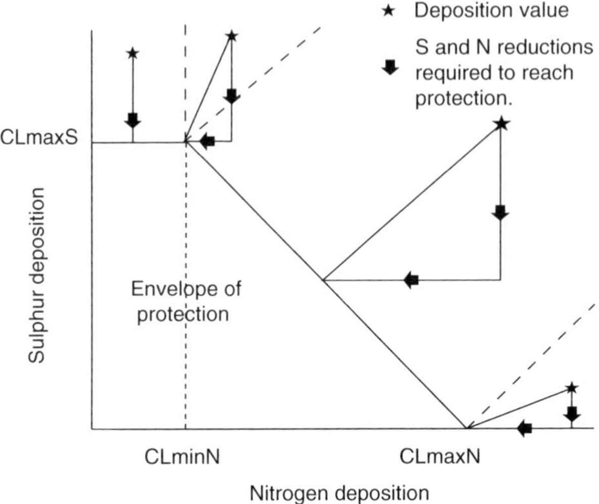

Figure 12.1 *The components of the critical load function (CLF). Note that where both sulphur and nitrogen deposition reductions are required, the exceedance must be calculated as the sum of both depositions and not simply the value equating to the length of the straight line from the deposition point to the CLF*

loads are calculated for the sensitive ecosystems defined above (i.e. not for the whole of the UK) (Hall *et al.*, 2001a). The total area of sensitive ecosystems in the UK is estimated as 95 782.4 km^2. It is these critical load values for each ecosystem that are used for this study, which total 328 713 records.

2.1.2 Deposition data

Deposition input data are a combination of wet, dry and cloud nitrogen, non-marine sulphur and ammonia deposition. Note that marine sources of sulphur are not included in the calculations of exceedance, since the aim here is to assess the impact of acidifying deposition from anthropogenic sources only. Deposition estimates for moorland were applied in the calculations for acid grassland, calcareous grassland and heathland; woodland deposition estimates were used for coniferous and deciduous woodland and mean deposition estimates for freshwaters.

Two different baseline deposition scenarios have been used to investigate the effect of deposition uncertainty on exceedance within the UK. The first related to the period 1995–97. This scenario was based on 5 km spatial resolution data derived from annual measured mean deposition values for the three-year period 1995–97. Maps of total deposition for each 5×5 km grid square of the UK were provided by the Centre for Ecology and Hydrology (CEH) Edinburgh for this work (Smith and Fowler, 2002; Smith *et al.*, 2000). Figure 12.2a shows sulphur deposition and Figure 12.2c shows the total nitrogen deposition for this scenario. This scenario is referred to elsewhere in this chapter as the 1995–97 baseline scenario.

The second scenario was the Gothenburg Protocol deposition scenario modelled from the Hull Acid Rain Model (HARM) (Metcalfe *et al.*, 2001) and the Fine Resolution Ammonia Model (FRAME) (Singles *et al.*, 1998; Sutton *et al.*, 1995)

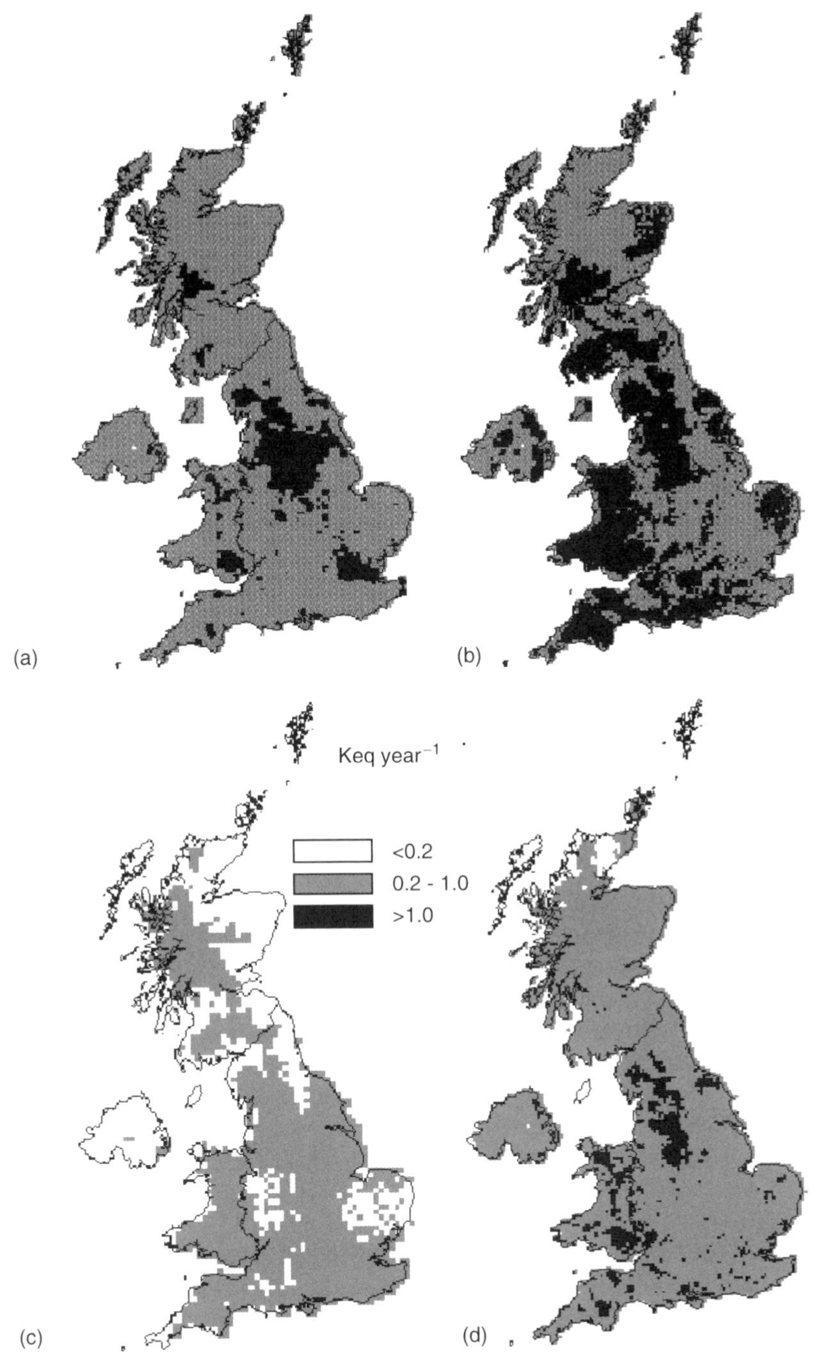

Figure 12.2 *Baseline deposition scenarios at a 5 km spatial resolution. (a) 1995–97 sulphur, (b) 2010 non-marine sulphur, (c) 1995–97 total nitrogen (nitrogen oxide + ammonia) and (d) 2010 total nitrogen (nitrogen oxide + ammonia). Note: there is currently no data available for the Isle of Man for this scenario. Nitrogen oxide is the collective term for nitric oxide and nitrogen dioxide. Sources: CEH Edinburgh, Edinburgh University and Lancaster University*

atmospheric deposition models. This is a deposition scenario for 2010, assuming the implementation of the recent United Nations Economic Committee for Europe Gothenburg Protocol (UN/ECE, 1999). HARM deposition was calculated at 10 km spatial resolution and FRAME at 5 km spatial resolution. HARM was converted to 5 km spatial resolution by sub-dividing the 10 km grid cells so, like the first scenario, it also had a 5 km cell size. The HARM outputs are used for sulphur and nitrogen oxide deposition while FRAME outputs are used for ammonia deposition. The deposition scenario is illustrated in Figures 12.2b for sulphur and 12.2d for nitrogen. This scenario is referred to elsewhere in this chapter as the 2010 baseline scenario.

2.2 The exceedance calculation

The acidity CLF can be used to calculate the total acid deposition reduction necessary to obtain ecosystem protection. Five different regions are defined on the CLF (Figure 12.1); one being the envelope of protection below the CLF and the other four defining combinations of sulphur and nitrogen deposition, where reductions are required to reach the CLF (Figure 12.1). Note that where both sulphur and nitrogen deposition reductions are required, the exceedance must be calculated as the sum of both depositions and not simply the value equating to the length of the straight line from the deposition point to the CLF. Elaboration on the calculation of the exceedance function for acidity critical loads as defined above can be found in Posch *et al.* (1999).

The total area of ecosystems exceeded (in km^2) in each 5 km square can be calculated and mapped. The AE values can be summed to give total AE values for each 5 km grid square. These values can then be summed for regions of interest. For this study, statistics for the whole UK were used to give an estimate of the effect of deposition uncertainty at the national level. Maps at a 5 km spatial resolution of the ecosystem areas exceeded and AE for the 1995–97 and 2010 baseline deposition scenarios are shown in Figure 12.3. The existing national methodology used to produce Figure 12.3 (Hall *et al.*, 2001b) is referred to as the deterministic approach throughout this chapter.

2.3 Uncertainty analyses

Uncertainty analysis involves the computation of uncertainty in the output induced by the quantified uncertainty in data inputs and model parameters. Two methods for examining uncertainty in deposition values were used: fixed value analysis and Monte Carlo analysis. The fixed value analysis, although an unrealistic method of simulating uncertainty, was used as an initial attempt to create bounds of uncertainty of the exceedance calculation.

2.3.1 The fixed value analysis

The estimates of total sulphur to a 20×20 km area were speculated to have an uncertainty of $\pm 40\%$ in central England increasing to $\pm 80\%$ in the west of Scotland

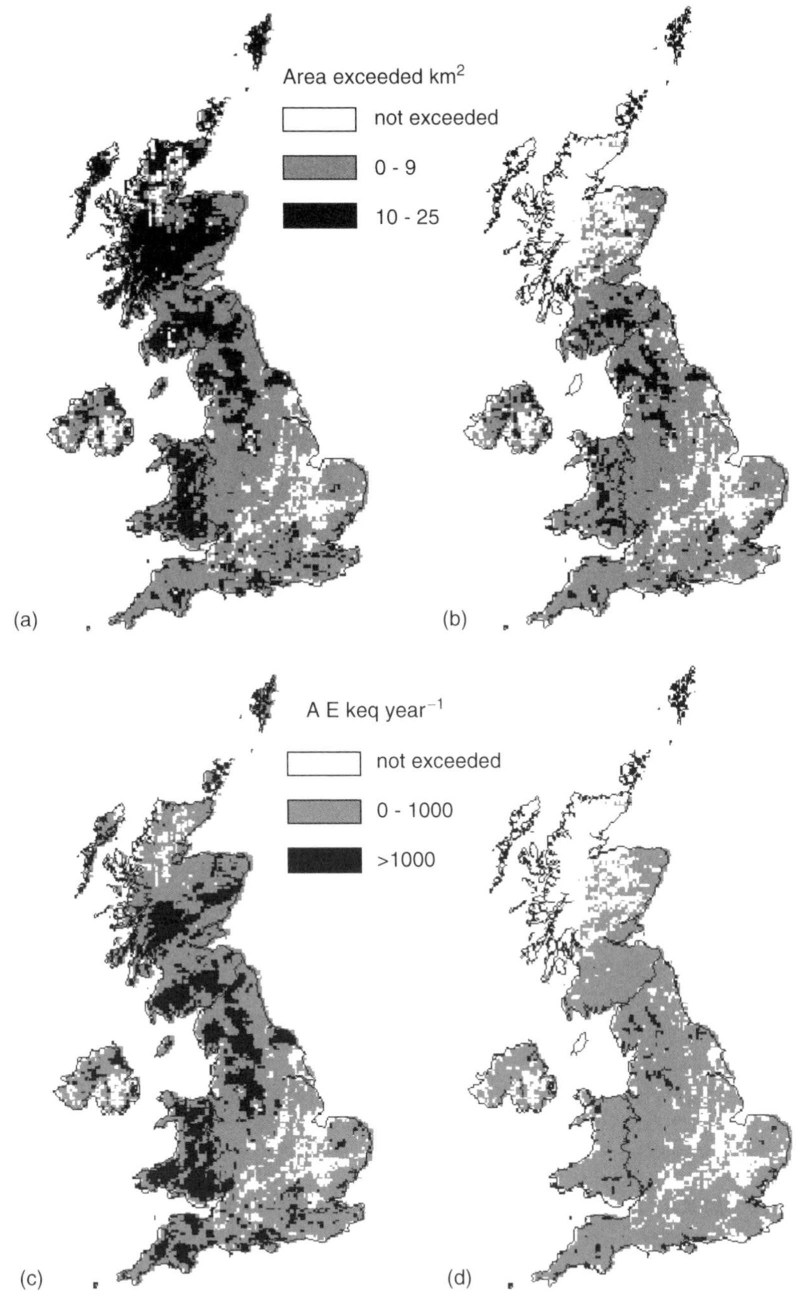

Figure 12.3 *Maps at a 5 km spatial resolution showing where acidity loads are exceeded by acid deposition for (a) area of ecosystems exceeded by 1995–97 baseline deposition, (b) area of ecosystems exceeded by 2010 baseline deposition, (c) accumulated exceedances for 1995–97 baseline deposition and (d) accumulated exceedances for 2010 baseline deposition. Source: CEH Monks Wood*

and in Wales. These values were derived from the work of Smith *et al.* (1995) who also noted that, as there are many uncertainties in the system, the estimates may be inaccurate. For the purposes of this study, the estimated total sulphur and nitrogen deposition to any 5×5 km area in the UK was assumed to have an uncertainty of $\pm 40\%$ and to be uniform across the country. It is recognized that further investigation into this limit is necessary but it was deemed a reasonable estimate.

Eight simulations were investigated, for this fixed values analysis, D1 to D8. These were designed to investigate variation in sulphur only, variation in nitrogen only and variation in sulphur and nitrogen together:

$$\text{(D1)} \quad \begin{aligned} N_{gen} &= N_{base} \\ S_{gen} &= S_{base} + 0.4 \cdot S_{base} \end{aligned}$$

$$\text{(D2)} \quad \begin{aligned} N_{gen} &= N_{base} \\ S_{gen} &= S_{base} - 0.4 \cdot S_{base} \end{aligned}$$

$$\text{(D3)} \quad \begin{aligned} N_{gen} &= N_{base} + 0.4 \cdot N_{base} \\ S_{gen} &= S_{base} \end{aligned}$$

$$\text{(D4)} \quad \begin{aligned} N_{gen} &= N_{base} - 0.4 \cdot N_{base} \\ S_{gen} &= S_{base} \end{aligned}$$

$$\text{(D5)} \quad \begin{aligned} N_{gen} &= N_{base} + 0.4 \cdot N_{base} \\ S_{gen} &= S_{base} + 0.4 \cdot S_{base} \end{aligned}$$

$$\text{(D6)} \quad \begin{aligned} N_{gen} &= N_{base} - 0.4 \cdot N_{base} \\ S_{gen} &= S_{base} - 0.4 \cdot S_{base} \end{aligned}$$

$$\text{(D7)} \quad \begin{aligned} N_{gen} &= N_{base} + 0.4 \cdot N_{base} \\ S_{gen} &= S_{base} - 0.4 \cdot S_{base} \end{aligned}$$

$$\text{(D8)} \quad \begin{aligned} N_{gen} &= N_{base} - 0.4 \cdot N_{base} \\ S_{gen} &= S_{base} + 0.4 \cdot S_{base} \end{aligned}$$

where N_{base} is the total (oxidized + reduced) nitrogen deposition value as estimated from a selected baseline deposition scenario (1995–97 or 2010) for nitrogen and S_{base} the corresponding value for sulphur while N_{gen} and S_{gen} represent the generated deposition scenario used in the simulation for nitrogen and sulphur respectively.

Scenarios D5 and D6 were run as worst and best case scenarios calculated by increasing (worst case) and reducing (best case) both nitrogen and sulphur deposition.

2.3.2 The Monte Carlo analysis

Monte Carlo simulation is a well-established technique for assessing the effects of uncertainty in inputs and model parameters. 'Monte Carlo simulations start with the sampling of parameter values from a known or suspected distribution. A set of distributions for each parameter is called a scenario. The scenario is used as input to the model which results in a distribution of output values. The chosen number of samples, and subsequent runs, depends on the accuracy required for the resulting output distribution' (Barkman *et al.*, 1995, p. 20).

The majority of previous studies have recognized that the underlying distribution of deposition values is largely unknown due to limited data availability (Jonsson *et al.*, 1995; Barkman *et al.*, 1995). Hence, previous studies have often assumed a uniform distribution between the upper and lower quoted deposition values. Since the shape of the deposition distributions was unknown it was decided to sample from both uniform (or rectangular) (D9–D11) and triangular (D12–D14) distributions and see how the different distributions affected the result. One of the attractive features of a triangular distribution appears to be that it exhibits a rough similarity to a normal distribution without tails, a seeming advantage if little is known about the distribution.

The following Monte Carlo simulations were used:

$$\text{(D9)}\quad \begin{aligned} N_{\text{gen}} &= N_{\text{base}} \\ S_{\text{gen}} &= S_{\text{base}} - r_{\text{s}} \cdot (S_1 - S_0) && \text{for } r_{\text{s}} \leq 0.5 \\ S_{\text{gen}} &= S_{\text{base}} + (1 - r_{\text{s}}) \cdot (S_1 - S_0) && \text{for } 0.5 < r \leq 1 \end{aligned}$$

$$\text{(D10)}\quad \begin{aligned} N_{\text{gen}} &= N_{\text{base}} - r_{\text{N}} \cdot (N_1 - N_0) && \text{for } r_{\text{N}} \leq 0.5 \\ N_{\text{gen}} &= N_{\text{base}} + (1 - r_{\text{N}}) \cdot (N_1 - N_0) && \text{for } 0.5 < r \leq 1 \\ S_{\text{gen}} &= S_{\text{base}} \end{aligned}$$

$$\text{(D11)}\quad \begin{aligned} N_{\text{gen}} &= N_{\text{base}} - r_{\text{N}} \cdot (N_1 - N_0) && \text{for } r_{\text{N}} \leq 0.5 \\ N_{\text{gen}} &= N_{\text{base}} + (1 - r_{\text{N}}) \cdot (N_1 - N_0) && \text{for } 0.5 < r_{\text{N}} \leq 1 \\ S_{\text{gen}} &= S_{\text{base}} - r_{\text{s}} \cdot (S_1 - S_0) && \text{for } r_{\text{s}} \leq 0.5 \\ S_{\text{gen}} &= S_{\text{base}} + (1 - r_{\text{s}}) \cdot (S_1 - S_0) && \text{for } 0.5 < r_{\text{s}} \leq 1 \end{aligned}$$

$$\text{(D12)}\quad \begin{aligned} N_{\text{gen}} &= N_{\text{base}} \\ S_{\text{gen}} &= S_{\text{base}} - \left((S_{\text{base}} - S_0) - \sqrt{r_{\text{s}} \cdot (S_1 - S_0) \cdot (S_{\text{base}} - S_0)} \right) \quad \text{for } r_{\text{s}} \leq 0.5 \\ S_{\text{gen}} &= S_{\text{base}} + \Big((S_1 - S_{\text{base}}) \\ &\quad - \sqrt{S_1^2 - S_1 \cdot (S_{\text{base}} + S_0) + S_0 \cdot S_{\text{base}} - r_{\text{s}} \cdot (S_1 - S_0) \cdot (S_1 - S_{\text{base}})} \Big) \\ &\quad \text{for } 0.5 < r_{\text{s}} \leq 1 \end{aligned}$$

$$\text{(D13)}\quad \begin{aligned} N_{\text{gen}} &= N_{\text{base}} - \left((N_{\text{base}} - N_0) - \sqrt{r_{\text{N}} \cdot (N_1 - N_0) \cdot (N_{\text{base}} - N_0)} \right) \\ &\quad \text{for } r_{\text{N}} \leq 0.5 \end{aligned}$$

$$N_{gen} = N_{base} + \Big((N_1 - N_{base}) - \sqrt{N_1^2 - N_1 \cdot (N_{base} + N_0) + N_0 \cdot N_{base} - r_N \cdot (N_1 - N_0) \cdot (N_1 - N_{base})}\Big)$$

for $0.5 < r_N \leq 1$

$$S_{gen} = S_{base}$$

(D14) $$N_{gen} = N_{base} - \Big((N_{base} - N_0) - \sqrt{r_N \cdot (N_1 - N_0) \cdot (N_{base} - N_0)}\Big) \text{ for } r_N \leq 0.5$$

$$N_{gen} = N_{base} + \Big((N_1 - N_{base}) - \sqrt{N_1^2 - N_1 \cdot (N_{base} + N_0) + N_0 \cdot N_{base} - r_N \cdot (N_1 - N_0) \cdot (N_1 - N_{base})}\Big)$$

for $0.5 < r_N \leq 1$

$$S_{gen} = S_{base} - \Big((S_{base} - S_0) - \sqrt{r_s \cdot (S_1 - S_0) \cdot (S_{base} - S_0)}\Big)$$

for $r_s \leq 0.5$

$$S_{gen} = S_{base} + \Big((S_1 - S_{base}) - \sqrt{S_1^2 - S_1 \cdot (S_{base} + S_0) + S_0 \cdot S_{base} - r_s \cdot (S_1 - S_0) \cdot (S_1 - S_{base})}\Big)$$

for $0.5 < r_s \leq 1$

where N_{gen} is the generated nitrogen deposition scenario used in the simulation, S_{gen} is the generated sulphur deposition scenario used in the simulation, N_0 is the lower limit for nitrogen deposition ($N_0 = N_{base} - 0.4 \cdot N_{base}$), S_0 is the lower limit for sulphur deposition ($S_0 = S_{base} - 0.4 \cdot S_{base}$), N_1 is the higher limit for nitrogen deposition ($N_1 = N_{base} + 0.4 \cdot N_{base}$), S_1 is the higher limit for nitrogen deposition ($S_1 = S_{base} + 0.4 \cdot S_{base}$), r_N is a random value between 0 and 1 and r_s is a random value between 0 and 1.

Monte Carlo simulations involve assumptions about the mutual independence of the input variables. It has, therefore, been assumed here that the nitrogen and sulphur depositions are independent. The validity of this assumption may be questioned as oxidized nitrogen emissions can also originate from the same sources as sulphur, from, for example, coal-burning power stations and other stationary sources, but additionally from vehicle exhaust fumes. These simulations have also ignored the spatial correlation of the model input data as a single value of r_s and r_N has been used for the whole map.

2.4 Calculated outputs

The outputs analysed are the exceeded area and AE. In the next section, these are mapped to give the spatial distribution of these variables in the UK using the deterministic approach. Frequency distributions of the exceedance values for each critical load value (totalling 328 713 for all ecosystems) calculated using the deterministic approach are displayed to show how areas of ecosystems are distributed around the exceeded/non-exceeded cut-off level. These are used to help explain how perturbations to the deposition scenarios affect the area exceeded and AE outputs. For the fixed value analysis the variation in uncertainty was quantified as upper and lower bounds of exceeded area and AE. For the Monte Carlo trial runs the results were presented in the form of frequency diagrams. The uncertainty is quantified both as ranges and confidence intervals for these trial runs.

3 Results

3.1 Deterministic

This section gives the exceedance results using the baseline exceedance calculations defined in Hall *et al.* (2001b), and summarized above, as well as the 1995–97 and 2010 baseline deposition scenarios. The critical loads data used were the data described in Hall *et al.* (2001a). The total area of each ecosystem for which the critical load was exceeded and the AE across the UK, not accounting for uncertainty, were calculated. The ecosystem area exceeded in the UK for 1995–97 was estimated to be 68 265 km^2 and for 2010 as 29 330 km^2, 71 % and 31 % of sensitive ecosystems respectively. The AE in the UK for 1995–97 was estimated to be 7 236 915 keq year^{-1} and for 2010 as 1 436 411 keq year^{-1}. These estimates are the deterministic baseline values to which the effect of uncertainty is compared in the next sections. The exceeded areas and AE have been mapped at a 5 km spatial resolution and are shown in Figure 12.3.

The frequency distribution of exceedance values for all ecosystems was formed for both 1995–97 and 2010 baseline deposition scenarios (Figure 12.4). The peak of the 1995–97 frequency distribution (Figure 12.4a) fell into the positive range indicating that deposition generally exceeded the critical load while the peak of the frequency distribution for 2010 was in the negative range (Figure 12.4b). This is to be expected, as the deposition values for 2010 were less than those for 1995–97 (Figure 12.2), so the critical loads were less likely to be exceeded, resulting in smaller areas of ecosystems exceeded.

3.2 Fixed value uncertainty analysis

The effects of fixed value uncertainty in deposition values on the total area of ecosystems exceeded and AE are displayed in Table 12.1 for all eight simulations (D1 to D8). Changes relative to the deterministic values are calculated using 68 265 km^2 and 7 236 915 keq year^{-1} for area exceeded and AE respectively for the 1995–97 simulations and 29 330 km^2 and 1 436 411 keq year^{-1} for the 2010 simulations.

3.2.1 Area exceeded

The worst case scenario for the 1995–97 period resulted in an increase in the area of critical load exceedance of 12 676 km^2, whereas the best case scenario resulted in a

Table 12.1 Change in area of critical load exceedance and AE from the deterministic case for 1995–97 and 2010

Simulation	Change in ecosystem area exceeded (km²)		Change in AE (keq year⁻¹)	
	1995–97	2010	1995–97	2010
D1 = base N, high S	+ 8 224	+ 3 165	+2 246 559	+ 309 284
D2 = base N, low S	−10 388	−3 572	−2 044 765	−277 427
D3 = high N, base S	+10 080	+11 071	+4 276 278	+1 867 068
D4 = low N, base S	−16 897	−17 170	−3 745 202	−1 157 104
D5 = high N, high S	+12 676	+14 065	+6 660 613	+2 277 535
D6 = low N, low S	−30 448	−20 767	−5 299 371	−1 264 338
D7 = high N, low S	+ 3 810	+ 8 548	+1 975 553	+1 484 453
D8 = low N, high S	− 4 957	−16 031	−1 851 343	−1 007 360

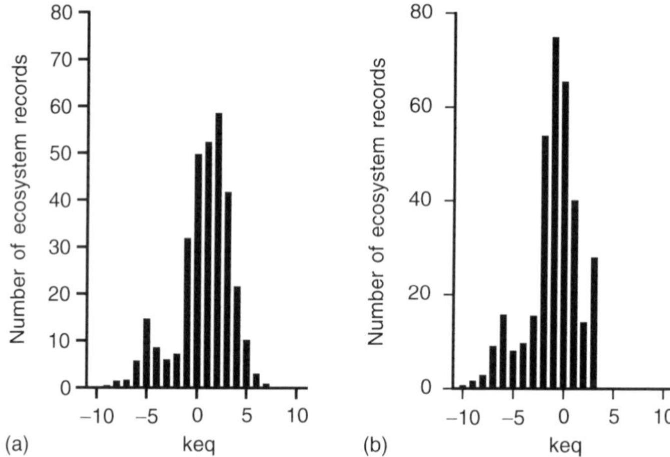

Figure 12.4 Frequency distributions showing the number of individual areas of ecosystem which comprise the UK that fall into 0.5 keq exceedance bins. (a) 1995–97 simulation and (b) 2010 simulation

decrease of 30 448 km^2. For the 2010 deposition the worst case scenario resulted in an increase in exceeded area of 14 065 km^2 and the best case scenario a decrease in area exceeded of 20 767 km^2.

Across the range of deposition uncertainty (−40% to +40% of the baseline deposition estimate) there was a non-linear relationship between deposition and area of critical load exceedance for both the 1995–97 and 2010 deposition scenarios. It was also apparent that a larger change in exceeded area was associated with nitrogen rather than sulphur deposition (Table 12.1).

3.2.2 *Accumulated exceedance*

The 1995–97 worst case scenario resulted in an increase in AE of 6 660 613 keq year^{-1}. The best case scenario resulted in a decrease in AE of 5 299 371 keq year^{-1}. The 2010 deposition worst case scenario resulted in an increase of AE of 2 277 535 keq year^{-1}. The best case scenario resulted in a decrease in AE of 1 264 338 keq year^{-1}.

Across the range of deposition uncertainty (−40% to +40% of the baseline deposition estimate) there was a non-linear relationship between deposition and AE for both the 1995–97 and 2010 deposition scenarios. As with the area exceeded, a larger change in AE was apparent for nitrogen rather than sulphur deposition (Table 12.1).

Although using the above worst and best case scenarios may not realistically simulate deposition uncertainty under field conditions, the results indicate the possibility of a substantial under-estimation of areas of critical load exceedance at the national scale.

3.3 Monte Carlo uncertainty analysis

The results of the Monte Carlo simulations are stochastic in nature so a result obtained from one run of the simulation will not necessarily be the same as from the next run. To check whether the number of calculations undertaken for the UK was sufficient to achieve reproducible results, the cumulative standard deviations were calculated from:

$$\sigma_x = \sqrt{\frac{n\Sigma x^2 - (\Sigma x)^2}{n(n-1)}} \tag{1}$$

where n = number of trials and x = result of Monte Carlo simulation. Each nitrogen and sulphur deposition value was simulated 250 times. Figure 12.5 displays the cumulative standard deviations for the 250 trials of each of the uniform deposition scenarios. Each graph, through its stabilization, indicates that sufficient trials were undertaken. The cumulative standard deviations for the triangular deposition scenarios also showed that 250 trials were adequate. Table 12.2 gives the range of outputs from the Monte Carlo simulations D9–D14 (i.e. the difference between the smallest and largest values of the Monte Carlo trials).

3.3.1 Area exceeded

The results from simulation D9–D11 are summarized in Table 12.2 and demonstrate that the uncertainty in sulphur deposition causes the least variation in area exceeded. Variation in exceeded area is around 25% of the deterministic value in both the 1995–97 and 2010 simulations. Table 12.2 also gives the range of Monte Carlo results using the triangular distribution. Table 12.3 compares the results from the rectangular and triangular deposition scenario results for area exceeded by giving the value for the following index:

$$\frac{\text{triangular range} - \text{rectangular range}}{\text{rectangular range}}$$

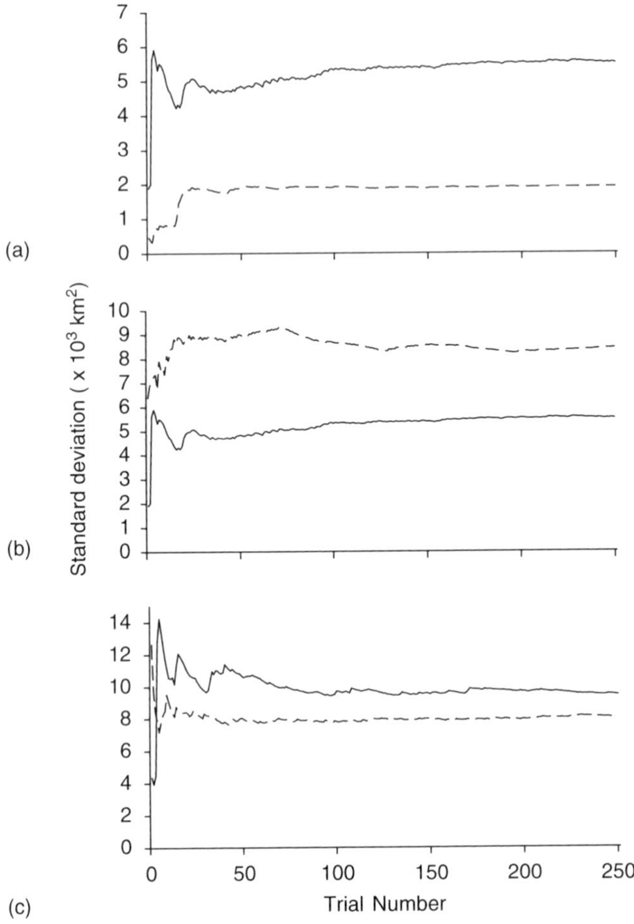

Figure 12.5 Change in cumulative standard deviation of area exceeded for 250 trial runs. 1995–97 acid deposition (solid line), 2010 acid deposition (short dash). (a) Variation in sulphur deposition only, (b) variation in nitrogen deposition only and (c) variation in both sulphur and nitrogen deposition

Table 12.2 Range of area exceeded and AE from 250 Monte Carlo trials for 1995–97 and 2010

Simulation	Range of area exceeded		Range of AE	
	1995–97	2010	1995–97	2010
Rectangular				
D9 = vary S	17 769	6 614	4 049 159	578 592
D10 = vary N	26 849	25 673	7 984 705	2 787 779
D11 = vary N + S	35 013	32 497	9 788 379	3 281 056
Triangular				
D12 = vary S	17 054	6 146	3 854 910	535 785
D13 = vary N	23 944	24 345	6 885 697	2 593 114
D14 = vary N + S	32 123	28 582	8 674 521	2 916 138

Table 12.3 Difference in range of area exceeded and AE from the Monte Carlo trials using rectangular(R) and triangular(T) distributions for the deposition uncertainty

	Simulation	Area exceeded $(T-R)/R$	AE $(T-R)/R$
1995–97	D9 with D12	−0.04	−0.05
	D10 with D13	−0.10	−0.14
	D11 with D14	−0.08	−0.11
2010	D9 with D12	−0.07	−0.07
	D10 with D13	−0.05	−0.06
	D11 with D14	−0.12	−0.11

The discrepancies are quite small, suggesting that in all cases the selection of a rectangular or triangular distribution makes little difference to the area exceeded results. The negative signs of these values show that the triangular distribution produces a narrower set of Monte Carlo results than the rectangular distribution. This is expected as the triangular distribution clusters the most likely values around the most expected value whereas the rectangular distribution is spread evenly between the two extremes.

3.3.2 *Accumulated exceedance*

The rectangular deposition scenario results in Table 12.2 show that the uncertainty in sulphur deposition caused the least variation in AE. Variation in AE was around 50 % of the deterministic value in both the 1995–97 and 2010 simulations. Uncertainty in nitrogen deposition resulted in variation that was over 100 % of the deterministic value for the 1995–97 and nearly 200 % of the deterministic value for the 2010 simulations. The uncertainty in sulphur and nitrogen deposition combined caused a variation in AE that was approximately 150 % of the deterministic value for the 1995–97 simulation and over 200 % of the deterministic value for the 2010 simulation.

Table 12.3 also gives a comparison of the Monte Carlo results for ranges of AE. Again the discrepancies are small and the ranges of values for the triangular distribution results are narrower than are those for the rectangular distribution.

3.3.3 *Frequency distributions*

Figure 12.6 gives the frequency distributions of ecosystem areas exceeded and AE generated in the trial runs using the rectangular distribution simulations.

As there was little difference between the results from the rectangular and triangular distribution simulation all subsequent analyses were based on the rectangular distribution.

The probabilistic exceedance plots shown in Figure 12.7 were formed from the output of the 250 Monte Carlo simulation using the rectangular distribution. The results for area exceeded and AE were ranked from largest to smallest. Each was assigned a probability according to the equation:

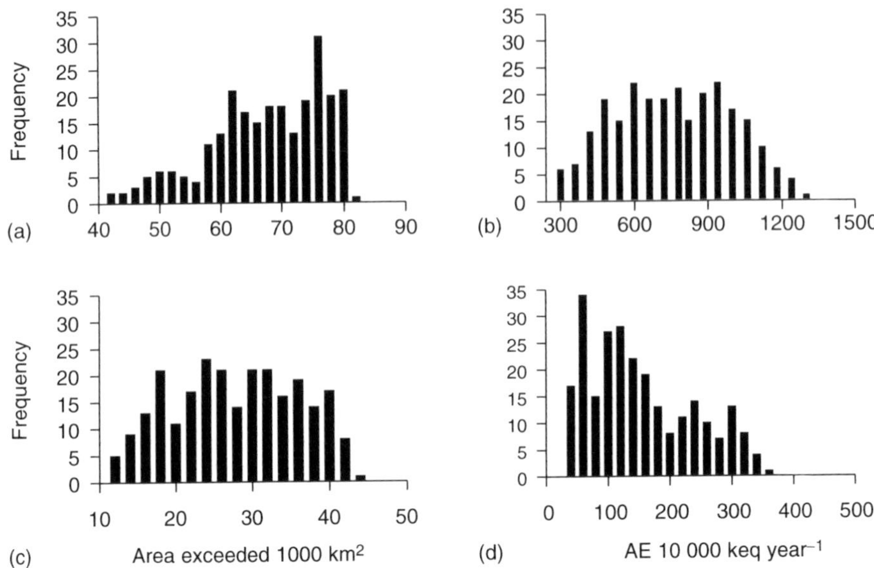

Figure 12.6 Results of the Monte Carlo approach presented in the form of an empirical histogram, developed by grouping the outputs of the 250 trial runs into suitable categories. (a) Area exceeded 1995–97 simulation, (b) accumulated exceedance 1995–97 simulation (c) area exceeded 2010 simulationand (d) accumulated exceedance 2010 simulation

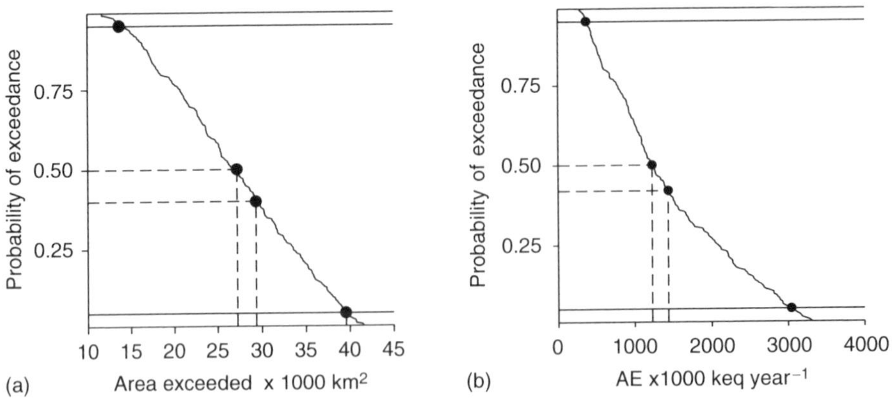

Figure 12.7 Probabilistic exceedance plots. (a) The estimates of ecosystem areas exceeded in the UK in 2010 with the Monte Carlo approach. (b) The estimates of accumulated exceedance in the UK in 2010 with the Monte Carlo approach. Confidence intervals are also calculated. Both figures also show the results of the deterministic approach. The area protection estimate of the deterministic approach corresponds to a 40 % probability of exceedance and the accumulated exceedance protection estimate of the deterministic approach corresponds to a 42 % probability of exceedance

$$P = \frac{r}{N+1} \tag{2}$$

where r = rank and N = number of trials. For example in Figure 12.7(a) the value 78 673 km^2 is the 13th largest out of 250 values. The probability of this being exceeded is therefore $13/250 \times 100 = 5.2\,\%$. Put another way, the probability of having protected all ecosystems at risk using the value 78 673 km^2 is 94.8 %.

4 Discussion

4.1 Fixed value analysis

4.1.1 Area exceeded

The deterministic estimate of areas of critical load exceedance for 2010 was 31 % of sensitive ecosystems. By comparison, the worst case scenario resulted in 45 % and the best case scenario 9 %. Note that this range is not equally spread around the deterministic value (i.e. it is non-linear (Table 12.1)). This was also seen to be the case for the 1995–97 simulation. Both the 1995–97 and 2010 simulations show a larger decrease in exceeded area than increase when fixed value deposition uncertainty is taken into account.

4.1.2 Accumulated exceedance

The range of AE from the best case to the worst case scenario was 172 073 to 3 713 946 keq year^{-1} for the 2010 deposition scenario.

Both the 1995–97 and 2010 simulations show a larger increase in AE than decrease when fixed value deposition uncertainty is taken into account. A numerical example can illustrate this point if it is assumed that the area of every ecosystem record is identical and assumed to be 1 ha for the purpose of this analysis. Calculating exceedances using an unperturbed deposition scenario, two ecosystem records have, for example, exceedances of –0.157 and 0.243 keq. This would make an AE of 0.243 keq as non-exceeded ecosystem records do not contribute to the AE. An increase of 40 % deposition increases these exceedances to 0.173 and 0.573 keq respectively. This would make an increase in AE of $0.173 + 0.573 - 0.243 = 0.503$ keq year^{-1}. Taking the same two ecosystem records as before and decreasing the deposition by 40 % gives exceedances of –0.486 and –0.086 respectively. This would make a decrease in AE of 0.243 keq year^{-1}. This argument holds true in most cases and explains the higher increase in AE than decrease.

4.1.3 Sensitivity to nitrogen and sulphur deposition

Nitrogen deposition uncertainty was shown to result in a wider output variation than sulphur deposition uncertainty (Table 12.1) for both area exceeded and AE. Investigation into why this was the case involved looking at how far away the nitrogen and sulphur deposition values were from the critical load and whether a reduction of nitrogen only, sulphur only or nitrogen and sulphur together was required to reach the critical load. It was found that in the vast majority of cases a combination of sulphur and nitrogen reductions was required. It was also found that the same reduction in nitrogen and sulphur was required. Since nitrogen deposition is larger than sulphur deposition (Figure 12.2) for both the 1995–97 and 2010 scenarios perturbing both deposition values by $\pm 40\,\%$ produced a larger change in deposition on the (larger) nitrogen deposition values than on the (smaller) sulphur deposition values. Hence the change in exceedance values is more sensitive to nitrogen deposition even though the relationship between the frequency distribution of nitrogen and sulphur deposition values and the critical load value was the same.

4.2 Monte Carlo uncertainty analysis

The area in the UK remaining at risk of acidification in 2010 was calculated to be 29 330 km^2 (i.e. 31 % of the total ecosystem area) using the deterministic approach. Figure 12.7a summarizes the estimates of protected areas of the deterministic approach and the Monte Carlo analysis. The estimated range of values the area exceeded takes was from approximately 10 000 km^2 to approximately 40 000 km^2. Taking a conservative interpretation (i.e. risk of exceedance less than 5 %), as it is desirable to be certain that all areas at risk of acidification are protected, gives an approximately 35 % increase (39 540 km^2) in the area exceeded above the deterministic approach. If the deposition uncertainty is taken into account, there is a 40 % probability of exceeding the critical loads for the 29 330 km^2 estimated to be exceeded by the deterministic approach. If a 50 % probability of exceeding were used as an indicator, the estimate of area at risk would be slightly more optimistic (27 200 km^2) than the deterministic approach. The probabilities of ecosystem exceedance decreased by 2010 because the deposition levels decreased and thus the critical load was less likely to be exceeded.

Figure 12.7b shows the probabilistic graph for AE. The values range from about 300 000 keq $year^{-1}$ to approximately 3 500 000 keq $year^{-1}$. A conservative interpretation (risk of exceedance less than 5 %) gives a greater than 100 % increase (3 040 000 keq $year^{-1}$) compared to the deterministic approach. When deposition uncertainty is taken into account, there is a 42 % probability of the accumulated exceedance being greater than the 1 436 411 keq $year^{-1}$ estimated by the deterministic approach. If a 50 % probability of exceeding were used as an indicator, the accumulated exceedance would be slightly more optimistic (1 230 000 keq $year^{-1}$) than the deterministic approach.

4.3 Comparison with previous work

Barkman *et al.* (1995) investigated the effects of uncertainties in both input data and model parameters in calculations of the critical load of acidity and its exceedance, for forest ecosystems in two studies using Monte Carlo simulations of the regional PROFILE model (Warfvinge and Sverdrup, 1992; Sverdrup and Warfvinge, 1991). For southern Sweden, Barkman *et al.* (1995) found that taking into account the uncertainties in critical load and deposition estimates and allowing a 5% exceedance probability resulted in more pessimistic estimates of protected areas than according to the deterministic approach. The Monte Carlo results for area exceeded are in broad agreement with Barkman *et al.* (1995) in that using a 5% exceedance probability indicated a more pessimistic result than the figures currently used by policy makers.

In this chapter, the variation in deposition gave a non-linear relationship with area exceeded. This is in disagreement with the results of Smith *et al.* (1995) who established that in the range of −40% to +40% of the current 20 km spatial resolution deposition estimate there was an almost linear relationship between deposition and area of critical load exceedance. These differences may be accounted for, in part, by the methods used, especially in the calculation of critical loads.

5 Conclusions

Uncertainties in deposition estimates have a substantial effect on exceedance calculations. The fixed value analysis suggests that by including uncertainty in the calculation of exceedance for the 2010 scenario, the uncertainty in area exceeded could be between +50 and −70% of the value derived by the deterministic method. The Monte Carlo analysis narrowed this range to +35 and −50% and gives a median of 27 200 km^2, a decrease of 7% of the value derived from the deterministic method. In relation to AE, the fixed value analysis suggests that the uncertainty may be even greater (e.g. between +160 and −90% of the value derived by the deterministic method). The Monte Carlo analysis again narrowed the range to +100 and −75% and gave a most likely estimate of 1 230 000 keq year^{-1}, a decrease of 14% of the value derived from the deterministic method.

The Monte Carlo analysis enabled the association of exceedance probabilities with a whole range of exceeded areas and AE. The fixed value analysis was useful in assessing bounds within which the value may fall but contained no information on how probable each value was. The Monte Carlo analysis was hence more helpful than the fixed value analysis for estimating the impacts of uncertainties in the calculation of exceedances in this study.

In the Monte Carlo analysis a uniform distribution was used to model the uncertainty of the deposition distributions. The decision to use a uniform distribution was felt to be justified by a lack of information about the form of the underlying nitrogen and sulphur deposition distribution. In reality, a uniform distribution may not be the most appropriate probability function for this variable. However a triangular distribution was also tested and made little difference to the results. Normal, log normal and other distribution types may also have been considered applicable in this study.

Studying both the areas exceeded and AE was interesting as they represented different ways of viewing potential 'harm' to the environment. However, AE was more useful as it takes account of both the area of ecosystem exceeded and the magnitude of the exceedance.

Current international emission abatement strategies are based on exceedance calculations without considering uncertainties. Syri *et al.* (2000, p. 274) state that

'For many countries achieving emission ceilings will require costly abatement installations with high unit reduction costs'. The Monte Carlo approach provides information about the confidence limits for the environmental impact estimates of emission reduction efforts. Emission reductions and their costs can be weighted against the risk of critical load exceedance, thus providing additional information for the decision-making process. The inclusion of uncertainties in critical load and exceedance calculations is going to become a key activity in the UK and the rest of Europe in preparation for reviews of international protocols.

The analyses in this chapter are based on examining uncertainty in one variable only, input deposition. Other variables affect the exceedance estimates and these also have uncertainties, which are important to analyse. Indeed, the critical loads values themselves are uncertain and thus could impact considerably on the results presented. Additional refinements for future studies could also include an allowance for non-homogeneity in grid squares and the problems associated with using data of differing spatial resolution.

Acknowledgements

The authors acknowledge the Department of Environment, Food, and Rural Affairs (contract EPG1/3/116) for funding to carry out this work, however, the views expressed are those of the authors. The authors would like to thank S. Metcalfe, R. Smith and M. Sutton for the provision of the deposition data.

References

Barkman, A., Warfinge, P. and Sverdrup, H., 1995, Regionalization of critical loads under uncertainty, *Water, Air and Soil Pollution*, **85**, 2515–20.

Hall, J., Ullyett, J., Hornung, M., Kennedy, F., Reynolds, B., Curtis, C., Langan, S. and Fowler, D., 2001a, *Status of UK Critical Loads and Exceedances, Part 1 – Critical Loads and Critical Loads Maps, Update to January 1998*, Report: February 2001. Report to the Department of Environment, Transport and the Regions. NERC/DETR Contract EPG1/3/116.

Hall, J., Broughton, R., Bull, K., Curtis, C., Fowler, D., Heywood, L., Hornung, M., Metcalfe, S., Reynolds, B., Ullyett, J. and Whyatt, D., 2001b, *Status of UK Critical Loads and Exceedances. Part 2 – Exceedances*. UK National Focal Centre, CEH Monks Wood. Report to the Department of Environment, Transport and the Regions. NERC/DETR Contract EPG1/3/116.

Hettelingh, J-P., Posch, M., de Smet, P. A. M. and Downing, R. J., 1995, The use of critical loads in emission reduction agreements in Europe, *Water, Air and Soil Pollution*, **85**, 2381–8.

Jonsson, C., Warfvinge, P. and Sverdrup, H. 1995, Uncertainty in predicting weathering rate and environmental stress factors with the PROFILE model, *Water, Air and Soil Pollution*, **81**, 1–23.

Metcalfe, S. E., Whyatt, J. D., Broughton, R., Derwent, R. G., Finnegan, D., Hall, J., Mineter, M., O'Donaghue, M. and Sutton, M. A., 2001, Developing the Hull Acid Rain Model: its validation and implications for policy makers, *Environment Science and Policy*, **4**, 25–37.

Nilsson, J. and Grennfelt, P. (eds), 1988, Critical Loads for Sulphur and Nitrogen. Report 1988:15 (Copenhagen: Nordic Council of Ministers)

Posch, M., 1995, Percentiles and protection isolines, in M. Posch, P. A. M. de Smet, J-P. Hettelingh and R. J. Downing (eds), *Calculation and Mapping of Critical Thresholds in Europe: Status Report 1995* (Bilthoven, The Netherlands: Coordination Centre for Effects, National Institute of Public Health and the Environment (RIVM)), pp. 43–7.

Posch, M. and Hettelingh, J-P., 1997, Remarks on critical load calculations, in M. Posch, P. A. M. de Smet, J-P. Hettelingh and R. J. Downing (eds), *Calculation and Mapping of Critical Thresholds in Europe: Status Report 1997* (Bilthoven, The Netherlands: Coordination Centre for Effects, National Institute of Public Health and the Environment (RIVM)), pp. 25–8.

Posch, M., de Smet, P. A. M. and Hettelingh, J-P., 1999, Critical loads and their exceedances in Europe: an overview, in M. Posch, P. A. M. de Smet, J-P. Hettelingh, and R. J. Downing (eds), *Calculation and Mapping of Critical Thresholds in Europe: Status Report 1999* (Bilthoven, The Netherlands: Coordination Centre for Effects, National Institute of Public Health and the Environment (RIVM)), pp. 3–11.

Rodhe, H., Grennfelt, P., Wisnieuski, J., Agren, C., Bergtssen, G., Johanssen, K., Kauppi, P., Kucera, V., Rasmussen, L., Rosseland, B., Schotte, L. and Selden, G. 1995, Acid Reign '95?-Conference Summary Statement *Water Air and Soil Pollution*, **85**, 1–14.

Singles, R., Sutton, M. A. and Weston, K., 1998, A multi-layer model to describe the atmospheric transport and deposition of ammonia in Great Britain, *Atmospheric Environment*, **32**, 393–9.

Smith, R. and Fowler, D., 2002, Uncertainty in estimation of wet deposition of sulphur, *Water, Air and Soil Pollution* (in press).

Smith, R., Hall, J. and Howard, D. C., 1995, Estimating uncertainty in the current critical loads exceedance models, *Water, Air and Soil Pollution*, **85**, 2503–8.

Smith, R., Fowler. D., Sutton, M., Flechard, C. and Coyle, M., 2000, Regional estimation of pollutant gas dry deposition in the UK: model description, sensitivity analyses and outputs, *Atmospheric Environment*, **34**, 3757–77.

Sutton, M. A., Schjørring, J. K. and Wyers, G. P., 1995, Plant–atmosphere exchange of ammonia, *Philosophical Transactions of the Royal Society, London*, Series A, **351**, 261–78.

Sverdrup, H. and Warfvinge, P., 1991, On the geochemistry of chemical weathering, in K. Rosen (ed.), *Chemical Weathering under Field Condition. Reports on Forest Ecology and Forest Soils* 63 (Uppsala, Sweden: Swedish University of Agricultural Science), p. 79.

Syri, S., Suutari, R. and Posch, M., 2000, From emission in Europe to critical load exceedances in Finland – uncertainty analysis of acidification integrated assessment, *Environmental Science and Policy*, **3**, 263–76.

UN/ECE, 1999, *Protocol to the Convention on Long-Range Transboundary Air Pollution to Abate Acidification, Eutrophication and Ground-Level Ozone* (New York: United Nations Commission for Europe).

Warfvinge, P. and Sverdrup, H., 1992, Calculating critical loads of acid deposition with PROFILE – a steady-state soil chemistry model, *Water, Air and Soil Pollution*, **63**, 119–43.

13

Vertical and Horizontal Spatial Variation of Geostatistical Prediction

A. Wameling

1 Introduction

Spatially related data are found in many environmental applications. Measurements are generally available at a limited number of points only, and due to financial restrictions this number is often small. However, subsequent use of a variable often necessitates its availability for every point in the region of interest. For example, assume that yearly average temperatures have been measured at several stations and that temperatures are used to decide which species or provenances are to be planted in a forest stand. For all stands where no measurement exists, temperature has to be predicted.

Geostatistical methods are popular for both the analysis and interpolation of spatially distributed data. The basic assumption is that the measured data $z(\mathbf{x}_1), \ldots, z(\mathbf{x}_n)$ form a sample from a single realization of a random process $\{Z(\mathbf{x}), \mathbf{x} \in D \subseteq IR^2\}$, where $Z(\mathbf{x})$ is a random function and D is the region of interest. Interpolation then means the prediction of process realization values $z(\mathbf{x})$ for arbitrary points $\mathbf{x} \in D$.

In contrast to non-statistical interpolation algorithms such as nearest neighbour interpolation or triangulation, geostatistical methods not only provide optimal predictions but also estimates of prediction precision. If the aim of prediction is a statement about the process realization at an individual point, then a combination of prediction and prediction precision is straightforward. For example, under the assumption of normally distributed prediction errors

$$\hat{Z}(\mathbf{x}) - Z(\mathbf{x}),$$

Uncertainty in Remote Sensing and GIS. Edited by G.M. Foody and P.M. Atkinson.
 ISBN: 0–470–84408–6

a confidence interval for the true realization value can be constructed. The errors of local predictions are denoted vertical errors in the following, as each prediction deviates from the true value in the vertical direction.

In environmental applications, statements about individual process values are normally not the primary concern of an analysis. The main purpose of many investigations is rather to make statements about regions. Quite often interest lies in the *locations* of points or regions having special characteristics concerning the process realization. Examples are the locations of maxima or minima and the locations of regions having a certain range $[c_1, c_2]$ of the variable under investigation. Expressed formally,

$$D_{[c_1,\, c_2]} = \{\mathbf{x} \in D\colon c_1 \leq Z(\mathbf{x}) \leq c_2\} \tag{1}$$

or

$$D_c = \{\mathbf{x} \in D\colon Z(\mathbf{x}) > c\}. \tag{2}$$

This chapter analyses the location of points with constant process values (i.e. isolines or contours)

$$I_c = \{\mathbf{x} \in D\colon Z(\mathbf{x}) = c\} \tag{3}$$

of a given process realization. For continuous process realizations, contours are the boundaries of regions as defined in equations (1) and (2). For the temperature example mentioned above, a possible task of a geostatistical analysis is to predict regions having a mean temperature above or below c. Such regions are bounded by I_c.

Areas and contours are usually derived from a combination of predictions made for many points $\mathbf{x} \in D$. Predictions of equations (1), (2) and (3) are then given by

$$\hat{D}_{[c_1,\, c_2]} = \{\mathbf{x} \in D\colon c_1 \leq \hat{Z}(\mathbf{x}) \leq c_2\},$$

$$\hat{D}_c = \{\mathbf{x} \in D\colon \hat{Z}(\mathbf{x}) > c\}$$

and

$$\hat{I}_c = \{\mathbf{x} \in D\colon \hat{Z}(\mathbf{x}) = c\}, \tag{4}$$

respectively, where $\hat{Z}(\cdot)$ denotes the interpolation process. Given that all individual values $\hat{Z}(\mathbf{x})$ are predictions with errors, it is obvious that the predicted regions equation (4) also have associated errors. The definition of these horizontal location errors is not as straightforward as the local vertical error concept described above. The uncertainty represented by imprecision in contour plots has so far been discussed in very few articles (e.g. Lindgren and Rychlik, 1995; Polfeldt, 1999). In many, if not most, investigations horizontal errors are completely disregarded. One common approach is a simultaneous display of contour maps of both the interpolated surface and the related kriging variances (e.g. Venkatram, 1988; Bruno and Capicotto, 1998). However, the information in these two maps is difficult to combine, as it is unclear how the horizontal spatial variation is related to the vertical local variability.

In sections 2 and 3, the concepts of vertical and horizontal variability are described. In section 4, an attempt is made to relate both concepts.

2 Local Accuracy of Kriging Prediction

Denote $\mathbf{Z} = [Z(\mathbf{x}_1), \ldots, Z(\mathbf{x}_n)]'$ and $\mathbf{z} = [z(\mathbf{x}_1), \ldots, z(\mathbf{x}_n)]'$. For an arbitrary point $\mathbf{x}_0$ a predictor $p(\mathbf{Z}; \mathbf{x}_0)$ is optimal in the least square sense, if it minimizes

$$E[Z(\mathbf{x}_0) - p(\mathbf{Z}; \mathbf{x}_0)]^2. \tag{5}$$

The solution to this minimization problem is given by the conditional expectation

$$E[Z(\mathbf{x}_0)|\mathbf{Z} = \mathbf{z}]$$

which is generally a non-linear and unknown function of the observed z_i. The conditional mean squared error reduces to

$$E\{[Z(\mathbf{x}_0) - E(Z(\mathbf{x}_0)|\mathbf{Z} = \mathbf{z})]^2|\mathbf{Z} = \mathbf{z}\} = Var[Z(\mathbf{x}_0)|\mathbf{Z} = \mathbf{z}]. \tag{6}$$

Instead of minimizing equation (5) over all possible predictors $p(\mathbf{Z}; \mathbf{x}_0)$, only the subgroup of all linear predictors is usually considered. The best linear predictor is known as the kriging predictor $\hat{p}(\mathbf{Z}; \mathbf{x}_0)$, with variations depending on whether or not the unconditional mean function

$$\mu(\mathbf{x}) = E\ [Z(\mathbf{x})]$$

is known. Simple kriging requires knowledge of $\mu(\mathbf{x}_1), \ldots, \mu(\mathbf{x}_n)$ and $\mu(\mathbf{x}_0)$, whereas in ordinary kriging a constant but unknown mean $\mu(\mathbf{x}) = \mu$ is assumed and estimated in the kriging process.

For both variants of kriging prediction, the prediction error decomposes into

$$Z(\mathbf{x}_0) - \hat{p}(\mathbf{Z}; \mathbf{x}_0) = \{Z(\mathbf{x}_0) - E[Z(\mathbf{x}_0)|\mathbf{Z}]\} + \{E[Z(\mathbf{x}_0)|\mathbf{Z}] - \hat{p}(\mathbf{Z}; \mathbf{x}_0)\} \tag{7}$$

(Cressie, 1991), and consequently, the conditional mean squared error is

$$E\{[Z(\mathbf{x}_0) - \hat{p}(\mathbf{Z}; \mathbf{x}_0)]^2|\mathbf{Z} = \mathbf{z}\} = Var[Z(\mathbf{x}_0)|\mathbf{Z} = \mathbf{z}] + \{E[Z(\mathbf{x}_0)|\mathbf{Z} = \mathbf{z}] - \hat{p}(\mathbf{z}; \mathbf{x}_0)\}^2, \tag{8}$$

which shows it to be composed of

(i) the conditional variance of the random variable $Z(\mathbf{x}_0)$ given the present data and
(ii) the squared error due to the replacement of the best predictor (the conditional mean) by the best linear predictor (simple or ordinary kriging).

In practical applications there is a third component resulting from the fact that the underlying covariance structure has to be estimated. In this chapter, the analysis is restricted to simulated examples where the structure is known and this error source can be disregarded.

If the underlying stochastic process is a Gaussian process, then the conditional expectation is linear. If, in addition, simple kriging is used, then equation (8) reduces to equation (6), that is, the error component (ii) is zero. This does not hold for ordinary kriging, where (ii) will be the imprecision due to the estimation of the stationary mean μ. For any process, the error component (ii) can be reduced by replacing the linear predictor by a more precise predictor whereas (i) can only be reduced by additional observations, that is, by increasing the number n of data values.

In practical kriging applications, local precision is measured by the kriging variance $\sigma_K^2(\mathbf{x})$, that is, the minimized unconditional mean squared error

$$E[Z(\mathbf{x}_0) - \hat{p}(\mathbf{Z}; \mathbf{x}_0)]^2, \tag{9}$$

which is a function of the position of the measuring points and of the underlying covariance model only, but which is independent of the data. Consequently, the kriging variance shows little variability over the region of interest, if the measuring points are spread regularly over D. The measuring points are, of course, an exception to this, where $\sigma_K^2(\mathbf{x}_i) = 0$ for $i = 1, \ldots, n$ by definition. The data independence is due to the fact that $\sigma_K^2(\mathbf{x})$ is the minimized *unconditional* mean squared error, whereas the conditional version

$$E[\{Z(\mathbf{x}_0) - \hat{p}(\mathbf{Z}; \mathbf{x}_0)\}^2 | \mathbf{Z} = \mathbf{z} \tag{10}$$

might be a more appropriate measure of local precision (Goovaerts, 1997; Yamamoto, 2000). The simple kriging predictor is identical to a Gaussian process. If ordinary kriging is used or if the process is not Gaussian, then equation (10) is data dependent with mean $\sigma_K^2(\mathbf{x})$. As a mean value, the kriging variance will be larger than equation (10) for some realizations and smaller for others.

Exact calculation of equation (10) requires knowledge of the multivariate distribution of $[Z(\mathbf{x}_0), Z(\mathbf{x}_1), \ldots, Z(\mathbf{x}_n)]$. Several methods have been proposed to estimate this term instead, including nonlinear kriging methods and conditional simulation. A detailed discussion would be beyond the scope of this chapter and has been given elsewhere (Goovaerts, 1997).

3 The Concept of Horizontal Spatial Variation

Let horizontal variation denote the variability of predictions concerning the location of process values, that is, predictions of sets defined in equations (1), (2) and (3). As mentioned above, contours (3) are boundaries of areas (1) and (2) for continuous process realizations. The subsequent analysis is, therefore, restricted to contour line variability.

The definition of horizontal prediction errors is not as straightforward as that of vertical errors because two-dimensional sets are analysed instead of one-dimensional values. Several concepts are conceivable, including

(i) areas,
(ii) maximal or average distance and
(iii) local distance

between predicted and true contours.

As described in section 1 the usual prediction of (3) is the isoline of the interpolation process. The interpolation is evaluated at a large number of points, usually located on a regular grid $D_G \subset D$. Then the prediction $\hat{I}_c = \{\mathbf{x} \in D: \hat{Z}(\mathbf{x}) = c\}$ of a contour I_c is approximated by

$$\hat{I}_c^{appr} = \{\mathbf{x} \in D: \hat{Z}_{appr}(\mathbf{x}) = c\},$$

where $\hat{Z}_{appr}(\cdot)$ is the interpolated surface constructed by kriging interpolation for $\mathbf{x} \in D_G$ and subsequent non-statistical interpolation for $\mathbf{x} \notin D_G$. Several non-statistical interpolation algorithms are used for the derivation of contours from grid data, including triangulation and polynomial interpolation. If rectangular grids are used then there is no unique interpolation method. For this reason, grids from equilateral triangles are often recommended. Linear interpolation is then possible within each triangle which is both simple and non-ambiguous. Then

$$\hat{Z}_{appr}(\mathbf{x}) = \begin{cases} \hat{Z}(\mathbf{x}), & \text{if } \mathbf{x} \in D_G \\ \sum_{i=1}^{3} w_i \hat{Z}_i & \text{else} \end{cases},$$

where $(\hat{Z}_1, \hat{Z}_2, \hat{Z}_3)$ are the kriging values evaluated at the three surrounding grid nodes and the weights w_i, $i = 1, 2, 3$ subject to the constraint $w_1 + w_2 + w_3 = 1$ are calculated from the position of point $\mathbf{x}$ within the triangle.

Given this two-step interpolation concept any horizontal error comprises three components, namely

A the horizontal variability of the conditional process
B the error due to the replacement of the conditional mean by the best linear predictor
C the second step interpolation error due to the replacement of $\hat{I}_c$ by $\hat{I}_c^{appr}$.

As in section 2, the choice of an appropriate interpolation method can reduce *B*. If the grid spacing (i.e. the length of each triangle side) is sufficiently small, then it is reasonable to expect that $\hat{Z}_{appr}(\mathbf{x}) \to \hat{Z}(\mathbf{x})$ and $\hat{I}_c^{appr} \to \hat{I}_c$ almost certainly and that, consequently, error component *C* can be neglected. Note that for a given grid spacing *d* this can be checked by calculation of both $\hat{Z}_{appr}(\mathbf{x})$ and $\hat{Z}(\mathbf{x})$ for some $\mathbf{x} \notin D_G$. The squared difference $[\hat{Z}_{appr}(\mathbf{x}) - \hat{Z}(\mathbf{x})]^2$ should be small compared to the local precision at point $\mathbf{x}$ (see section 2). To evaluate the horizontal prediction error the local distance concept (iii) is regarded, which reduces the two-dimensional prediction error to a one-dimensional prediction error.

Let $\hat{\mathbf{x}} \in \hat{I}_c^{appr}$ be an arbitrary point of a predicted contour. To analyse the distance of $\hat{\mathbf{x}}$ from the true contour, this distance is computed along a transect $T_\gamma(\hat{\mathbf{x}})$ in *D* through $\hat{\mathbf{x}}$ and forming an angle γ with the observed contour (Figure 13.1).

The variability of the directional distances depends on the slope of the process realization, (i.e. larger variability is expected in regions with small slope than in regions where the slope is steep). Therefore, the transect with the steepest gradient will be considered, which corresponds to $\gamma = 90°$ (cf. Watson, 1992). Denote this transect by $T_{90}(\hat{\mathbf{x}}) := T(\hat{\mathbf{x}})$.

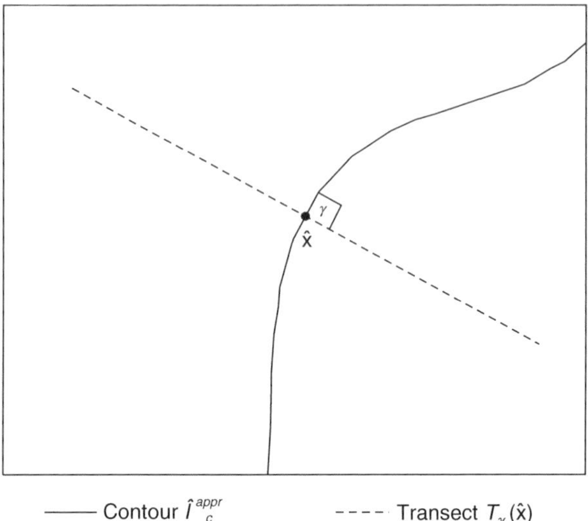

Figure 13.1 Restriction of process to a transect

If the transect is followed in one defined direction, an intersection with the contour of a specific realization occurs if the realization upcrosses (or downcrosses, depending on the direction) level c. Lindgren and Rychlik (1995) present a method to calculate confidence sets for the true level upcrossing (downcrossing) using the upcrossing (downcrossing) intensity. This method is restricted to Gaussian processes. Crossing intensities can then be derived by a generalization of Rice's formula (Cramér and Leadbetter, 1967).

An alternative to the above analytical construction method applicable to any spatial process is given in Wameling and Saborowski (2001). This method is based on the theory of conditional simulation of spatial processes. The idea is to generate a large number of realizations of the conditional process

$$\{Z(\mathbf{x})|Z(\mathbf{x}_i),\ i = 1,\ \ldots,\ n,\ \mathbf{x} \in D\} \tag{11}$$

for a large number of points $\mathbf{x} \in T(\hat{x})$, and to record the upcrossing nearest to $\hat{\mathbf{x}}$ for each of the M generated realizations. Denote these upcrossings $\mathbf{x}_{\min}^{(l)}$, $l = 1,\ \ldots,\ M$. Using a suitable simulation algorithm, the simulated upcrossings can be used to predict a simulated confidence interval.

For $\tau \in IR$ define

$$t(\tau) := \hat{\mathbf{x}} + \tau(\cos\hat{\theta},\ \sin\hat{\theta})',$$

where $\hat{\theta}$ is the angle of the transect with the horizontal (i.e. x-) axis. Then the transect is given by

$$T(\hat{\mathbf{x}}) := \{t(\tau),\ \tau \in IR\}.$$

Alternatively, the simulated upcrossings $\mathbf{x}_{\min}^{(l)}$ can be represented by $\tau_{\min}^{(l)}$, where

$$\mathbf{x}_{\min}^{(l)} = \hat{\mathbf{x}} + \tau_{\min}^{(l)}(\cos\hat{\theta}, \sin\hat{\theta})'.$$

The variability of the upcrossing point along the transect is estimated by

$$\frac{1}{M}\sum_{l=1}^{M}|\hat{\mathbf{x}} - \mathbf{x}_{\min}^{(l)}|^2 = \frac{1}{M}\sum_{l=1}^{M}(\tau_{\min}^{(l)})^2,$$

and an empirical confidence interval is constructed by

$$[\tau_{\alpha/2}, \tau_{1-\alpha/2}],$$

where $\tau_{\alpha/2}$ and $\tau_{1-\alpha/2}$ denote the $\alpha/2$- and $(1-\alpha/2)$-quantiles of $\{\tau_{\min}^{(l)}, l = 1, \ldots, M\}$.

Confidence intervals should be simulated for many points $\hat{\mathbf{x}} \in \hat{I}_c^{appr}$. The intervals can be included in a contour plot. They convey information about the precision of the contour. This information should be accounted for in the decision process (e.g. the choice of species or provenances mentioned in the introduction). Imprecise contours should be used with care.

A large number of simulation algorithms have been proposed (Chilès and Delfiner, 1999). Many algorithms rely on the assumption of Gaussian processes (e.g. Cholesky decomposition of the covariance matrix or sequential Gaussian simulation). If simple global kriging is utilized, the sequential algorithm is equivalent to the Cholesky method and simulation produces realizations of the conditional process (11). In practical applications of conditional simulation, several approximations are made. These include the use of local instead of global and ordinary instead of simple kriging in the sequential algorithm and the application of normal score transforms to the data with subsequent Gaussian simulation. Advantages and disadvantages of these approximations and guidelines for the choice of a suitable algorithm are frequently discussed (Deutsch, 1994; Goovaerts, 1997).

The construction method described above is applicable to any simulation technique, even if it only approximates (11). It is a flexible tool which allows the utilization of realizations generated by any algorithm to assess contour line variability, provided that the realizations are reasonable images of the process. So non-Gaussian algorithms like sequential indicator simulation (Journel, 1989) can also be employed if they are considered to be appropriate for a special application.

4 Attempt to Relate Vertical and Horizontal Variation

It is obvious from the previous sections that there is no simple analytic relation between horizontal and vertical variation. This is partly due to the fact that the vertical variability is locally restricted to one point in D whereas horizontal variation by definition relates to infinitely many points (in our definition all points on a transect). Moreover, it has been shown that even in the convenient case of a Gaussian process upcrossing intensities are rather complex terms (Lindgren and Rychlik, 1995).

The following example has been constructed to clarify these difficulties. An artificial data set of $n = 50$ points in a squared area of 20×20 units was generated.

A realization of a Gaussian process was generated using standard normal marginal distributions and a spherical covariance function with range 8.0 units (Wameling and Saborowski, 2001). Figure 13.2 presents a single contour ($c = 2.0$) from a global ordinary kriging interpolation together with the locations of the measuring points and the measured values.

Figure 13.3 shows a contour plot of the kriging variances in a neighbourhood of the contour $c = 2.0$. Local confidence intervals constructed from 1000 simulations on 120 transects across the contour are superimposed on the plot. For this example,

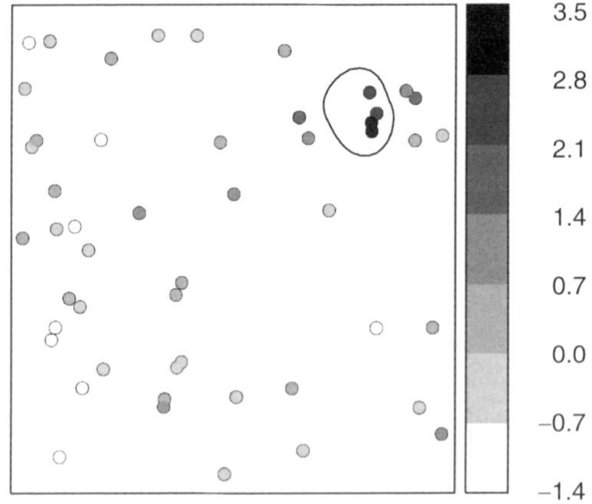

Figure 13.2 *Locations of measuring points and contour* $c = 2.0$

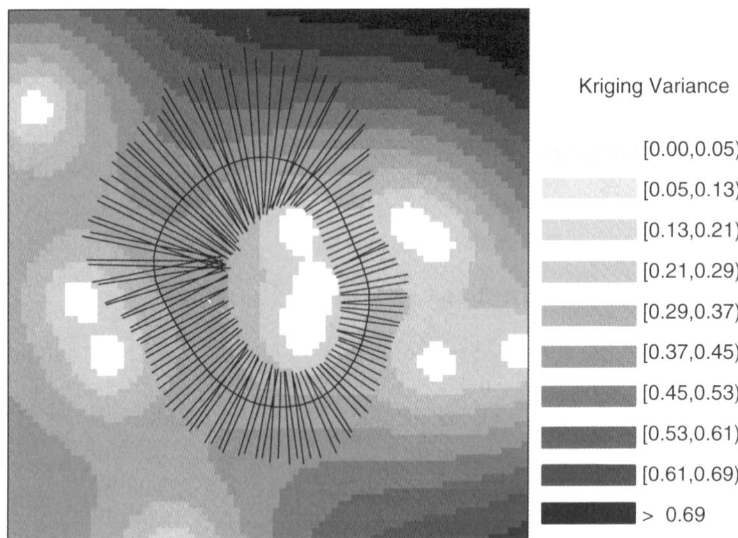

Figure 13.3 *Contour plot of kriging variances and local confidence intervals constructed from 1000 points*

simulation was carried out using Cholesky decomposition of the covariance matrix as the underlying process was known to be Gaussian.

The simulated confidence intervals are small in the right part of the contour line where kriging variances are also small. They are large in the upper part where kriging variances are larger. However, it is difficult to draw any general conclusion from Figure 13.3 about an explicit relationship between vertical and horizontal variability.

The relationship between vertical and horizontal variability can be approximated using the construction method of the predicted contours. Even though approximated, this relation can be helpful for the design of spatial investigations, if the aim is to keep horizontal variation as small as possible for a given effort. Kriging variances do not depend on the observed values but only on the positions of the measuring points, whereas the construction method for confidence intervals described in section 3 requires actual data. So this method is not available in advance of a survey.

Point predictions are calculated for all nodes $\mathbf{x} \in D_G$ of a triangular grid (see section 3). Contours are extracted by subsequent linear interpolation between the grid nodes, and a contour $\hat{I}_c^{appr}$ is consequently predicted as a polygon of line-segments (Figure 13.4a). Figure 13.4b shows a single triangle containing a line-segment.

Let P1, P2, P3 denote the vertices of the triangle, such that one end of the line-segment is in $\overline{\text{P1 P2}}$ and one is in $\overline{\text{P1 P3}}$. Let $\hat{T}$ be the segment endpoint in $\overline{\text{P1 P2}}$. Figure 13.5 illustrates how the existence of local, vertical uncertainty in P1 and P2 influences the location of the crossing $\hat{T}$.

Without loss of generality let $\text{P1} = (0, 0)'$ and $\text{P2} = (d, 0)'$, where d is the grid spacing. Then $\|\hat{T} - \text{P1}\| = \hat{T}.\hat{T}$ is calculated by linear interpolation between the kriging values at P1 and P2, that is

$$\hat{T} = \frac{c - \hat{Z}_1}{\hat{Z}_2 - \hat{Z}_1} d.$$

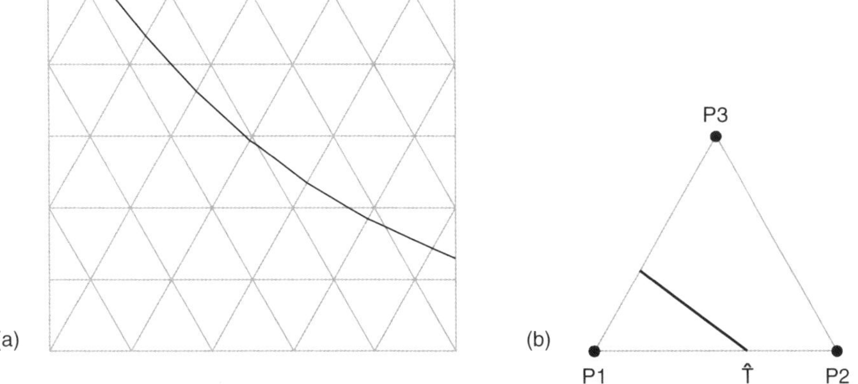

Figure 13.4 *Schematic diagram showing (a) Contour line contruction on triangular grids; (b) line-segment in a triangle*

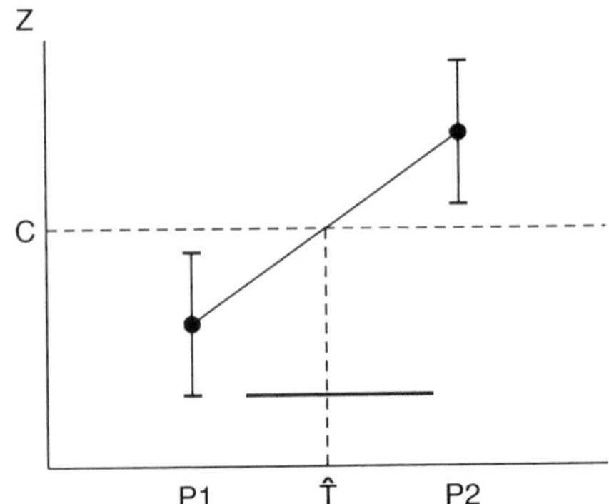

Figure 13.5 Vertical and horizontal prediction errors

For the conditional process there is a vertical error

$$R_i = Z_i - \hat{Z}_i,$$

Compare this result to equation (7). Provided that there is a crossing in $\overline{\text{P1 P2}}$ in a realization, this crossing is approximated by

$$\mathrm{T} = \frac{c - Z_1}{Z_2 - Z_1} d \tag{12}$$

and the horizontal error is approximated by

$$\begin{aligned} E := \mathrm{T} - \hat{\mathrm{T}} &= \left(\frac{c - Z_1}{Z_2 - Z_1} - \frac{c - \hat{Z}_1}{\hat{Z}_2 - \hat{Z}_1} \right) d \\ &= \left(\frac{c - (\hat{Z}_1 + R_1)}{(\hat{Z}_2 + R_2) - (\hat{Z}_1 + R_1)} - \frac{c - \hat{Z}_1}{\hat{Z}_2 - \hat{Z}_1} \right) d \\ &= \left(\frac{c(R_1 - R_2) + \hat{Z}_1 R_2 - \hat{Z}_2 R_1}{(\hat{Z}_2 - \hat{Z}_1)^2 + (\hat{Z}_2 - \hat{Z}_1)(R_2 - R_1)} \right) d. \end{aligned}$$

Via the error propagation law (Cameron, 1982)

$$Var[f(R_1, R_2)] \approx \left[\frac{\partial f}{\partial R_1}(\vec{\mu}_R) \right]^2 \sigma_1^2 + \left[\frac{\partial f}{\partial R_2}(\vec{\mu}_R) \right]^2 \sigma_2^2 + 2 \cdot \left[\frac{\partial f}{\partial R_1}(\vec{\mu}_R) \right] \cdot \left[\frac{\partial f}{\partial R_2}(\vec{\mu}_R) \right] \sigma_{12}$$

the variance $Var[E]$ is approximated by

$$Var[E] \approx (\hat{Z}_2 - \hat{Z}_1)^{-4} [(\hat{Z}_2 - c)^2 \sigma_1^2 + (\hat{Z}_1 - c)^2 \sigma_1^2 + 2 \cdot (\hat{Z}_2 - c)(c - \hat{Z}_1) \sigma_{12}], \tag{13}$$

where $\sigma_i^2 := Var[R_i]$ and $\sigma_{12} := Cov[R_1, R_2]$. The latter term is derived in a manner analogous to the derivation of the kriging variance as

$$\sigma_{12} = C(\|\mathrm{P1} - \mathrm{P2}\|) - \boldsymbol{\lambda}_1' \mathbf{c}_1 - \boldsymbol{\lambda}_2' \mathbf{c}_2 + \boldsymbol{\lambda}_1' \boldsymbol{\Sigma} \boldsymbol{\lambda}_2,$$

where $C(\cdot)$ is the covariance function of the random process for which second-order stationarity is assumed; λ_1 and λ_2 are the vectors of kriging weights at P1 and P2, respectively. Moreover,

$$\mathbf{c}_1 = [C(\|\mathrm{P1} - \mathbf{x}_1\|), \ \ldots, \ C(\|\mathrm{P1} - \mathbf{x}_n\|)]' \text{and } \mathbf{c}_2 = [C(\|\mathrm{P2} - \mathbf{x}_1\|), \ \ldots, \ C(\|\mathrm{P2} - \mathbf{x}_n\|)]'$$

are the vectors of covariances of the process values at P1 and P2 and the n measuring points, respectively, and

$$\mathbf{\Sigma} = \left(C(\|\mathbf{x}_i - \mathbf{x}_j\|)\right)_{\substack{i=1,\ \ldots n \\ j=1,\ \ldots n}}$$

is the covariance matrix of the process values at the measuring points.

Note that equation (13) is the variability of the horizontal error in the direction of the grid line. Variances are large where the angle between the contour segment and grid line is small. The most extreme conceivable case occurs when the grid line and contour segment coincide. Then $\hat{Z}_2 - \hat{Z}_1 = 0$ and, consequently, equation (13) is not defined. To make variances independent of the angle between the segment and grid line and thus comparable among different segments, E is projected onto the gradient of the interpolation within the triangle, which is perpendicular to the contour line segment. Let φ be the angle between the gradient and $\overline{\mathrm{P1}\,\mathrm{P2}}$. The projection E' of E onto the gradient is

$$E' = E \cdot \cos(\varphi),$$

and consequently

$$\begin{aligned} Var[E'] &= [\cos(\varphi)]^2 \cdot Var[E] \\ &\approx \frac{3}{4}\frac{d^2}{(\hat{Z}_2 - \hat{Z}_1)^2} \cdot \frac{(\hat{Z}_2 - c)^2\sigma_1^2 + (\hat{Z}_1 - c)^2\sigma_2^2 + 2\cdot(\hat{Z}_2 - c)(c - \hat{Z}_1)\sigma_{12}}{\hat{Z}_1^2 + \hat{Z}_2^2 + \hat{Z}_3^2 - \hat{Z}_1\hat{Z}_2 - \hat{Z}_1\hat{Z}_3 - \hat{Z}_2\hat{Z}_3}. \end{aligned} \quad (14)$$

This formula is approximate in several respects. The main drawback is that equation (12) is incorrect whenever there is no crossing in $\overline{\mathrm{P1}\,\mathrm{P2}}$ for certain realizations, and this is very likely if the grid spacing d is small or vertical variances are large. However, if d is small, neither the kriging estimates $\hat{Z}_i$ nor the kriging variances σ_i^2 will differ too seriously in a neighbourhood of P1 and P2. So, it is likely that equation (14) can be used to describe horizontal variability where the variances are small, whereas the formula will fail to model the relation of horizontal and vertical variation where variances are large.

To check this presumption, 50 unconditional realizations of the process described above were generated. For each realization, global ordinary kriging was used to interpolate on a triangular grid of spacing $d = 0.2$. Contours were extracted for level $c = 2.0$. From the contours of each realization, five segments were chosen randomly. For each segment, a confidence interval was simulated as described in section 3 and the horizontal variance was estimated by equation (14). Figure 13.6 shows a scatter-plot of the interval length and the square roots of the estimated variances.

There is a linear relationship between the length of the simulated confidence intervals and the estimated standard deviation of the horizontal error. However,

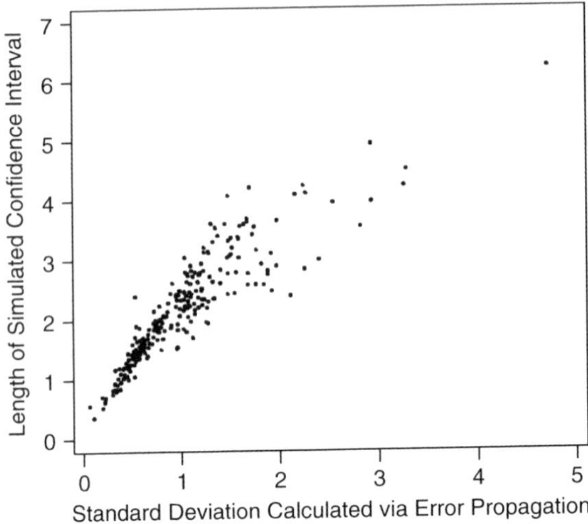

Figure 13.6 *Scatterplot of simulated confidence interval length and standard deviation of horizontal error predicted by error propagation*

this relationship only holds where both measures are small. Large estimated standard deviations can only partly explain confidence interval lengths. Note that scatterplots for other levels c show very similar relationships.

Now, suppose that it is possible to take some additional samples. If interest lies in a particular contour, the kriging variances in a neighbourhood of that contour can be decreased if the additional measuring points are chosen carefully. Suppose that a layout is found which reduces kriging variances and kriging covariances locally by half on average. By equation (14), $Var[E']$ will also decrease by half and the estimated standard deviation by around 30%. Fitting a regression line without intercept to Figure 13.6, the reduction of confidence interval length will also be about 30%. These calculations are only approximate ones and will not hold for large variances. This lessens the practical value, because a researcher will usually be interested in decreasing the lengths of large confidence intervals. However, if all vertical errors are reduced, this will also decrease the horizontal error, and this is also true for large errors.

5 Discussion

Two concepts relating to the precision of geostatistical interpolation have been described. Vertical precision is widely discussed in the literature and can easily be estimated by the kriging variance or other measures, whereas the concept of horizontal precision has often been disregarded. In many applications, point predictions are made for a large number of points and much effort is made to evaluate the associated precision. However, once the predictions are displayed in a map, this map is accepted without recognizing the inherent inaccuracy. Whenever an interpolated surface is used to draw conclusions about positions of points having certain characteristics, the associated horizontal prediction precision should be analysed.

Inaccurate contours or maps should not be used for subsequent inference. An alternative to interpolating and mapping the variable of interest is the indicator approach suggested by Journel (1982). Probabilities $P[Z(\mathbf{x}) > c_k]$ are predicted for given thresholds c_k via simple or ordinary kriging of indicator transformed data. From the predicted probabilities for $\mathbf{x} \in D_G$ sets of points having a high probability of exceeding the threshold can be derived. For example, consider the temperature example discussed above. Without direct prediction of the temperature at point $\mathbf{x}$, a possible decision could be to plant a certain provenance on subareas having a probability of at least 0.9 that the average temperature is above a threshold c. Predictions should then be made for subareas instead of individual points via block indicator kriging.

Conditional simulation allows an estimation of contour line precision by construction of confidence intervals along transects perpendicular to the line. However, the described construction method is computationally demanding and it can, moreover, only be performed for a given sample. An attempt was made to describe horizontal variability as a function of local prediction variance. Knowledge of a functional relationship, even if it is only approximate, can help to evaluate the effect of a reduction of vertical variability on the horizontal precision. Vertical variability can be reduced by additional samples and by the use of a more precise predictor and the reduction can be calculated in advance.

In this chapter, simple error propagation was used to derive an approximate relationship. The resulting formula (equation (14)) is independent of any distributional assumptions, but holds only for transects having small horizontal errors. This is due to the fact that the formula models horizontal variation only by the vertical errors in a neighbourhood to the predicted contour. The true relationship is much more complex as was shown in Lindgren and Rychlik (1995) for the special case of Gaussian processes.

The formula is based on only one possible concept of horizontal precision, namely the local distance concept. Other concepts are conceivable, including maximal or minimal distances between true and observed lines. Polfeldt (1999) analysed the quality of a contour map using the probability that a point is located on the correct side of a given contour. For these concepts, the relationship to vertical accuracy is certainly different and has not been investigated yet. Nevertheless, the derived formula can help to obtain an approximate and easy to compute impression of the precision of contour sections. Additional sampling effort should then be concentrated to areas where greater precision is required.

Acknowledgement

This work was part of a project ‘Spatial Accuracy of Geostatistical Interpolation Methods for Forest Sample Data’, funded by the Deutsche Forschungsgemeinschaft (DFG).

References

Bruno, R. and Capicotto, B. M., 1998, Geostatistical analysis of pluviometric data: IRF-K approach, *Journal of Geographic Information and Decision Analysis*, **2**, 137–51.

Cameron, J. M., 1982, Error analysis, in S. Kotz (ed.), *Encyclopedia of Statistical Sciences*, **2**, 545–51.

Cramér, H. and Leadbetter, M. R., 1967, *Stationary and Related Stochastic Processes* (New York: Wiley).

Cressie, N. A., 1991, *Statistics for Spatial Data* (New York: Wiley).

Deutsch, C. V., 1994, Algorithmically defined random function models, in R. Dimitrakopoulos (ed.), *Geostatistics for the Next Century* (Dordrecht: Kluwer Academic).

Goovaerts, P., 1997, *Geostatistics for Natural Resources Evaluation* (New York: Oxford University Press).

Journel, A. G., 1982, The indicator approach to estimation of spatial distributions, in T. B. Johnson and R. J. Barnes (eds), *Proceedings of the 17th APCOM International Symposium*, New York, April 1982 (New York: Society of Mining Engineers of the AIME), pp. 793–806.

Journel, A. G., 1989, Imaging of spatial uncertainty: a non-Gaussian approach, in B. E. Buxton (ed.), *Geostatistical, Sensitivity and Uncertainty Methods for Ground-Water Flow and Radionuclide Transport Modelling* (Columbus, OH: Battelle Press), pp. 585–99.

Lindgren, G. and Rychlik, I., 1995, How reliable are contour curves? Confidence sets for level contours, *Bernoulli*, **1**, 301–19.

Polfeldt, T., 1999, On the quality of contour maps, *Environmetrics*, **10**, 785–90.

Venkatram, A., 1988, On the use of kriging in the spatial analysis of acid precipitation data, *Atmospheric Environment*, **22**, 1963–75.

Wameling, A. and Saborowski, J., 2001, Construction of local confidence intervals for contour lines, in K. Rennolls (ed.), *Proceedings of IUFRO4.11 Conference 'Forest Biometry, Modelling and Information Science'* (Greenwich: University of Greenwich), http://cms1.gr.ac.uk/conferences/iufro/proceedings/.

Watson, D. F, 1992, *Contouring – A Guide to the Analysis and Display of Spatial Data* (Oxford: Pergamon Press).

Yamamoto, J. K., 2000, An alternative measure of the reliability of ordinary kriging estimates, *Mathematical Geology*, **32**, 489–509.

14

Geostatistical Prediction and Simulation of the Lateral and Vertical Extent of Soil Horizons

Benjamin Warr, Inakwu O. A. Odeh and Margaret A. Oliver

1 Introduction

Concerns about global climate change require estimates of various components of atmospheric and terrestrial carbon. This is particularly important with respect to terrestrial inorganic carbon, which is a major repository of global carbon dioxide (CO_2). Global and regional estimates of inorganic carbon, particularly carbonates, are unreliable and in many cases do not take into account the spatial variability of the occurrence and amount of carbonates in the soil horizons or examine the spatial uncertainty of the predictions. Gypsum, on the other hand, is important for soil management. Worldwide estimates of the extent of soil containing gypsum range from 90 million ha (FAO, 1993) to 207 million ha (Eswaran and Zi-Tong, 1991). These estimates are unreliable as is the case with the terrestrial carbonates. There is, therefore, the need to design appropriate approaches to provide an inventory of soil carbonate and gypsum, particularly at the regional level, which could be aggregated to give estimates at the national or continental scale. It is increasingly necessary to provide uncertainty information that is appropriate to the scale of spatial aggregation that is propagated through soil, crop and climate models (Addiscott, 1993).

Uncertainty arises in any situation when there is a lack of complete knowledge. Spatial prediciton uncertainty is caused both by a lack of precision when collecting data and incomplete knowledge, a characteristic of sparsely sampled point data sets. Both sources of uncertainty may lead to the introduction of error when choosing and

Uncertainty in Remote Sensing and GIS. Edited by G.M. Foody and P.M. Atkinson.
 ISBN: 0–470–84408–6

parametizing a spatial model, thereby limiting the accuracy with which spatial predictions can be made.

Accurate spatial prediction of soil horizon thickness implicitly requires accurate prediction of the locations where a particular horizon is present. Soil horizon thickness can be expected to lie within a limited range of values, whereas its true lateral extent can differ considerably from that indicated by the samples collected only at a few points across a region. Therefore, the uncertainty associated with the prediction of the horizon's lateral extent is likely to be far greater than that of predicted horizon thickness. Geostatistical interpolation and simulation techniques can be used to provide spatially distributed predictions and uncertainty measures, but the choice of which technique to use should be defined by our understanding of the pedology of the region and the data themselves.

Soil horizon thickness is a continuous variable often having a positively skewed distribution. This is caused by many zero values, where the horizon of interest is absent, and a few extreme values. Ordinary kriged (OK) predictions of positively skewed data are often biased, because of the underestimation of large values and overestimation of small ones. Several geostatistical techniques have been developed to overcome the prediction bias caused by positive skew; these include multi-Gaussian kriging (MG) and indicator kriging (IK) (Saito and Goovaerts, 2000). We shall propose a third method that uses the kriged prediction of an indicator of horizon presence to delineate regions where the soil horizon is most likely to exist, and which can be used to mask the continuous estimate.

Any statistical prediction is subject to uncertainty, but this is exaggerated when the number of sample locations is small and the distance between sites is large; a feature typical of regional scale reconnaissance soil surveys. By recreating sample variability, simulation techniques can describe the *spatial* uncertainty of the attributes across the study area; information that is additional to the *local* kriging variances. Again, there are many geostatistical algorithms available to simulate both continuous and categorical variables and their results can differ (Journel, 1996; Goovaerts, 2001).

Our aim here is to (i) compare methods to estimate soil horizon attribute thickness and lateral extent using simulated data that share the same statistical and spatial properties exhibited by real data, (ii) to apply the method that performed most accurately to the real *in-situ* data and (iii) then to investigate the uncertainty associated with the prediction of the soil horizon's lateral and vertical extent by geostatistical simulation.

2 Theory

Geostatistics is no longer new to the geographical information systems (GIS) and remote sensing communities. However, a brief treatment of the theory is given here to provide the links of basic geostatistics to the more recent methods of simulation, and in particular to emphasize the different goals when kriging and simulating continuous and indicator variables.

2.1 Predicting soil horizon attributes

Kriging is a geostatistical technique for optimal estimation that uses the variogram model parameters, which describe the form and degree of spatial continuity (Burgess and Webster, 1980a, b). The most commonly applied technique to estimate continuous soil properties is ordinary kriging (OK). However, the spatial distribution of a soil horizon is only continuous within regions where the soil horizon exists. Zero values of the continuous variable delineate the region where no horizon is present. At the regional scale of mapping, and typically within GIS, the boundaries between regions of horizon presence and absence represent discontinuities. Therefore, a sensible approach for predicting local soil horizon thickness would comprise an initial prediction of the regions where the soil horizon is present, and within which it can be assumed to be continuous in space, followed by an optimal estimation of the continuous variable within these zones.

2.1.1 Indicator coding

Indicator variables, $\mathbf{I}(\mathbf{x}_\alpha\,; z_k)$, are generated by applying an indicator transform to a continuous variable, $z(\mathbf{x}_\alpha)$, at selected thresholds, z_k which are often chosen to be equal to the decile values of the cumulative distribution,

$$\mathbf{I}(\mathbf{x}_\alpha\,; z_k) = \begin{cases} 1 & \text{if } z(\mathbf{x}_\alpha) \geq z_k \\ 0 & \text{otherwise} \end{cases} \quad k = 1, \ldots, K. \tag{1}$$

The indicator-transformed values describe the spatial distribution of sets with random location and shape (Bierkens and Burrough, 1993a). Therefore, the lateral extent of a soil horizon can be described by a *categorical indicator* variable defined at the threshold $z_k = 0$. The result is a categorical indicator variable equal to one if the horizon s_k, is present and zero if not,

$$\mathbf{I}(\mathbf{x}_\alpha\,; s_k) = \begin{cases} 1 & \text{if } s(\mathbf{x}_\alpha) = s_k \\ 0 & \text{otherwise} \end{cases} \quad k = 1, \ldots, K \tag{2}$$

2.1.2 Variography

For both continuous and (categorical) indicator variables we use the variogram as it describes the spatial pattern of an attribute and is used subsequently for prediction and simulation. The experimental variogram of a continuous variable, $\gamma^*(\mathbf{h})$, is calculated as half the average squared difference between data pairs,

$$\gamma^*(\mathbf{h}) = \frac{1}{2N(\mathbf{h})} \sum_{\alpha=1}^{N(\mathbf{h})} [z(\mathbf{x}_\alpha) - z(\mathbf{x} + \mathbf{h})]^2 \tag{3}$$

where $N(\mathbf{h})$ is the number of data pairs within a given class of distance and direction.

The accuracy with which the variogram is estimated, and hence the variable predicted, can be increased by reducing the effect of a few large values of semivariance

(Goovaerts, 1997). This can be achieved by transforming the original z-data into normal score y-values,

$$y(\mathbf{x}_\alpha) = \phi(z(\mathbf{x}_\alpha)) \tag{4}$$

where ϕ is a normal score transform that relates the p-quantiles z_p and y_p of the two distributions. The variogram of the $y(\mathbf{x}_\alpha)$ can then be recalculated using equation (2). Typically the $y(\mathbf{x}_\alpha)$ variogram model $\gamma_Y(\mathrm{h})$, has a smaller relative nugget effect and greater spatial continuity over short lag distances than its $\gamma_Z(\mathrm{h})$ equivalent.

The experimental variogram of an (categorical) indicator variable $\mathbf{I}(\mathbf{x};\, z_k)$ is computed in exactly the same way as that for continuous variables as

$$\gamma_I{}^*(\mathbf{h}) = \frac{1}{2}\mathrm{E}[\{\mathbf{I}(\mathbf{x}_\alpha + \mathbf{h}\,; s_k) - \mathbf{I}(\mathbf{x}_\alpha\,; s_k)\}^2] \tag{5}$$

It describes the frequency with which two locations a vector, **h**, apart belong to different categories – $k = 1, \ldots, K$. This is also the probability of passing from one category to another for a given separation.

2.1.3 *Kriging continuous attributes with positive skew*

Kriging can be used to predict a continuous soil attribute z, at unsampled locations $\mathbf{x}$ using sparse sample data $\{z(\mathbf{u}_\alpha),\ \alpha = 1, \ldots, n\}$ (Burgess and Webster, 1980a, b). All linear kriging systems are based on the same basic estimator $Z^*\,(\mathbf{x})$,

$$Z^*(\mathbf{x}) - m(\mathbf{x}) = \sum_{\alpha}^{n} \lambda_\alpha(\mathbf{x})[Z(\mathbf{x}_\alpha) - m(\mathbf{x}_\alpha)] \tag{6}$$

where $\lambda_\alpha(\mathbf{x})$ is the weight assigned to the data $z(\mathbf{x}_\alpha)$ located within a given neighbourhood $W(\mathbf{x})$. The weights are chosen so as to minimize the estimation variance $\sigma_E^2 = \mathrm{Var}\{Z^*(\mathbf{x}) - Z(\mathbf{x})\}$ under the constraint that the estimator is unbiased.

The ordinary kriging (OK) system assumes that the mean $m(\mathbf{x}_\alpha)$ is constant, but unknown within a local neighbourhood $W(\mathbf{x})$. Therefore, the local mean must be estimated for each $z(\mathbf{x}_\alpha)$. This estimate is not robust to the presence of extreme values. A single extreme value can impact the estimate of the local mean and cause under- and overestimation of the tails of the $z(\mathbf{x})$ histogram, generating systematic estimation bias that can be severe for skewed data. This is known as the smoothing or information effect. Smoothing increases with increasing relative nugget effect and sample skew, and with decreasing sample density (Goovaerts, 1997).

Multi-Gaussian kriging (MG) and indicator kriging (IK) have been developed to reduce the systematic prediction bias. Both require a non-linear (back) transform of the data prior to and after kriging. The first step in multi-Gaussian kriging is to normalize the data, $z(\mathbf{x}_\alpha)$ (equation (5)) to give a Gaussian distribution. This facilitates inference of the mean when kriging. The kriging estimates $Y^*_{SK}(\mathbf{x}_\alpha)$ are then back-transformed to provide the estimate $Z^*_{\mathrm{MG}}(\mathbf{x})$.

The estimate $Z_{IK}{}^*(\mathbf{x})$ is obtained by multiplying the estimate at each threshold z_k by the threshold value and summing the results for all K. Nevertheless, because kriging is a form of weighted spatial averaging OK, MG and IK can all be expected to produce a smoothing effect of the true $Z(\mathbf{x})$ histogram. None of these methods can be used to classify zero and non-zero zones based on the predicted values. However, this is what is required for a correct prediction of a horizon's lateral extent (and subsequently local thickness).

2.1.4 *Kriging categorical attributes to delineate zones of presence*

Consider a categorical indicator defined by presence or absence. Its sample mean is a prior estimate of the probability of the category's presence. The kriged estimate of the categorical indicator represents its conditional expectation, which in turn equals the conditional probability that the category s_k is present,

$$E(\mathbf{I}(\mathbf{x}\,;s_k)|n) = P(\mathbf{I}(\mathbf{x}\,;s_k)|n) \tag{7}$$

where $|n$ represents the conditioning of the prediction by the surrounding n data surrounding a location $\mathbf{x}$ (Bierkens and Burrough, 1993a, b). Since the estimate is a probability thresholds can be selected and the kriged estimate, coded as an indicator

$$\mathbf{I}(\mathbf{x}\,;s_{cIK}) = \begin{cases} 1 & \text{if } [\mathbf{I}(\mathbf{x}\,;s_k)]^* \geq z_k \\ 0 & \text{otherwise} \end{cases} \tag{8}$$

The resulting indicator variable delineates the most likely zones of horizon presence and absence. The value at which we apply a cut-off can be chosen on the basis of ancillary knowledge or with the intention of respecting certain properties of the sample data. For example, in the absence of prediction bias a reasonable classification criterion would ensure that the estimated lateral extent should equal the global sample proportion. For an unbiased estimate, values smaller than 0.5 indicate a greater likelihood of absence than presence and *vice versa*.

2.2 Examining uncertainty using simulation

Constrained spatial simulation techniques that reconstruct the variability of sample data can be used to investigate uncertainty in two ways that linear kriging cannot: (a) to create exhaustive 'realistic' datasets, which can be used to test various prediction techniques; and (b) to characterize *spatial* uncertainty. The simple kriging estimation variance σ^2 and predicted value of multi-Gaussian variables can be combined to provide conditional probability distributions $F(\mathbf{x}\,;z|(n))$ that describe the *local* uncertainty of the estimate at location $\mathbf{x}$, using the surrounding sample data, n (Goovaerts, 1994). It is then possible to simulate local uncertainty by drawing random numbers from the conditional distribution. No measure of the spatial uncertainty for a series of M locations can be derived from the single-point ccdfs $F(\mathbf{x}_m\,;z|(n))$, $j = 1, \ldots, M$. Each estimate when considered individually is best (in the least-squares sense), but the map of the local predictions may not be best in a wider sense (Goovaerts, 1997), and importantly, the estimation variance does not describe the spatial uncertainty associated with possible patterns of values.

When simulating, information on uncertainty is provided by the repeated simulation of many equi-probable realizations. In contrast to kriging, simulation selects randomly a value $z(\mathbf{x}_j^{'})$, $j = 1, \ldots, N$ at location $\mathbf{x}$, from the N-point conditional cumulative distribution function (ccdf) that models the joint (hence *spatial*) uncertainty at the N locations surrounding the simulated point,

$$F(\mathbf{x}_1', \ldots, \mathbf{x}_N'; z_1, \ldots, z_N|(n)) = \mathrm{Prob}\{Z(\mathbf{x}_1') = z_1, \ldots, Z(\mathbf{x}_N') = z_N(n)\} \quad (9)$$

This process is repeated for each location $\mathbf{x}_\alpha'$, under constraints that include reproduction of the variogram and sample histogram and that the simulated values at sample points are equal to the true sample values. The realizations can then be analysed and patterns that exist in many are considered probable and *vice versa*. Sequential Gaussian (sGs) and sequential indicator simulation (sis) can simulate both continuous and categorical (indicator) variables (Journel, 1996; Goovaerts, 1997). In contrast to kriged estimates of continuous variables, thresholds can be applied to the realizations of a continuous variable because truncation of a single realization of a random function generates a random set described by an indicator. Consequently sGs can provide, as can sis, uncertainty information about both horizon attributes; lateral extent and thickness. This enabled a comparison of uncertainty of the lateral extent provided by parametric (sGs) and non-parametric (sis) techniques.

3 The Study Area

The Bourke region is approximately 700 km northwest of Sydney, Australia (Figure 14.1). On the surface, the soils are distinguished primarily by their colour. The most common Munsell colour (hue, value, chroma) assignments were red (2.5 YR), grey (10 YR 5/2) and light grey (5 YR). Although the relative elevation is small the red soil is predominantly on the raised levees or swale dunal formations. These are surrounded by the low-lying alluvial fans, which are dominated by the grey soil. In contrast to gypsum horizons, carbonate horizons are more prevalent in the red and transitional soils and uncommon in the grey soils. The origin of the raised levees is not known, but we can speculate that their formation is probably due to several mechanisms given the variation in soil texture (Chittleborough and Oades, 1980). Glassford and Sonenvik (1995) suggest that the surficial deposits of much of southern Australia, which have subsequently undergone varied pedogenesis, are mostly desert aeolian sediments composed of aeolian and fluvial sandy clayrock and altered aeolian sands and dusts (Harden *et al.*, 1991). Certain red soils are almost uniform sands. This suggests a relatively young ($\cong$ 5000 year BP) dunal parent material (Urushibara-Yoshino, 1996). Others are duplex soil types, suggesting that there has been illuviation of clay from the upper horizons, leaving behind fine sands coated with resistant iron oxide minerals. These may simply be older or have a distinct genesis, perhaps as ancient fluvial levees or alluvial terraces having undergone dust accession (Nettleton, 1991).

4 Methods and Analysis

We stated earlier that a sensible approach to estimating local horizon thickness should involve two steps: an initial delineation of the zones where a horizon is

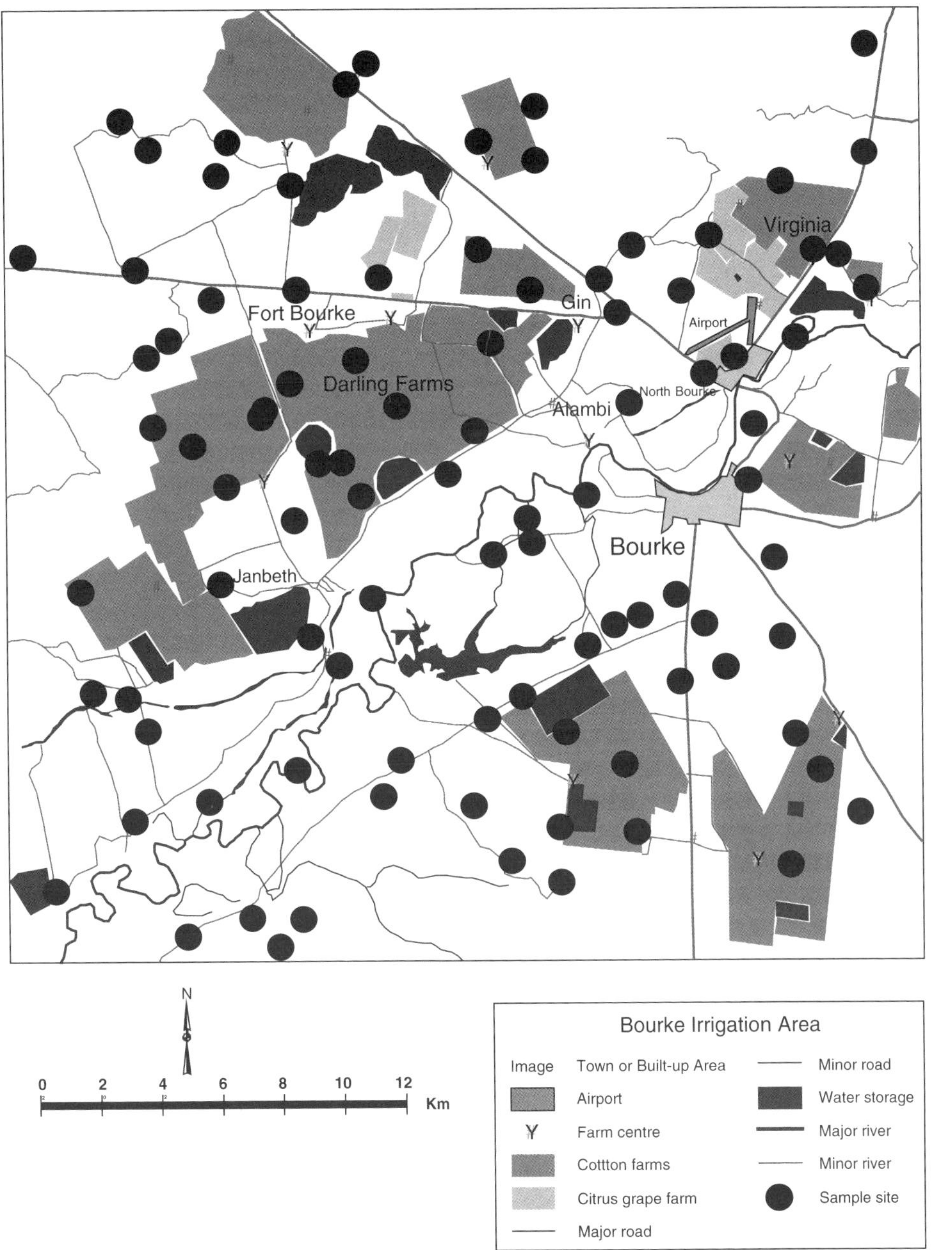

Figure 14.1 Bourke irrigation area. Origin (bottom left): 337500 E, 6655000 N

most likely to be present and an optimal kriging of horizon thickness within these zones. We tested this hypothesis using simulated data because the *in-situ* data were at a few sparsely located samples. Comparing methods of prediction on simulated data with similar statistical properties to those of the real data provided a useful alternative to cross-validation, for hypothesis testing and investigating uncertainty caused by the choice of technique. Creation of the simulated data started with a detailed analysis of the true sample data to ensure that they had similar statistical and spatial properties.

The second part of the analysis compared the tools available to describe spatial uncertainty. First, we evaluated and compared the measures provided by kriging and simulation methods. Second, we focused on the difference between the measures of uncertainty of horizon lateral extent provided by parametric sGs and non-parametric sis algorithms, showing that the results they provide are not equivalent.

4.1 Soil data

Three binary indicator variables describing the presence of diffuse and nodular calcium carbonate, and gypsum, were defined from observations of 1.5 m cores collected at 101 sites over a 30 km by 30 km region surrounding Bourke. We used a stratified random sampling design (Figure 14.1). The sampling density was approximately 0.156 samples per km^2. The upper and lower extents of the horizon were used to calculate the horizon thickness, which was then used to define the indicator variable. Carbonates ($CaCO_3$) were identified using 0.1 M HCl and gypsum accumulations, indicating a content greater than 10% (Verheye and Boyadgiev, 1997), were recognized by visual identification.

4.2 Structural analysis of in-situ data

The *in-situ* variables used were presence (or absence) of: (1) diffuse CO_3^{2+} (D), (2) nodular CO_3^{2+} (N) and (3) gypsum (G). We refer to the categorical indicator variables as $\mathbf{I}(\mathbf{x};D)$ for the diffuse CO_3^{2+}, $\mathbf{I}(\mathbf{x};N)$ for the nodular CO_3^{2+} and $\mathbf{I}(\mathbf{x};G)$ for gypsum. The horizon thickness values are D_T, or N_T, or G_T, respectively, with subscript T indicating the thickness. The sample data indicate that the diffuse carbonate horizon is more prevalent than horizons with carbonate nodules. Gypsum horizons are relatively uncommon, being present at only 20% of the sites. However, the lack of an explicitly defined test method and the reliance solely on visual recognition of its presence as crystal coatings within the profile probably led to an underestimation of gypsum. The mean depths of the upper depth limits followed the expected pattern for diffuse carbonates, typically occurring closer to the surface than nodular carbonates and the latter forming above gypsum concretions. The diffuse carbonate horizon often extends into the nodular carbonate horizon as expected. The correlation between indicators suggests that nodular calcium is often co-located with diffuse carbonates ($\rho = 0.45$). Their spatial co-dependence is weak and the intrinsic model provides an acceptable fit. Gypsum has a small negative correlation with diffuse and nodular carbonates ($\rho = -0.09$). The histograms of the three horizon thickness values were positively skewed. This skew was caused by a few extreme values and many zero values (Figure 14.2).

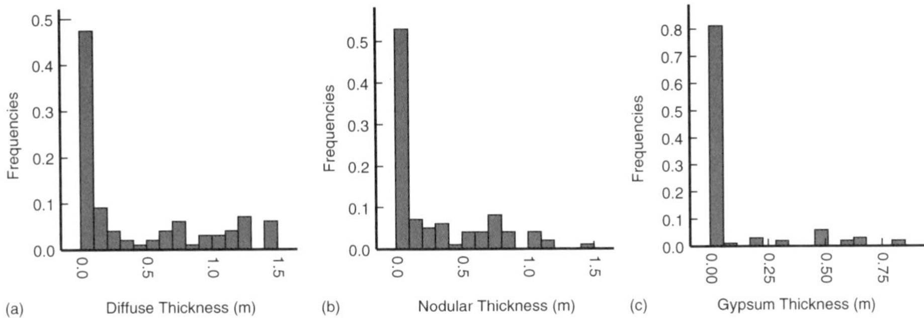

Figure 14.2 *Histograms of horizon thickness: (a) diffuse carbonates, (b) nodular carbonates and (c) gypsum*

4.3 Variogram modelling

Authorized variogram models were fitted by eye to the experimental variograms of the variables of horizon thickness and horizon presence for each type of horizon. Variogram models of the thickness variables for each horizon are shown in Figure 14.3. The best-fitting model was an exponential function. The nugget effect accounted for over half of the sill variance. It was largest for the nodular horizon and smallest for the diffuse carbonate horizon. The spatial structures for gypsum and diffuse carbonate horizons had similar ranges of 15.3 km and 24.28 km, respectively. In contrast, the range for nodular carbonates was approximately half this distance.

For multi-Gaussian kriging and simulation, each thickness variable was transformed to normal scores (equation (4)). The experimental variograms of the normal scores were computed and modelled. The resulting models had smaller relative nugget effects and greater spatial continuity, shown by their larger ranges. Variograms were also fitted to the (categorical) indicator variables of each horizon. Again the exponential model provided the best fit. Interestingly the spatial ranges of both diffuse carbonate and gypsum indicators were about half those of the corresponding thickness variables, while that of the nodular carbonate indicator was twice the length of its thickness counterpart. For each variable, the variograms of indicators defined at increasingly large decile values showed a typical decrease in the structure (a destructuration effect). For a cut-off greater than 1 m the variograms were almost pure nugget.

4.4 Simulating data with realistic characteristics

Thirty values were drawn at random from a Gaussian distribution N (0, 1) at equidistant locations along a hypothetical 25 km long transect. The variogram model, $\gamma_{D_{Tg}}(\mathbf{h})$, of the normal-score transformed *in-situ* thickness variable was then used to simulate a variable $T_{SIM\ g}$ at 500 points along the transect conditioned by the 30 samples. The values were back-transformed. A single realization T_{SIM}, was

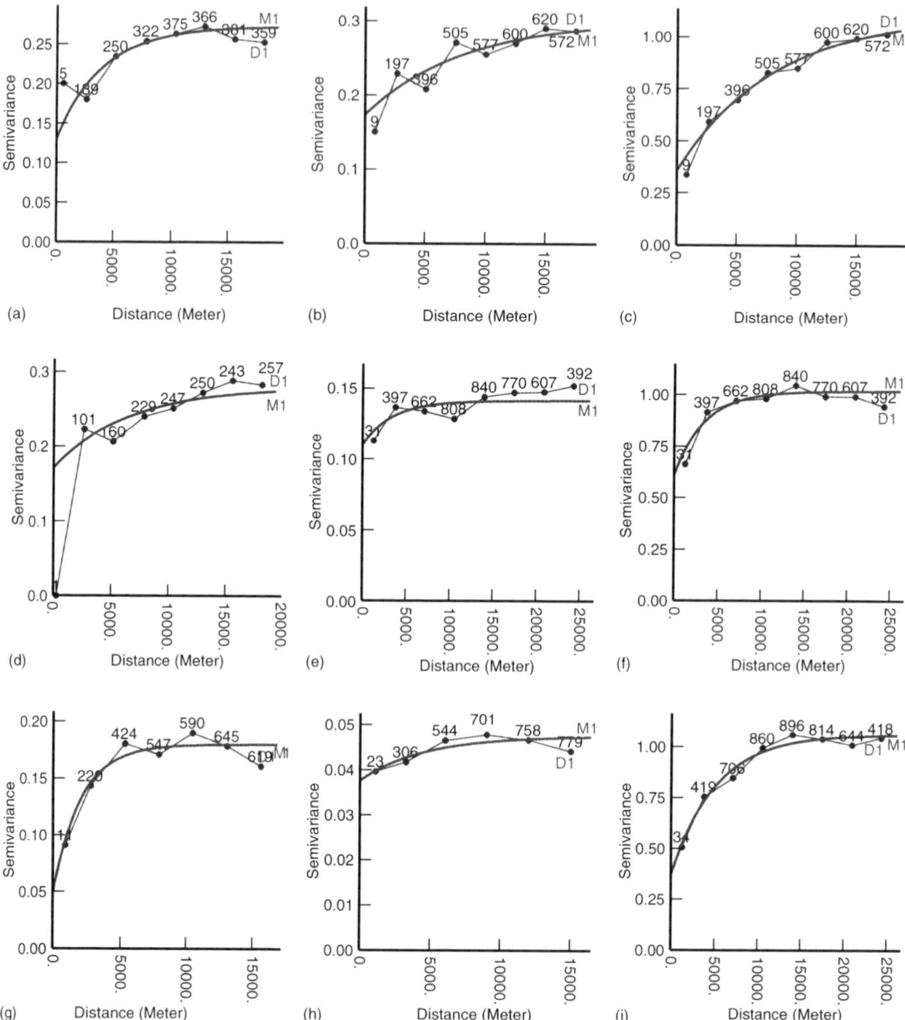

Figure 14.3 Experimental (dots) and modelled (thick solid line) variograms for (a) diffuse carbonate indicator, (b) diffuse carbonate thickness, (c) diffuse carbonate thickness normal scores, (d) nodular carbonate indicator, (e) nodular carbonate thickness, (f) nodular carbonate thickness normal scores, (g) gypsum indicator, (h) gypsum thickness, (i) gypsum thickness normal scores

selected that has a similar sample variogram and histogram as the diffuse carbonate horizon (Figure 14.4a).

For estimation the variogram, $\gamma_{T_{SIM}}(\mathbf{h})$, was adjusted to the chosen realization for ordinary kriging. By using the exhaustive dataset uncertainty associated with estimation of the variogram was not a concern. A second variogram, $\gamma_{T_{SIM\,g}}(\mathbf{h})$, was fitted to the Gaussian transform of the same realization, for multi-Gaussian kriging. Thickness, T_{SIM}, was also transformed to an indicator at a cut-off corresponding to the deciles of the cumulative distribution. Variograms were fitted to each indicator for

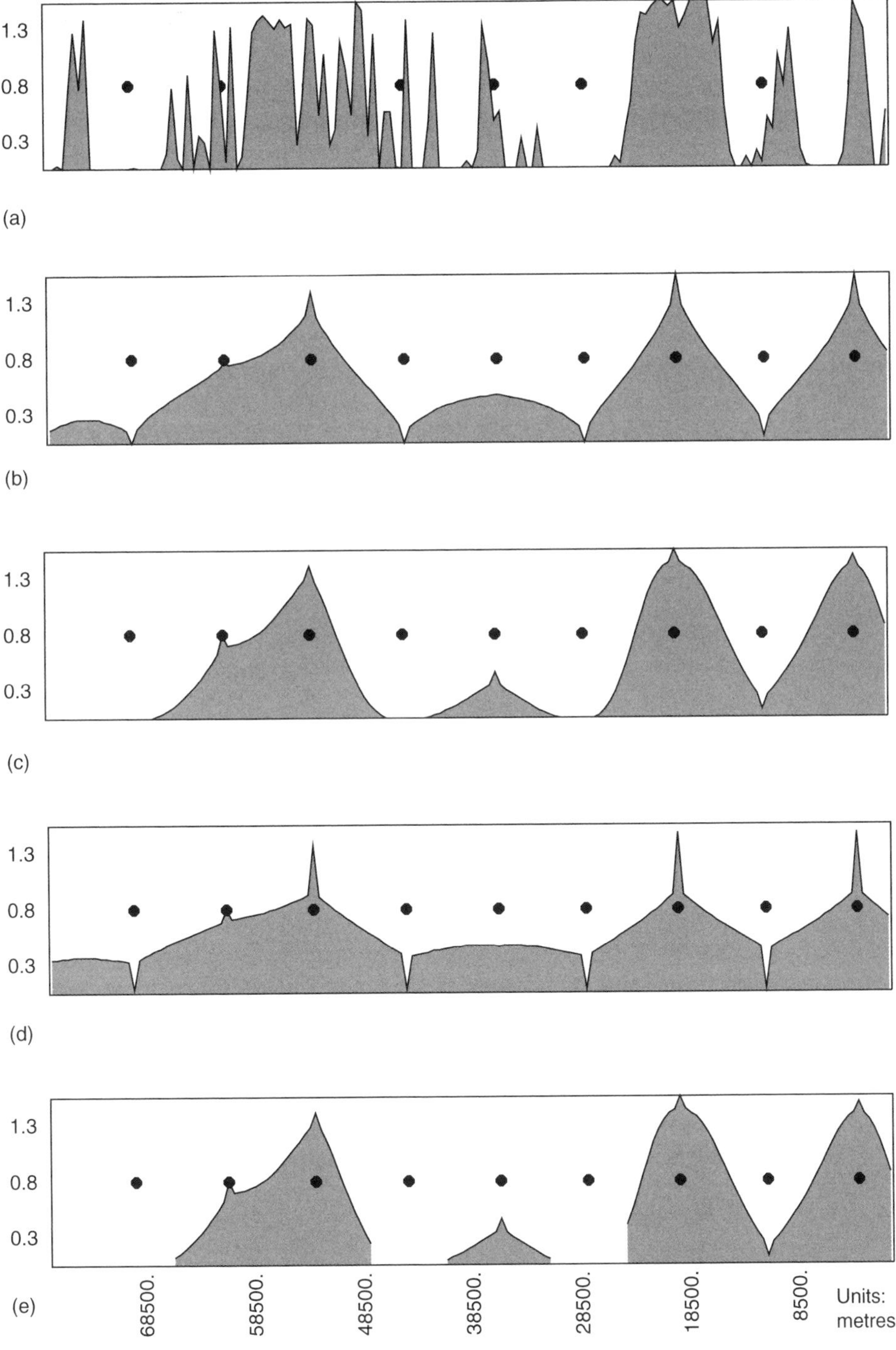

Figure 14.4 *Simulated transect: (a) true simulated thickness, T_{SIM}, (b) $T^*_{SIM\ OK}$, (c) $T^*_{SIM\ MG}$, (d) $T^*_{SIM\ IK}$, (e) $T^*_{SIM\ OPT}$*

indicator kriging. Finally a categorical indicator was defined to create the simulated horizon lateral extent P_{SIM},

$$P_{SIM} = \mathbf{I}(\mathbf{x}\,; T_{SIM} > 0) = \begin{cases} 1 & \text{if } T_{SIM}(\mathbf{x}) > 0 \\ 0 & \text{otherwise} \end{cases}, \tag{10}$$

and a variogram model, $\gamma_{P_{SIM}}(\mathbf{h})$, was adjusted to this variable for kriging the indicator of presence/absence (categorical indicator kriging). From the 500 simulated values 30 were selected as data for kriging. The remaining 470 were used for accuracy assessment.

4.5 Comparing prediction methods

As a first step, OK, MG and IK were used to predict values of simulated horizon thickness, T_{SIM}. The predictions $T^*_{SIMOK,\ MG,\ IK}$, were compared against the accuracy assessment data to determine which showed the least systematic bias. The results were also used to examine the error that accrues if any of these methods is used to classify zeros and non-zeros.

In the second step, the simulated horizon lateral extent P_{SIM}, was estimated by kriging the indicator of horizon presence $\mathbf{I}(\mathbf{x}\,; T_{SIM} > 0)$. A cut-off was applied to the conditional probability of horizon presence provided by the kriged categorical indicator $[\mathbf{I}(\mathbf{x}\,; T_{SIM} > 0)]^*_{cIK}$, to provide an estimate of the simulated horizon lateral extent,

$$[\mathbf{I}(\mathbf{x}\,; P_{SIM})]^* = \begin{cases} 1 & \text{if } [\mathbf{I}(\mathbf{x}\,; Z_{SIM} > 0)]^*_{\text{cIK}} \geq 0.5 \\ 0 & \text{otherwise} \end{cases} \tag{11}$$

This simulated value was used to mask the kriged estimate of simulated horizon thickness by calculating the product of the thickness estimate $T_{SIM}{}^*(\mathbf{x})$ and the indicator coded estimate of horizon presence $[\mathbf{I}(\mathbf{x}\,; P_{SIM})]^*$,

$$T^*_{OPT}(\mathbf{x}) = T^*_{SIM}(\mathbf{x}).[\mathbf{I}(\mathbf{x}\,; P_{SIM})]^*. \tag{12}$$

The quality of the estimate of T_{SIM} obtained using the estimate of P_{SIM} to delineate the zones of horizon presence was compared using the histograms of errors calculated from the exhaustive simulated data set.

4.6 Simulating the in-situ variables to characterize spatial uncertainty

Two runs, each consisting of 500 simulation realizations, generated: (a) conditional sGs of soil horizon thickness following a normal score transform, and (b) conditional sis of horizon presence. The local neighbourhood was constrained to include no more than 25% simulated values when drawing new simulated values. The results were then post-processed. The mean and standard deviation were calculated for each node across the realizations (*E-type* estimate). In addition, each realization of horizon thickness, simulated using sGs, was indicator coded according to equation (10). This provided a second set of realizations of the indicator of horizon presence, which was subsequently post-processed to find the *E-type* uncertainty measures.

5 Results

5.1 Comparing prediction bias using a simulated dataset

The kriged transects (Figure 14.4b–e) are less variable than reality (Figure 14.4a). The transects and error histograms (Figure 14.5a–d) show that this results primarily from the overestimation of zero values, where no horizon is present, and to a lesser extent by the underestimation of large values. Of the OK, IK and MG estimates both OK and IK predicted the presence of a horizon along the entire transect. The MG estimate predicted a horizon presence along 90% of the transect. In reality the horizon is present along only 59% of the transect. As expected the OK estimate, with no prior transformation of the data, was the most biased. Surprisingly the IK estimate was equally biased. It is suggested that a lack of non-zero indicator values for certain thresholds was responsible (Saito and Goovaerts, 2000).

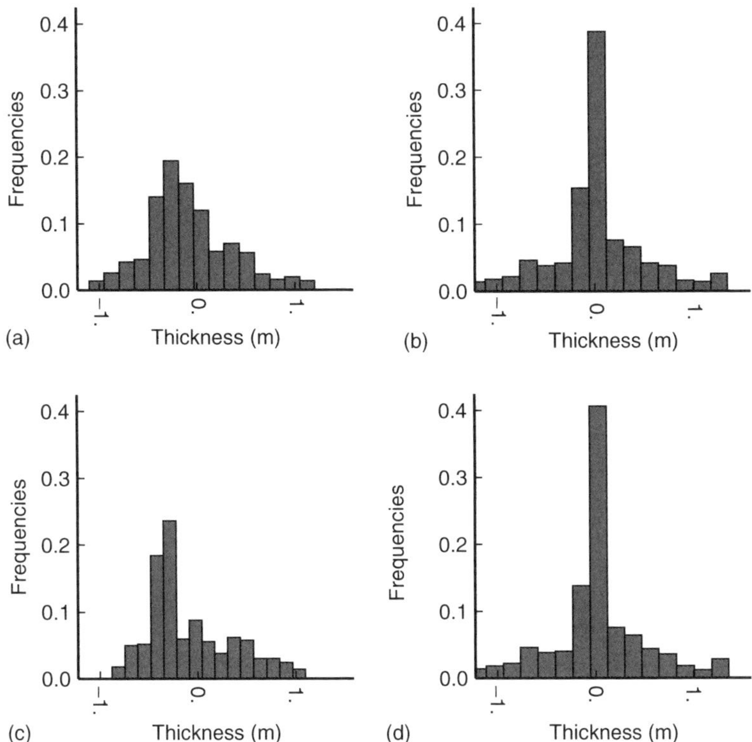

Figure 14.5 Simulation experiment prediction error histograms: (a) $T^*_{SIM\ OK}$, (b) $T^*_{SIM\ MG}$, (c) $T^*_{SIM\ IK}$, (d) $T^*_{SIM\ OPT}$

Clearly, where no mask was imposed on the thickness estimates, the lateral extent of the horizon P_{SIM} and local thickness T_{SIM} were markedly overestimated. Figure 14.4e shows the optimal estimate obtained by masking the least biased of the horizon thickness estimates, provided by MG kriging. A comparison of Figure 14.5c and 14.5d shows that imposing the mask has reduced the slight overestimation of true zero values. As a result the estimate of the simulated horizon lateral extent $[\mathbf{I}(\mathbf{x}; P_{SIM})]^*$, 59% matches exactly the true lateral extent P_{SIM}. Moreover we can see from Figure 14.4e that the location of the zones of horizon absence approximate well the reality (Figure 14.4a).

The validation results were conclusive:

- MG kriging was the least systematically biased of the three continuous estimators.
- OK, MG and IK should *not* be used to classify zeros and non-zeros.
- Kriging the conditional probability of horizon presence, thresholding the result and using this to mask the kriged thickness reduced the estimation bias and delineated zones of horizon absence.
- In the absence of other information the classification criterion should be the reproduction of sample proportions. In the absence of prediction bias, this can be achieved by applying a threshold to the categorical indicator estimates at a value $z_k = 0.5$.

5.2 In-situ prediction using the 'optimal' method

We applied the optimal two-step technique to predict the lateral extent and local horizon thickness for the *in-situ* data described above. Multi-Gaussian kriging was used to estimate the horizon thickness and categorical IK (cIK) to predict the conditional probability of horizon presence. Thresholding the cIK estimates at a value of 0.5 did not reproduce the lateral extent of horizon thickness. Therefore, the categorical indicators of horizon presence were thresholded at quantile values equal to 1 – *sample proportion*, before multiplying the two predictions (equation (12)). Maps of the predictions (Figure 14.7a–c) show (in black) the regions where no horizon is expected to exist. Predicted horizon thickness is typically greatest towards the centre of zones of presence. However, because the mask is estimated independently from the thickness estimate, boundaries of the zones of presence exist side-by-side with horizon estimates.

The small estimation values show that kriging has greatly smoothed the thickness estimate histogram. The small values of the mean thickness also result from the underestimation of large values and the overestimation of zero values. This effect is best illustrated by scatterplots of cross-validation estimates for each sample location (Figure 14.6) and the estimate of the thickness of nodular carbonate thickness. The latter thickness had the largest relative nugget effect, arising from isolated large thickness values surrounded by zero values, which were severely underestimated by kriging. The smoothing effect was also severe for the thickness of the gypsum layer because the sample was dominated by zero values (sample proportion 20%).

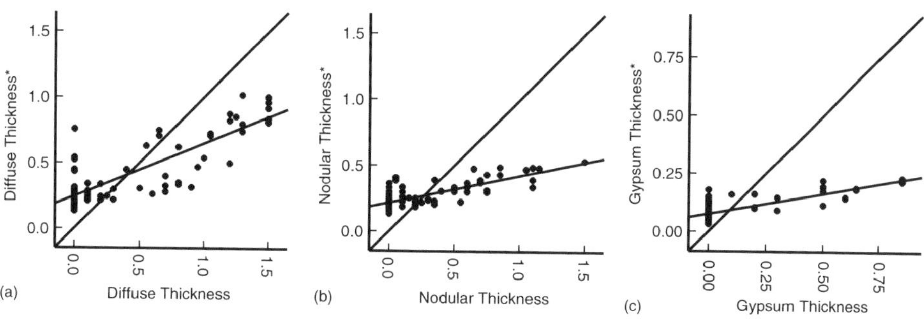

Figure 14.6 *Scatterplots of cross-validation estimates versus sample values of (a) diffuse carbonates, (b) nodular carbonates and (c) gypsum horizon thickness*

The maps of the predictions confirmed our initial hypotheses of the spatial distribution of carbonate and gypsum horizons. The diffuse and nodular carbonate horizons are more common than that of gypsum. The thickest nodular and diffuse carbonate horizons are co-located to the northeast and west of the study area in red soil on raised levees. In contrast, the gypsum horizon is limited to the grey soils, near the centre of the study area, used for cotton production. Nevertheless, the maps of prediction suggest that in the north of the study area isolated gypsum accumulations exist within larger regions of carbonate accumulations.

5.3 Kriging and simulation measures of uncertainty

Figures 14.7d–i and Figures 14.8a–f show the MG standard error and simulation standard deviation for each horizon type and each simulation method. The maps of prediction precision provided by simulation differ markedly from those provided by kriging. The MG estimation standard deviation maps (Figure 14.7a–c) show a similar spatial pattern. They do not indicate zones where uncertainty in the estimate is more prevalent, only that caused by the sample geometry, the relative nugget effect and form of the chosen variogram model.

In contrast, the E-type maps (Figure 14.7g–l) derived from the simulated realizations show patterns that are specific to each variable because they reflect the likelihood of finding particular spatial patterns of values. Hence, in zones where the sample values were homogeneously small (zero) or large the spatial uncertainty is low. For zones where the sample values are both small and large, the predicted uncertainties are larger. Hence we can delineate regions based on two criteria, (a) whether we expect there to be a horizon present, using the kriging/simulation conditional probability and (b) the degree of certainty in the estimate, via the simulation standard deviation.

5.3.1 Comparing sis and sGs measures of the uncertainty of lateral extent

Despite the similarities between the results from each simulation method, subtle differences exist. For example, the E-type simulated mean lateral extents of the sis realizations approximated well the sample estimates. However, the means of the

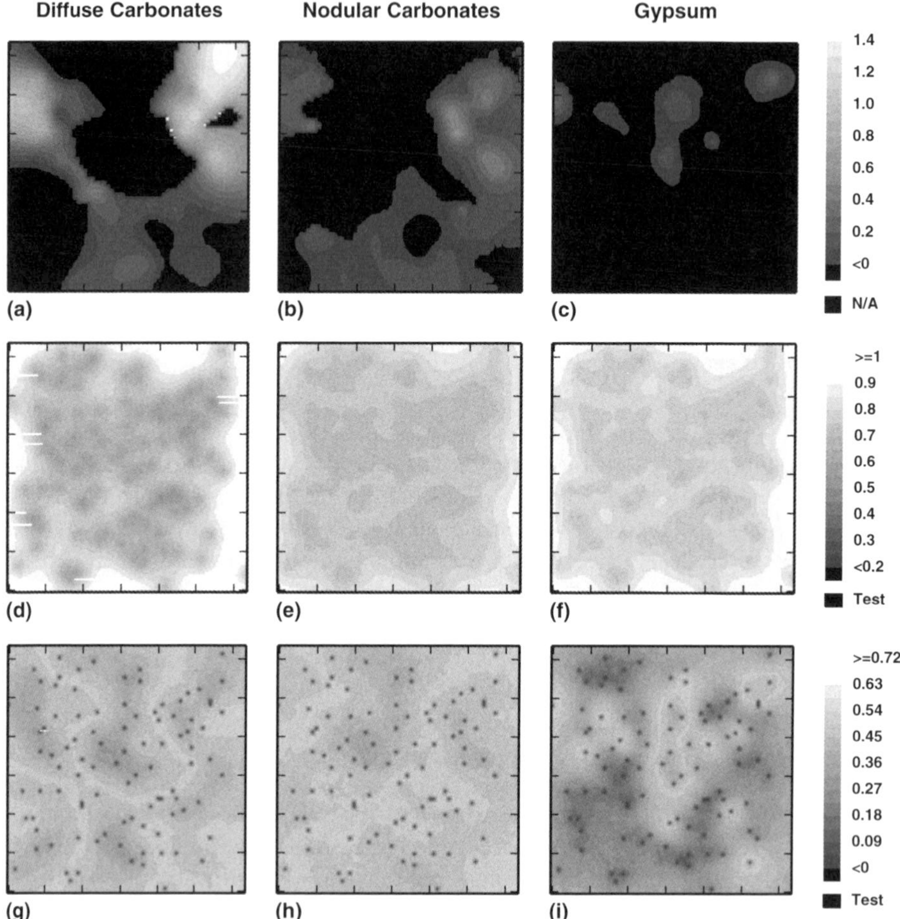

Figure 14.7 *Images showing (a–c) Optimal two-step kriging estimates of horizon thickness, (d–f) MG kriging estimation variance and (g–i) sGs thickness E-type mean, for each horizon type (Units are metres)*

indicator-coded sGs realizations, for each horizon type, were consistently 2–5% larger than the corresponding sample estimates. Also the 5% and 95% quantile values were systematically larger than those provided by sis. However, the simulated range of uncertainty was systematically smaller when using sGs to simulate lateral extent (−4.73 to 7.05% of the total area).

These differences are caused not just by differences in the variogram models used, but by the effect of the sample values when sequentially and randomly drawing values using the simulation model of spatial uncertainty (equation (9)). For example for diffuse carbonates, but representative of all three horizon types, there is a non-linear correlation between the sGs simulation of the thickness mean and standard deviation. By contrast, the sis indicator conditional probability is not correlated with the estimated value of thickness. The thickness

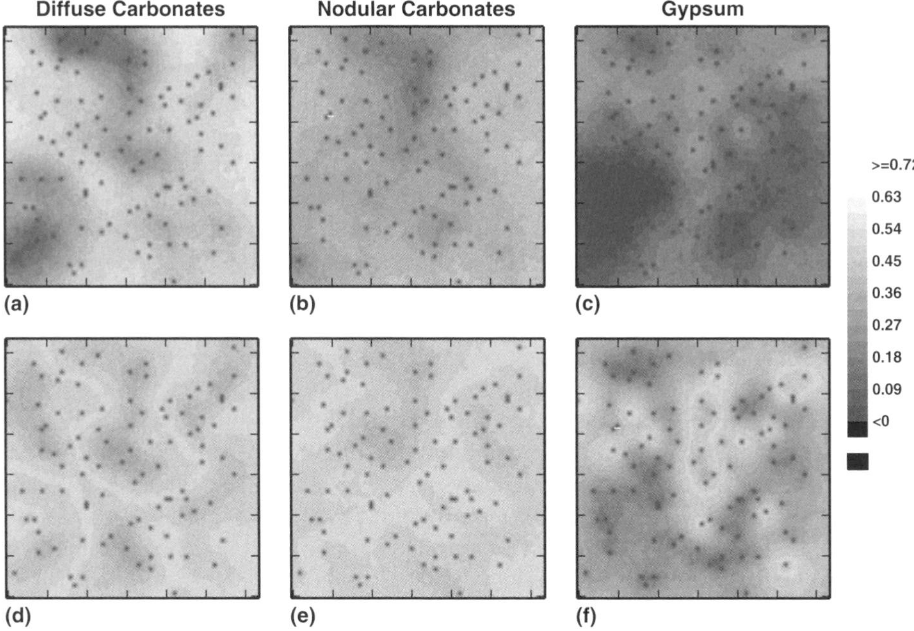

Figure 14.8 Simulation standard deviation for each horizon type. (a–c) sGs indicator simulation standard deviation, (d–f) sis simulation standard deviation

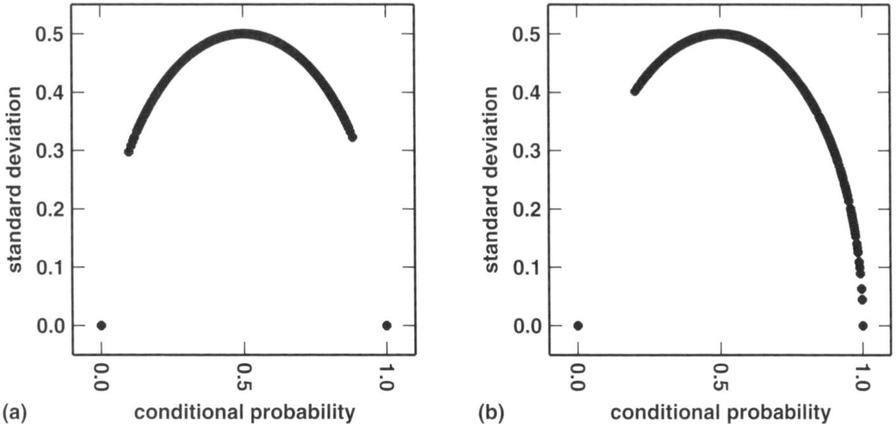

Figure 14.9 Scatterplots of simulation post-processing results for the diffuse carbonate horizon. (a) sis mean versus sis standard deviation, (b) sGs indicator coded mean versus sGs indicator coded standard deviation

value is not involved when simulating an indicator (and when estimating the indicator variogram) and has no effect on the conditional probability and the standard deviation generated by post-processing. This is not the case when indicators are provided by simulating thickness. The thickness value then plays a role when simulating the local thickness and thus the lateral extent. Put simply, large values are simulated more often in regions where the sample values are large. This effect is carried through to the estimated conditional probabilities and standard deviations of the indicator variable. Comparing Figures 14.9a and 14.9b shows that the maximum conditional probabilities are greater for the sGs method. Points with the largest conditional probabilities, with values close to one, have the smallest standard deviations almost equal to zero. The points with the largest conditional probabilities are in regions of homogeneous thick horizons. This reflects the degree of certainty of the horizon being present. Yet this effect is not expressed equally for zones of horizon absence. The conditional probabilities are not equivalent and the simulation local standard deviations are not as small (conditional probability $\cong 0.2$, standard deviation $\cong 0.4$). Yet these regions are just as homogeneous as their counterparts. Smaller uncertainty in regions of homogeneous thick horizon presence seems logical. However, this must be equalled by a similar reduction in uncertainty for regions of homogeneous horizon absence. The sis minimum conditional probability, in regions of homogeneous horizon absence, is associated with a smaller standard deviation than that provided by the sGs result (conditional probability $\cong 0.1$, standard deviation $\cong 0.3$). This is a value equivalent to that of equally homogeneous regions of horizon presence (conditional probability $\cong 0.9$, standard deviation $\cong 0.3$).

6 Conclusions

Simulation provides a useful tool with which to validate the choice of prediction method, in particular when the sample data are sparse. Our test validated the choice of a two-step kriging method for the reasons outlined in section 5.1. Nevertheless, the *in-situ* predicted thickness values were systematically biased and the estimates of both thickness and lateral extent subject to much uncertainty given the scarcity of data. By simulating the lateral and vertical extent we were able to generate maps of simulation standard deviation that described the spatial uncertainty, which was not the case for kriging. Both sequential simulation algorithms tested produced realizations that reflected the likelihood of finding areas of large and small values, irrespective of the sample geometry. Importantly, we have shown that the results provided by each are not equivocal. A choice between which to use should be made on the basis of pedological interpretations derived from the sample data and those made in the field. In the absence of additional information sis represents the more robust choice because the sample statistics were reproduced with less bias and the range of predicted uncertainty was more conservative.

Acknowledgements

This work was completed as part of a Leverhulme Study Abroad Studentship granted to Benjamin Warr. We would also like to thank Professor McBratney and Dr Balwant Singh of the Agricultural Chemistry and Soil Science Department, University of Sydney for having provided the opportunity to work in Australia. All geostatistical analysis was performed using ISATIS® kindly provided by Geovariances®, Fontainebleau.

References

Addiscott, T. M., 1993, Simulation modelling, *Geoderma*, **60**, 15–40.

Bierkens, M. F. P and Burrough, P. A., 1993a, The indicator approach to categorical soil data: I. Theory, *Journal of Soil Science*, **44**, 361–68.

Bierkens, M. F. P. and Burrough, P. A., 1993b, The indicator approach to categorical soil data: II. Application to mapping and land use suitability analysis, *Journal of Soil Science*, **44**, 369–81.

Burgess, T. M. and Webster, R., 1980a, Optimal interpolation and isarithmic mapping of soil properties. I. The semi-variogram and punctual kriging, *Journal of Soil Science*, **31**, 315–31.

Burgess, T. M. and Webster, R., 1980b, Optimal interpolation and isarithmic mapping of soil properties. II. Block kriging, *Journal of Soil Science*, **31**, 333–41.

Chittleborough, D. J. and Oades, J. M., 1980, The development of red-brown earth. III. The degree of weathering and translocation of clay, *Australian Journal of Soil Research*, **18**, 383–93.

Eswaran, H. and Zi-Tong, Gong, 1991, Properties, genesis, classification and distribution of soils with gypsum, in W.D. Nettleton (ed.), *Occurrence, Characteristics, and Genesis of Carbonate, Gypsum, and Silica Accumulations in Soils*, SSSA Special Publication Number 26.

FAO, 1993, World Soil Resources. An explanatory note on the FAO World Soil Resources Map at scale 1:25 000 000, *FAO World Soil Resources Report*, **66**, (Rome: FAO), 64 pp.

Glassford, D. K. and Sonenvik, V., 1995, Desert-aeolian origin of the Late Cenozoic regolith in arid and semi-arid Southwestern Australia, *Palaeogeography, Palaeoclimate, Palaeoecology*, **114**, 131–66.

Goovaerts, P., 1994, Comparison of coIK, IK and mIK performances for modeling conditional probabilities of categorical variables, in R. Dimitrakopoulos (ed.), *Geostatistics for the Next Century* (Dordrecht: Kluwer Academic), pp. 18–29.

Goovaerts, P., 1997, *Geostatistics for Natural Resources Evaluation* (New York: Oxford University Press).

Goovaerts, P., 2001, Geostatistical modeling of uncertainty in soil science, *Geoderma*, **103**, 3–26.

Harden, J. W., Taylor, E. M., Reheis, M. C., and McFadden, L. D., 1991, Calcic, gypsic and siliceous soil chronosequences in arid and semiarid environments, in W. D. Nettleton (ed.), *Occurrence, Characteristics, and Genesis of Carbonate, Gypsum, and Silica Accumulations in Soils*, SSSA Special Publication Number 26.

Journel, A. G., 1996, Modelling uncertainty and spatial dependence: stochastic imaging, *International Journal Geographical Information Systems*, **10**, 517–22.

Nettleton, W. D. (ed.), 1991, *Occurrence, Characteristics, and Genesis of Carbonate, Gypsum, and Silica Accumulations in Soils* (Madison, WI: Soil Science Society of America).

Saito, H. and Goovaerts, P., 2000, Geostatistical interpolation of positively skewed and censored data in a dioxin contaminated site, *Environmental Science and Technology*, **34**, 4228–35.

Soil Survey Staff, 1975, *Soil Taxonomy: A Basic System for Making and Interpreting Soil Surveys. USDA-SCS Handbook 436* (Washington, DC: U.S. Government Print Office).

Urushibara-Yoshino, K., 1996, The soils on the calcareous sand dunes southeast of South Australia, *Environmental Geology*, **28**, 154–60.

Verheye, W. H. and Boyadgiev, T. G., 1997, Evaluating the land use potential of gypsiferous soils from field pedogenic characteristics, *Soil Use and Management*, **13**, 97–103.

15

Increasing the Accuracy of Predictions of Monthly Precipitation in Great Britain using Kriging with an External Drift

Christopher D. Lloyd

1 Introduction

Maps of precipitation are used in a wide range of disciplines and in many different applications. Partly as a result of this, researchers have spent much effort to derive and test appropriate techniques for using data from rainfall monitoring networks to predict accurately precipitation at locations for which there are no observations. There is also a need to quantify the uncertainty associated with such predictions. Geostatistics has been shown to be a suitable set of tools both for predicting the spatial distribution of precipitation and for assessing uncertainty in predictions. Uncertainty in predictions of precipitation is a function of several factors. As well as spatial and temporal variation in precipitation, these include measurement error and sampling density. If uncertainty is not quantified effectively then applications in which the resulting predictions are used may be flawed in that the assumed accuracy of the precipitation values may be spurious. Some relevant case studies are listed below.

The focus of this work was twofold. Firstly, the relation between elevation and precipitation was assessed. Secondly, the effect of using this relation to inform geostatistical prediction was examined. Three techniques were used: (i) inverse distance weighting (IDW); (ii) ordinary kriging (OK) and (iii) kriging with an external drift (KED, a form of kriging with a trend model, KT). Of these, IDW and OK make no use

Uncertainty in Remote Sensing and GIS. Edited by G.M. Foody and P.M. Atkinson.

of secondary data. If we have secondary data that are (i) available at the primary data locations as well as at all locations for which predictions are desired and (ii) are linearly related to the primary variable, then these data may be used to inform the predictions using KED. Each prediction using KED is a function of the form of the variogram model, the precipitation data near to the target point and the (modelled) relation between the primary variable and the secondary variable locally.

The general increase in precipitation in Great Britain with elevation is well known, this is due to the fact that hills are barriers to moist airstreams, forcing the airstreams to rise and they act as high-level heat sources on sunny days. The latter causes convective clouds to form over them preferentially, resulting in showery precipitation (Atkinson and Smithson, 1976). The relation between precipitation and elevation locally in Great Britain has been explored in detail by Brunsdon *et al.* (2001).

Several studies have compared different algorithms for deriving predictions of precipitation from point data in conjunction with secondary data (for example, Hevesi *et al.*, 1992a, 1992b; Goovaerts, 2000; Gòmez-Hernàndez *et al.*, 2001; Deraisme *et al.*, 2001). Goovaerts (2000) compared IDW, OK, KED, simple kriging with locally varying means (SKlm), ordinary co-located cokriging (OCK) and regression. The focus in that study was on a region in the south of Portugal, and elevation was used as a secondary variable. Goovaerts reported that the techniques that used elevation outperformed OK where the correlation coefficient was larger than 0.75. Here, the aim was to determine whether the incorporation of elevation could increase the accuracy of precipitation predictions in Great Britain.

Predictions of monthly precipitation for Great Britain were made using the techniques above and they were compared using a visual examination of the maps derived through prediction and a cross-validation procedure. This chapter shows that when elevation was used as a secondary variable the accuracy of predictions of precipitation was increased.

2 Techniques

The techniques have been described in detail elsewhere, and appropriate references are given below. The IDW predictor is a well-known method and it is described by Burrough and McDonnell (1998). The fundamentals of geostatistics are given by Matheron (1971). Several more recent geostatistical textbooks are available (for example, Isaaks and Srivastava, 1989; Cressie, 1991; Goovaerts, 1997; Armstrong, 1998; Chilès and Delfiner, 1999; Webster and Oliver, 2000) that describe the techniques used in this chapter. A brief overview of the most salient aspects is given below.

2.1 Inverse distance weighting

IDW is well known and is defined merely for consistency. The IDW predictor can be given as

$$\hat{z}(\mathbf{x}_0) = \frac{\sum_{\alpha=1}^{n} z(\mathbf{x}_\alpha) \cdot d_{\alpha 0}^{-2}}{\sum_{\alpha=1}^{n} d_{\alpha 0}^{-2}} \tag{1}$$

where the prediction is made to the location $\mathbf{x}_0$ with the α observations, $z(\mathbf{x}_\alpha)$, and d is the distance by which the location $\mathbf{x}_0$ and each observation are separated. In this chapter, the exponent was set to 2.

2.2 The variogram

The central tool in geostatistical analysis is the variogram. The variogram describes the degree and form of spatial dependence, which is a measure of similarity between values separated by a given distance (and direction if anisotropic). The experimental variogram can be estimated for the $p(\mathbf{h})$ paired observations, $z(\mathbf{x}_\alpha)$, $z(\mathbf{x}_\alpha + \mathbf{h})$, $\alpha = 1, 2, \ldots p(\mathbf{h})$

$$\hat{\gamma}(\mathbf{h}) = \frac{1}{2p(\mathbf{h})} \sum_{\alpha=1}^{p(\mathbf{h})} \{z(\mathbf{x}_\alpha) - z(\mathbf{x}_\alpha + \mathbf{h})\}^2 \tag{2}$$

A mathematical model may then be fitted to the experimental variogram. The coefficients of this model are used with the data for kriging. A variogram model must be conditional negative semidefinite (CNSD, Webster and Oliver, 2000). In practice, a model is usually selected from one of a set of what are termed authorized models (see McBratney and Webster, 1986). Other models are described by Chilès and Delfiner (1999).

There are two principal classes of variogram model. Transitive (bounded) models have a sill (finite variance), and indicate a second-order stationary process. Unbounded models do not reach an upper bound, they are intrinsic only (McBratney and Webster, 1986). For a process to be second-order stationary the expected value of the variable should not depend on the location $\mathbf{x}$ and the covariance between locations separated by lag $\mathbf{h}$ should depend only on the lag, and not on $\mathbf{x}$. Intrinsic stationarity is less strict; the expected value of the difference between values at location $\mathbf{x}$ and $\mathbf{x} + \mathbf{h}$ should be zero for all $\mathbf{x}$ and $\mathbf{h}$, and the variance of the increments should be finite (that is, the semivariance should depend only on $\mathbf{h}$) (Myers, 1989).

The model used here was a spherical function defined by

$$\gamma(h) = \begin{cases} c \cdot [1.5\frac{h}{a} - 0.5(\frac{h}{a})^3] & \text{if } h \leq a \\ c & \text{if } h > a \end{cases} \tag{3}$$

where a is the range (the limit of spatial dependence). The sill of the variogram comprises a nugget effect (c_0) and one or more structured components, ($c_1, \ldots, _n$).

2.3 Ordinary kriging

OK has been called 'the anchor algorithm of geostatistics' (Deutsch and Journel, 1998, p. 66). Its predictions are, like those of IDW, weighted averages of the available data. The main benefit of OK is that the data are used to describe the degree and form of spatial dependence whereas for IDW the weights are assigned arbitrarily.

The weights define the Best Linear Unbiased Predictor. Goovaerts (1997) provides a more detailed account of OK and related algorithms.

The OK prediction, $\hat{z}_{\mathrm{OK}}(\mathbf{x}_0)$, is a linear weighted moving average of the available n observations defined as

$$\hat{z}_{\mathrm{OK}}(\mathbf{x}_0) = \sum_{\alpha=1}^{n} \lambda_\alpha^{\mathrm{OK}} z(\mathbf{x}_\alpha) \tag{4}$$

with the constraint that the weights, $\lambda_\alpha^{\mathrm{OK}}$, sum to 1 to ensure an unbiased prediction

$$\sum_{\alpha=1}^{n} \lambda_\alpha^{\mathrm{OK}} = 1 \tag{5}$$

The prediction error must have an expected value of 0

$$E\{\hat{Z}_{\mathrm{OK}}(\mathbf{x}_0) - z(\mathbf{x}_0)\} = 0 \tag{6}$$

The kriging (or prediction) variance, σ^2_{OK}, is expressed as

$$\begin{aligned}\hat{\sigma}^2_{\mathrm{OK}}(\mathbf{x}_0) &= E[\{\hat{Z}_{\mathrm{OK}}(\mathbf{x}_0) - Z(\mathbf{x}_0)\}^2] \\ &= -\gamma(0) - \sum_{\alpha=1}^{n}\sum_{\beta=1}^{n} \lambda_\alpha^{\mathrm{OK}} \lambda_\beta^{\mathrm{OK}} \gamma(\mathbf{x}_\alpha - \mathbf{x}_\beta) + 2\sum_{\alpha=1}^{n} \lambda_\alpha^{\mathrm{OK}} \gamma(\mathbf{x}_\alpha - \mathbf{x}_0)\end{aligned} \tag{7}$$

That is, we seek the values of $\lambda_1, \ldots, \lambda_n$ that minimize this expression with the constraint that the weights sum to 1 (equation (5)).

This minimization is achieved through Lagrange multipliers. The conditions for the minimization are given by the OK system comprising $n + 1$ equations and $n + 1$ unknowns

$$\begin{cases} \sum_{\beta=1}^{n} \lambda_\beta^{\mathrm{OK}} \gamma(\mathbf{x}_\alpha - \mathbf{x}_\beta) + \psi_{\mathrm{OK}} = \gamma(\mathbf{x}_\alpha - \mathbf{x}_0) & \alpha = 1, \ldots, n \\ \sum_{\beta=1}^{n} \lambda_\beta^{\mathrm{OK}} = 1 \end{cases} \tag{8}$$

where ψ_{OK} is a Lagrange multiplier.

Knowing ψ_{OK}, the prediction variance of OK can be given as

$$\hat{\sigma}^2_{\mathrm{OK}} = \psi_{\mathrm{OK}} - \gamma(0) + \sum_{\alpha=1}^{n} \lambda_\alpha^{\mathrm{OK}} \gamma(\mathbf{x}_\alpha - \mathbf{x}_0) \tag{9}$$

With OK, the mean is predicted within a moving window neighbourhood. If the mean is unlikely to be constant within the window then a generalized form of OK, KT, may be adopted. The following section discusses KED, which is a particular case of KT.

2.4 Kriging with an external drift

As noted above, KED exploits relations between primary and secondary variables. It is essential that secondary data are available at the locations of the primary data as well as at all locations for which predictions are required and that they are linearly related to the primary variable. The KED predictions are a function of (i) the form of the variogram model, (ii) the neighbouring precipitation data, and (iii) the (modelled) relationship between the primary variable (precipitation) and the secondary variable (elevation) locally. In KED, the secondary data act as a shape function. The function describes the average shape of the primary variable (Wackernagel, 1998). The local mean of the primary variable is derived using the secondary information and simple kriging (SK) is carried out on the residuals from the local mean. Wackernagel (1998) notes that cross-validation with KED can be a useful means to assess the influence of the values of the drift function on the KED predictor within a particular neighbourhood.

The KED prediction is given by

$$\hat{z}_{\mathrm{KED}}(\mathbf{x}_0) = \sum_{\alpha=1}^{n} \lambda_{\alpha}^{\mathrm{KED}} z(\mathbf{x}_{\alpha}) \tag{10}$$

The weights are determined through the KED system of $n + 2$ linear equations (Goovaerts 1997; Deutsch and Journel, 1998)

$$\begin{cases} \sum_{\beta=1}^{n} \lambda_{\beta}^{\mathrm{KED}} \gamma(\mathbf{x}_{\alpha} - \mathbf{x}_{\beta}) + \psi_0^{\mathrm{KED}}(\mathbf{x}) + \psi_1^{\mathrm{KED}}(\mathbf{x}) y(\mathbf{x}_{\alpha}) = \gamma(\mathbf{x}_{\alpha} - \mathbf{x}_0) & \alpha = 1, \ldots, n \\ \sum_{\beta=1}^{n} \lambda_{\beta}^{\mathrm{KED}}(\mathbf{x}) = 1 & \\ \sum_{\beta=1}^{n} \lambda_{\beta}^{\mathrm{KED}}(\mathbf{x}) y(\mathbf{x}_{\beta}) = y(\mathbf{x}) & \end{cases} \tag{11}$$

where $y(\mathbf{x})$ are the secondary (elevation) data.

A major problem with KT and KED is that the underlying (trend-free) variogram is assumed known. A potential solution is to infer the trend-free variogram from paired data that are largely unaffected by any trend (Goovaerts, 1997; Wackernagel, 1998). Hudson and Wackernagel (1994), in an application concerned with mapping mean monthly temperature in Scotland, achieved this by estimating directional variograms and retained the variogram for the direction that showed minimal evidence of trend. Assuming that the trend-free variogram was isotropic the variogram for the direction selected was used for kriging.

3 Study Area and Data

The data used in the analysis were ground precipitation data measured across Great Britain under the direction of the UK Meteorological Office as part of the national rain gauge network. The data were obtained from the British Atmospheric Data Centre (BADC) web site (http://www.badc.rl.ac.uk). Daily and monthly data were obtained and combined into a single monthly data set. Only data for complete months were used. Summary statistics for the data for the selected months of January 1999 and July 1999 are given in Table 15.1. These months were selected to enable assessment of the performance of different prediction techniques for different

Table 15.1 Summary statistics for the January 1999 and July 1999 monthly precipitation data

Month in 1999	Number	Mean (mm)	Standard deviation (mm)	Skewness	Minimum (mm)	Maximum (mm)
January	3099	162.247	104.952	1.650	23.0	793.50
July	3037	38.727	37.157	2.269	0.0	319.00

seasons. The locations of the observations are shown in Figure 15.1 (January) and Figure 15.2 (July).

The relevant section of the global 30 arc-second (GTOPO 30) Digital Elevation Model (DEM) (http://edcdaac.usgs.gov/gtopo30/gtopo30.html) was used as the secondary data for KED. After conversion, using a nearest-neighbour algorithm, from geographic co-ordinates to British National Grid the spatial resolution of the DEM (Figure 15.3) was 661.1 m.

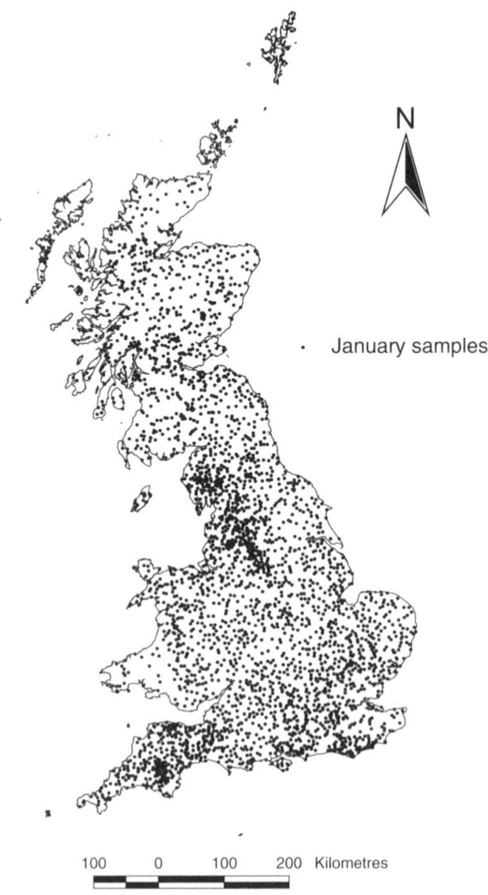

Figure 15.1 Location of measurements of January 1999 precipitation

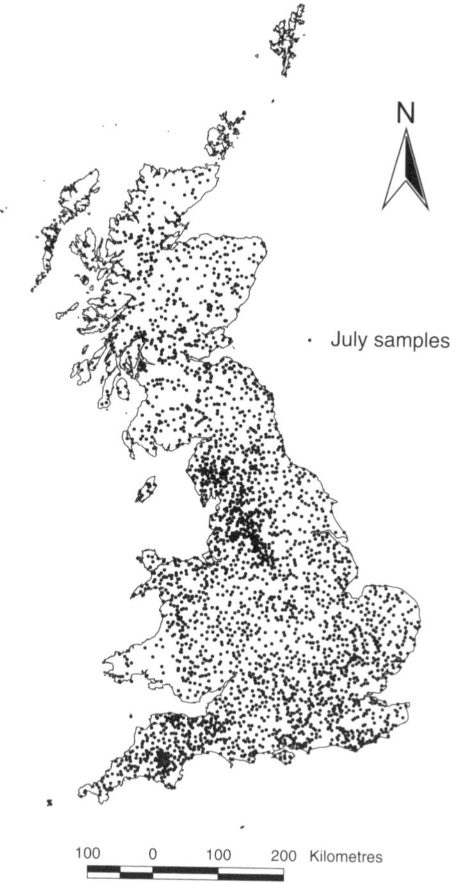

Figure 15.2 Location of measurements of July 1999 precipitation

4 Analysis

The results obtained using IDW, OK and KED were compared visually with maps (derived using a neighbourhood of 16 observations) and by cross-validation. With cross-validation, each observation is removed temporarily from the data set and predicted using the remaining data. The variograms were estimated and models were fitted using the Gstat package (Pebesma and Wesseling, 1998). All kriging was carried out using code provided in the Geostatistical Software Library (GSLIB, Deutsch and Journel, 1998).

Plots of precipitation against elevation are given in Figure 15.4 (January) and Figure 15.5 (July). The coefficient of determination, r^2, for the January data (0.278) was larger than that for the July data (0.111). This suggested that KED would be more useful for predicting using the January data than the July data. Neither of the plots nor values of r^2 suggested a clear relation between precipitation and elevation across Great Britain as a whole. However, since KED uses the local correlation

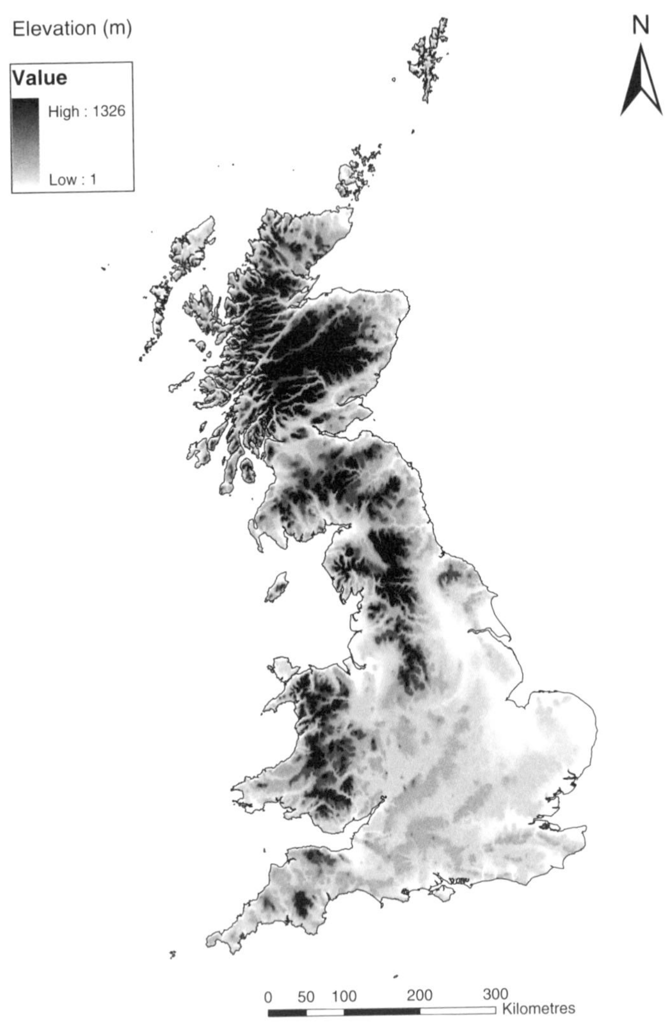

Figure 15.3 GTOPO 30 DEM of Great Britain. The elevation scale is in metres

coefficient it is possible that this is larger. Indeed, previous work has shown clearly that the relation between precipitation and elevation varies across Great Britain (Brunsdon *et al.*, 2001).

Omnidirectional variograms were estimated from the January (Figure 15.6) and the July data (Figure 15.7). There is evidence of periodicity in both of the variograms. Both variograms were fitted with a nugget effect and two spherical structures using weighted least squares fitting in Gstat. The models fitted to both the January and July variograms clearly reached a sill. The coefficients of the models fitted are given on the figures and they were used for OK.

Directional variograms were also estimated to assess the degree to which the variogram was affected by trend in different directions. There is a large scale trend

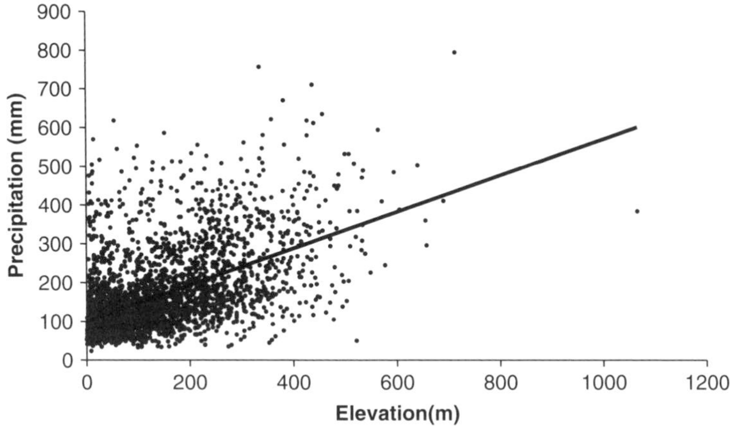

Figure 15.4 *Monthly rainfall against elevation for January 1999, $r^2 = 0.278$*

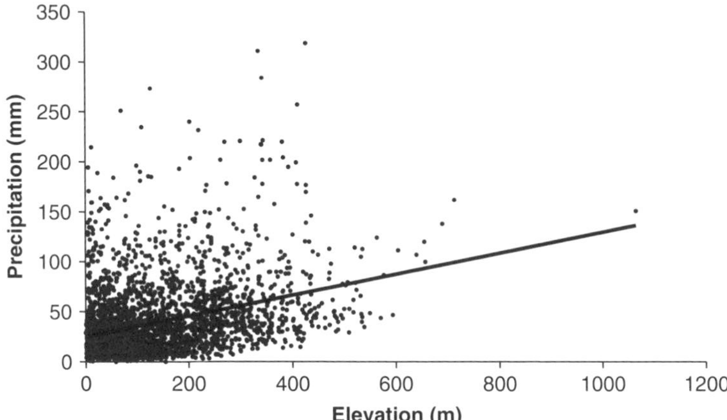

Figure 15.5 *Monthly rainfall against elevation for July 1999, $r^2 = 0.111$*

Table 15.2 *Precipitation amount against x and y: values of R^2 for polynomials of order 1, 2 and 3*

Month	Order 1	Order 2	Order 3
January	0.377	0.437	0.460
July	0.433	0.553	0.567

in precipitation in Great Britain for both January and July and so variation in the variogram for different directions was expected. First-, second- and third-order polynomials were fitted to the data for both months. The R^2 values for each model are given in Table 15.2 and they indicate that quite a large proportion of the variance in precipitation is accounted for by the global polynomials, although this is more clearly the case for July than for January.

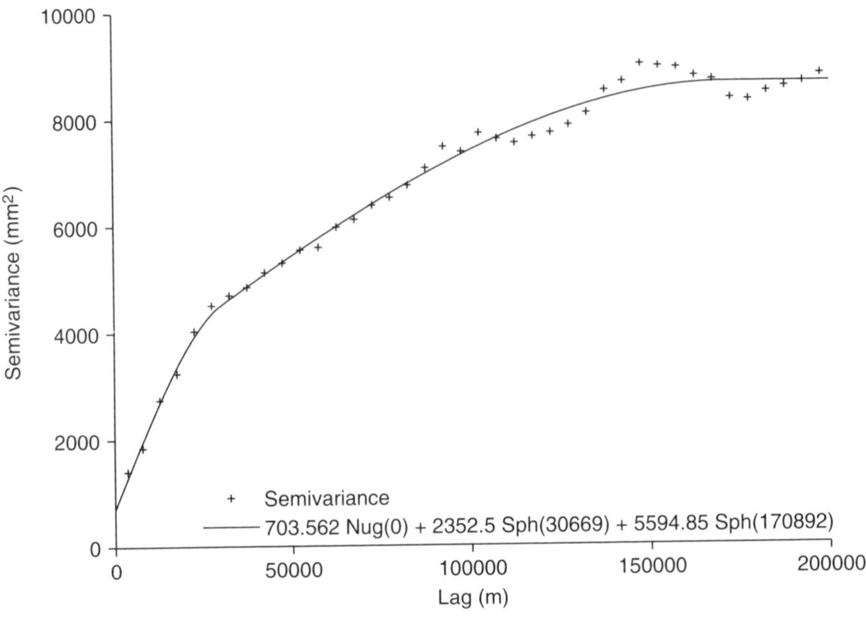

Figure 15.6 *Experimental variogram for January 1999 precipitation, with fitted model*

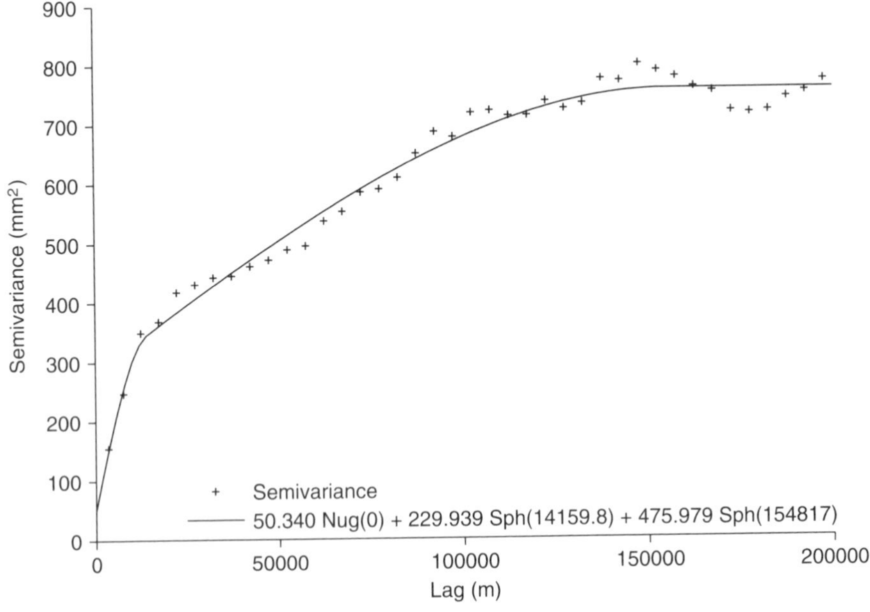

Figure 15.7 *Experimental variogram for July 1999 precipitation, with fitted model*

Directional variograms were estimated for January (Figure 15.8) and July (Figure 15.9). They were estimated at 22.5 decimal degree intervals within a tolerance of 22.5 decimal degrees. Directions in all cases refer to degrees clockwise from north. As expected following the trend surface analysis, there was marked change in the form of the variogram in different directions for both months. There was evidence that the variogram in some directions was more affected by trend than was the variogram in other directions. This was apparent for variograms of both months.

The variogram for 0 degrees for the January data was accepted as being less affected by trend than the variogram at 90 degrees and it was assumed that this variogram was suitable for prediction in all directions. That is, the underlying (trend-free) variogram was considered to be omnidirectional. A model was fitted to the variogram at 0 degrees (Figure 15.10) and it was used for predicting precipitation with KED. For the July data, the variogram at 45 degrees was accepted as the variogram least affected by trend and a model was fitted to it (Figure 15.11) and used in KED.

5 Results and Discussion

Initially, IDW, OK and KED maps were compared to assess the effect on predictions of using secondary data in KED. The IDW and OK maps (Figures 15.12 and 15.13 respectively) were more smooth than that for KED (Figure 15.14) which shows

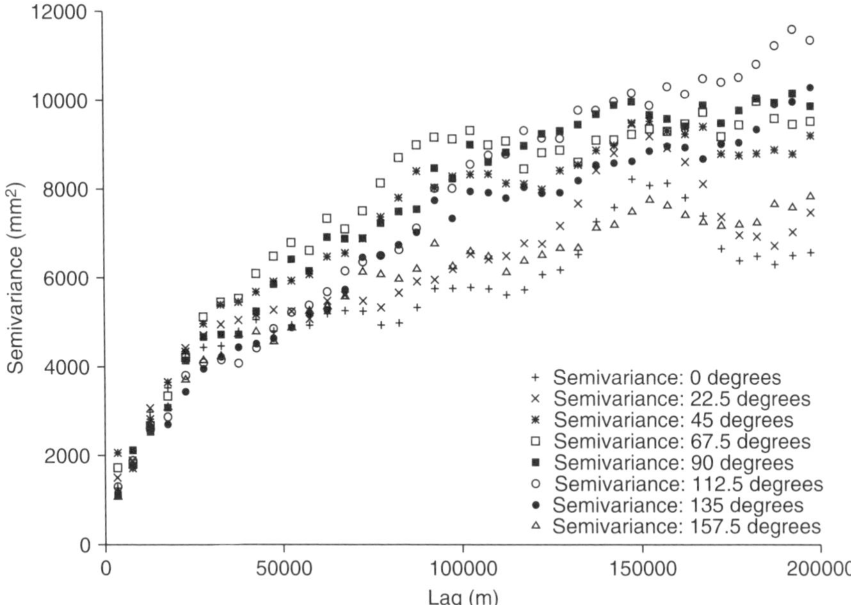

Figure 15.8 *Experimental variogram for January 1999 precipitation for eight directions. Directions are shown as decimal degrees clockwise from north (22.5 decimal degree tolerance)*

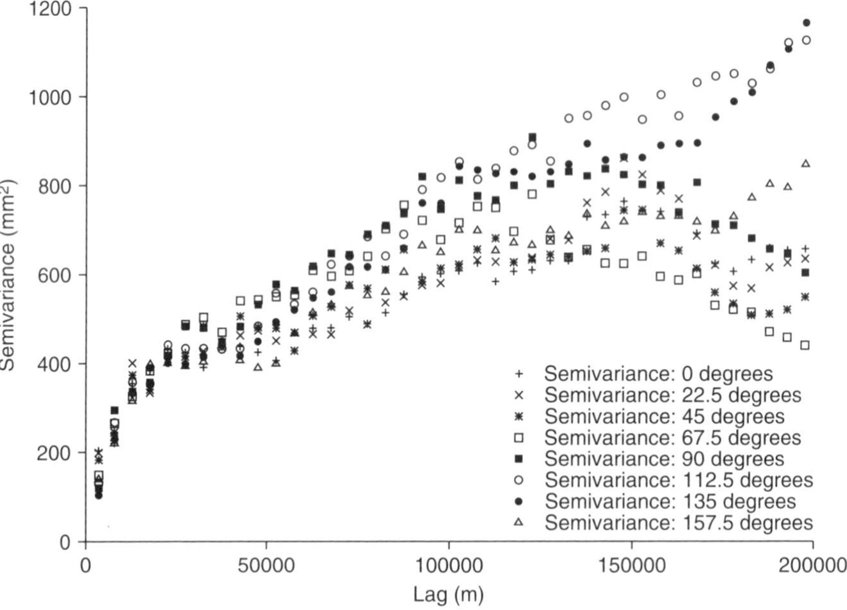

Figure 15.9 Experimental variogram for July 1999 precipitation for eight directions. Directions are shown as decimal degrees clockwise from north (22.5 decimal degree tolerance)

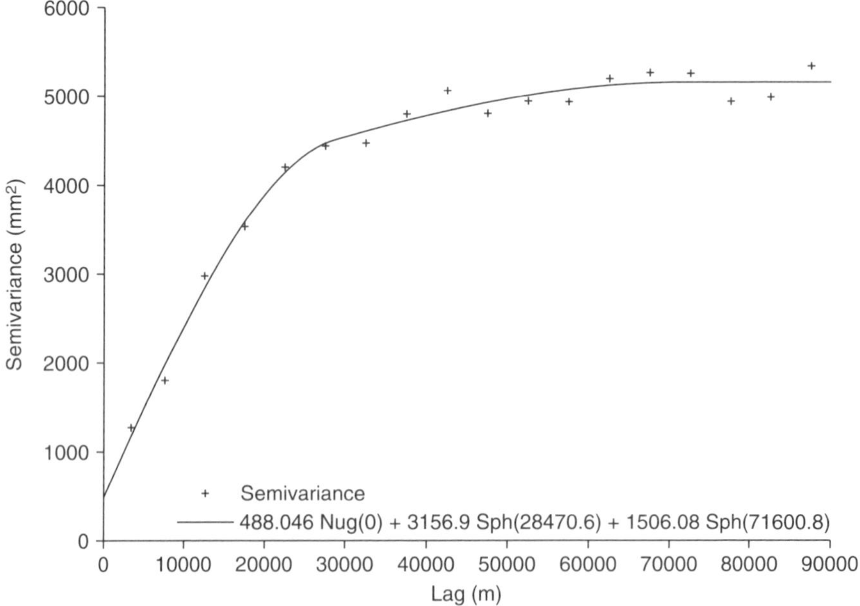

Figure 15.10 Experimental variogram for January 1999 precipitation for 0 degrees clockwise from north (22.5 degree tolerance)

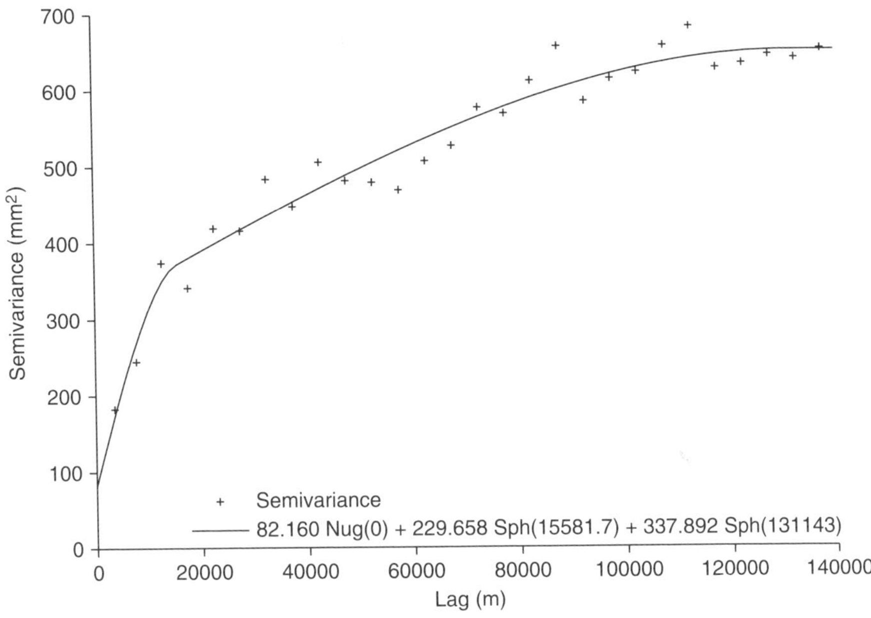

Figure 15.11 *Experimental variogram for July 1999 precipitation for 45 degrees clockwise from north (22.5 degree tolerance)*

clearly the form of the terrain (particularly in Scotland). A small number of very large precipitation values were predicted by KED although the distribution of precipitation predicted by OK is similar to that for KED. As Wackernagel (1998) has shown, KED can extrapolate outside the range of the data where values of the exhaustively sampled secondary variable (the DEM) are outside the range of the values of the secondary variable at locations where the primary variable has also been sampled. The map of OK standard error for January is shown in Figure 15.15 and that for KED in Figure 15.16. Again, the form of the terrain is evident in the KED map. The largest standard errors for KED are larger than those of OK with the largest values being in central Scotland.

The IDW and OK maps for July (Figures 15.17 and 15.18), as for January, contrast clearly with the KED map for the same month (Figure 15.19). As for the KED map of January precipitation, there are a small number of very large values in the KED map of July precipitation. The OK standard error map for July is given in Figure 15.20. The minimum OK standard error is smaller than the minimum KED standard error for July (Figure 15.21), and the largest KED standard error is more than the maximum OK standard error.

The results of IDW, OK and KED were also compared by cross-validation. Summary statistics for the cross-validation errors using the three techniques are given in Table 15.3 (January) and Table 15.4 (July). Cross-validation was done for neighbourhoods of 8, 16, 32, 64 and 128 observations to assess the effect on predictions of changing the number of data used in prediction.

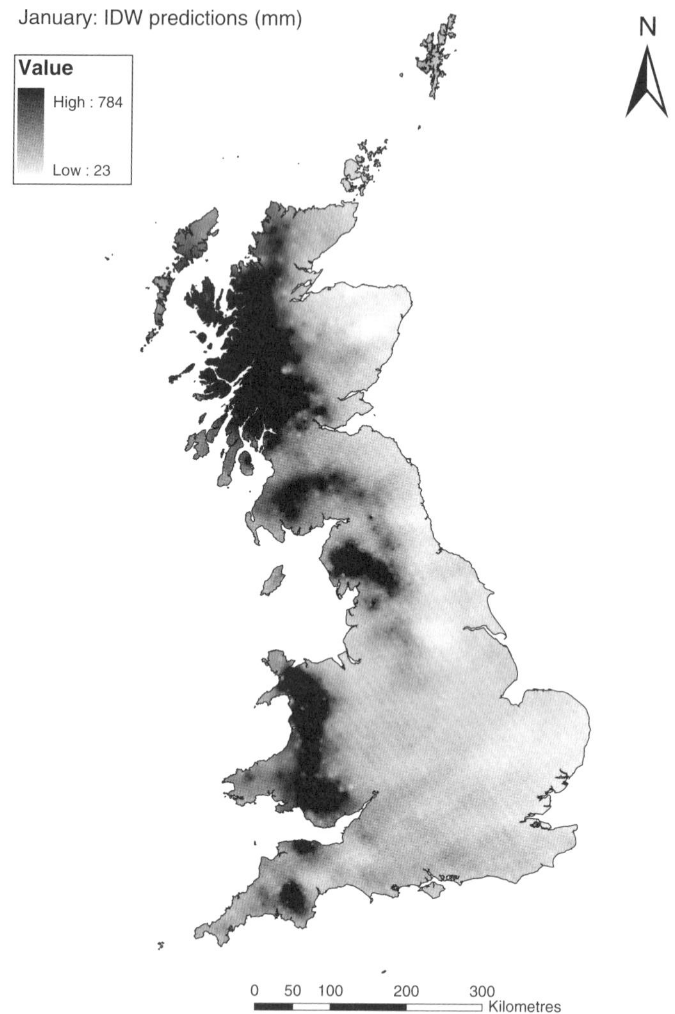

Figure 15.12 Map of precipitation derived using IDW, January 1999

The root mean squared error (RMSE) was used as a guide to the accuracy of the predictions. For January, the smallest RMSE for IDW was 32.811 mm for 18 observations, while for OK it was 30.606 mm for 16 observations and for KED it was 31.329 mm for 128 observations. When the single largest KED cross-validation error was removed the RMSE for KED was 30.205 mm for 16 observations. This was observed at the data location with the largest elevation in the data set (1065 m, which shows as an obvious outlier on the plot in Figure 15.4). This elevation value was much larger than those around it (the next largest elevation value in the data set was 713 m). For IDW, the RMSE values become larger as the number of observations

Table 15.3 Summary statistics for cross-validation errors for January 1999 monthly precipitation data. The summary statistics for KED are also provided after removal of the single largest cross-validation error

Method	N. data	Mean error (mm)	Std. dev. (mm)	Max. neg. error (mm)	Max. pos. error (mm)	RMSE (mm)
IDW	8	2.784	32.692	−313.321	322.311	32.811
IDW	16	3.543	33.413	−288.263	306.698	33.600
IDW	32	4.350	34.612	−285.741	295.895	34.884
IDW	64	4.996	36.206	−299.417	286.880	36.549
IDW	128	5.432	37.814	−309.258	279.646	38.202
OK	8	0.935	30.724	−301.628	223.284	30.738
OK	16	0.411	30.603	−294.378	220.008	30.606
OK	32	0.348	30.620	−293.927	220.715	30.622
OK	64	0.517	30.635	−292.867	218.895	30.639
OK	128	0.393	30.713	−291.482	218.917	30.716
KED	8	0.736	33.236	−308.454	607.645	33.244
KED	16	0.707	32.007	−316.758	591.510	32.015
KED	32	0.576	31.843	−307.996	540.131	31.848
KED	64	0.729	31.580	−312.527	478.712	31.588
KED	128	0.584	31.323	−314.552	448.059	31.329
KED	8	0.540	31.407	−308.454	237.400	31.406
KED	16	0.516	30.205	−316.758	223.490	30.205
KED	32	0.402	30.341	−307.996	215.850	30.339
KED	64	0.575	30.400	−312.527	216.200	30.400
KED	128	0.440	30.284	−314.552	216.480	30.282

Table 15.4 Summary statistics for cross-validation errors for July 1999 monthly precipitation data

Method	N. data	Mean error (mm)	Std. Dev. (mm)	Max. neg. error (mm)	Max. pos. error (mm)	RMSE (mm)
IDW	8	0.708	14.508	−169.355	109.295	14.525
IDW	16	0.863	14.479	−167.763	98.807	14.505
IDW	32	1.033	14.574	−164.748	91.857	14.611
IDW	64	1.099	14.810	−165.834	87.903	14.851
IDW	128	1.098	15.082	−168.942	86.382	15.122
OK	8	0.315	14.013	−173.714	91.643	14.017
OK	16	0.295	13.878	−168.544	92.235	13.881
OK	32	0.260	13.798	−169.548	93.489	13.801
OK	64	0.212	13.804	−169.518	93.320	13.805
OK	128	0.193	13.780	−169.059	93.964	13.781
KED	8	0.116	14.014	−162.579	88.635	14.015
KED	16	0.119	13.527	−138.897	86.234	13.528
KED	32	0.143	13.014	−123.399	85.681	13.014
KED	64	0.110	12.973	−133.962	92.204	12.973
KED	128	0.110	12.862	−133.786	91.139	12.863

used for prediction increase. For OK and KED the RMSE values differ for the number of data used, but not in a systematic way.

Bias in predictions is also of interest, and the mean error of cross-validation predictions was also explored. The IDW mean error closest to zero was 2.784 mm

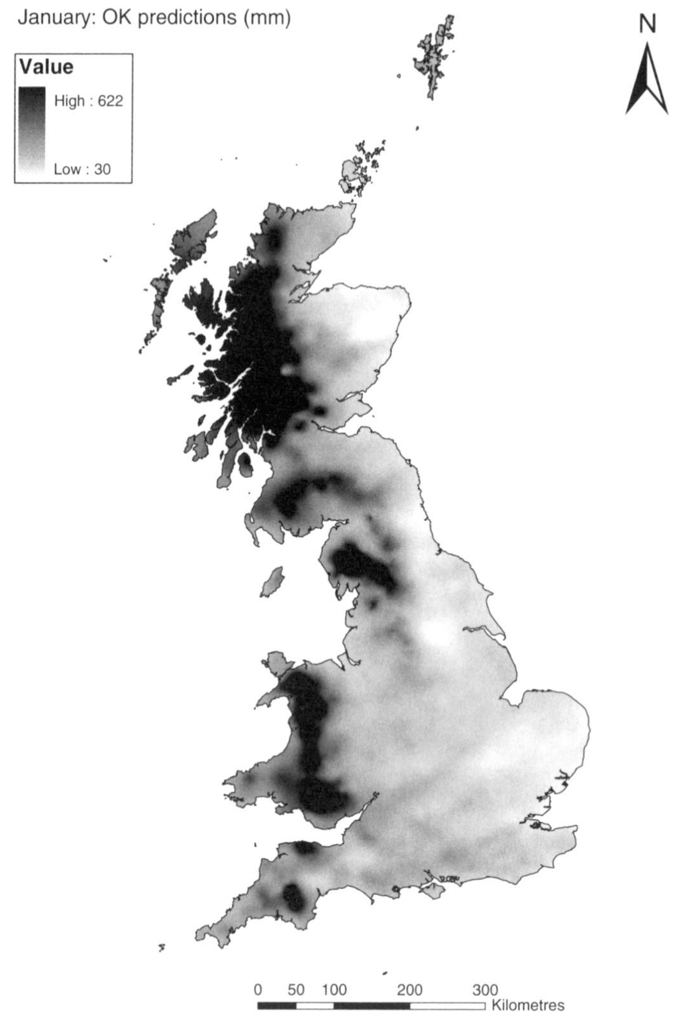

Figure 15.13 Map of precipitation derived using OK, January 1999

for 8 observations, while for OK it was 0.348 mm for 32 observations and for KED it was 0.576 mm for 32 observations, reducing to 0.402 mm when the largest single KED cross-validation error was removed.

There were larger differences in cross-validation errors for the July data than for the January data. The smallest RMSE for IDW was 14.505 mm for 16 observations, for OK it was 13.781 mm for 128 observations and for KED it was 12.863 mm for 128 observations. Therefore, the smallest KED RMSE was about 6.5% smaller than the equivalent for OK. In this case, the OK and KED RMS errors tend to decrease as the number of observations used in prediction increase.

The IDW mean error closest to zero was 0.708 mm for 8 observations, for OK it was 0.193 mm for 128 observations and for KED it was 0.110 mm for 64 and 128 observations. For these data, there appear to be benefits in using KED.

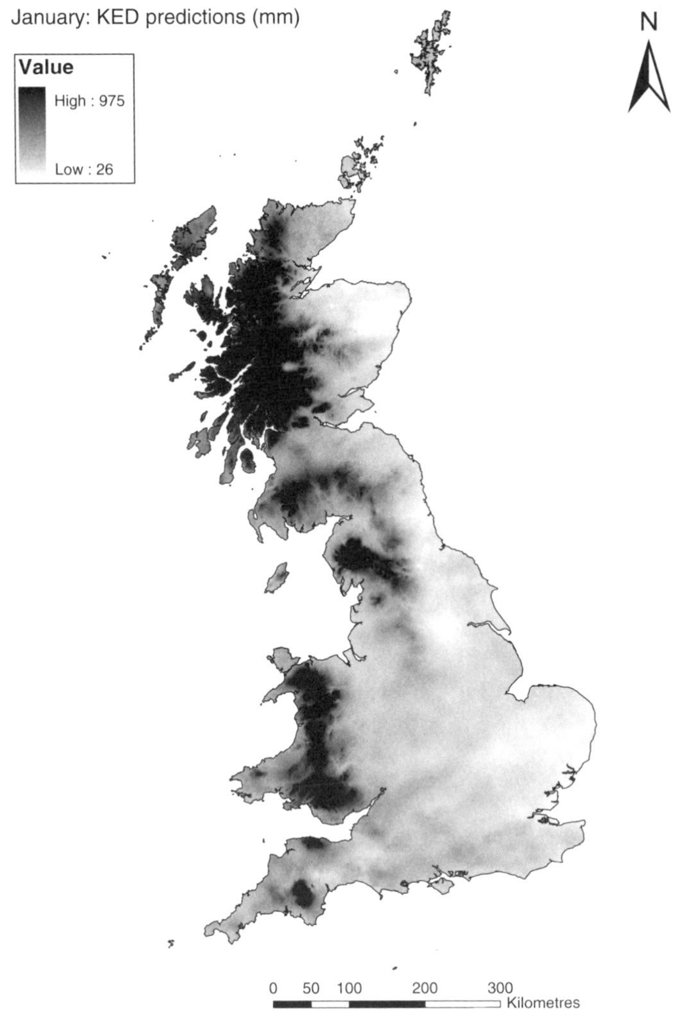

Figure 15.14 Map of precipitation derived using KED, January 1999

6 Conclusions

The results of KED suggest that it provides more accurate predictions than IDW and OK. The degree to which KED gives more accurate predictions than IDW or OK is a function of the relation between precipitation and elevation locally at a given time. Here, the benefits of using OK or KED over IDW were clear for both January and July. However, the benefits of using KED compared with OK were more obvious for July than for January. The correlation between precipitation and elevation was greater for January than for July, but since KED models this relationship locally the global relationship may not be an accurate guide to the local accuracy of KED.

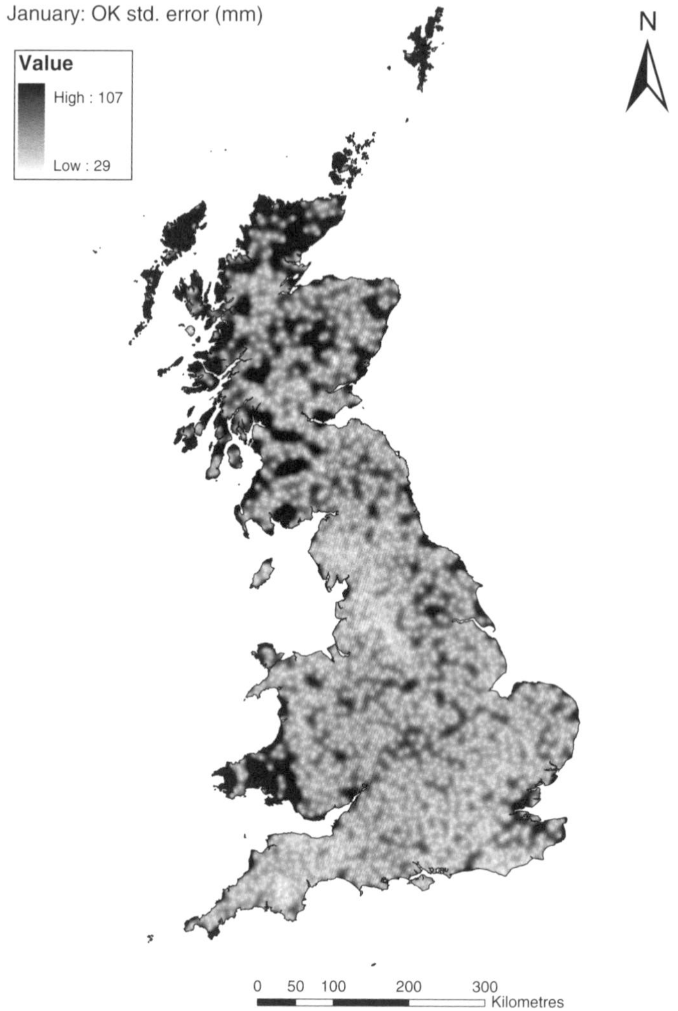

Figure 15.15 Map of OK standard error, January 1999

To understand the results further it will be necessary to examine the relations between the accuracy of predictions and the strength of the relationship between precipitation and elevation locally. Brunsdon *et al.* (2001) noted that elevation and precipitation were more clearly related in Scotland than, for example, in the south of England. KED might be expected, therefore, to provide greater benefits in some areas than in others. Additionally, there may be some benefit in stratifying the data and estimating variograms for these subsets.

The accuracy of predictions of precipitation varies markedly with season and there are many reasons why this may be the case. For example, as was shown in this chapter, the spatial variation of precipitation varies from month to month; some months are more 'reliably wet' than others (Atkinson and Smithson, 1976).

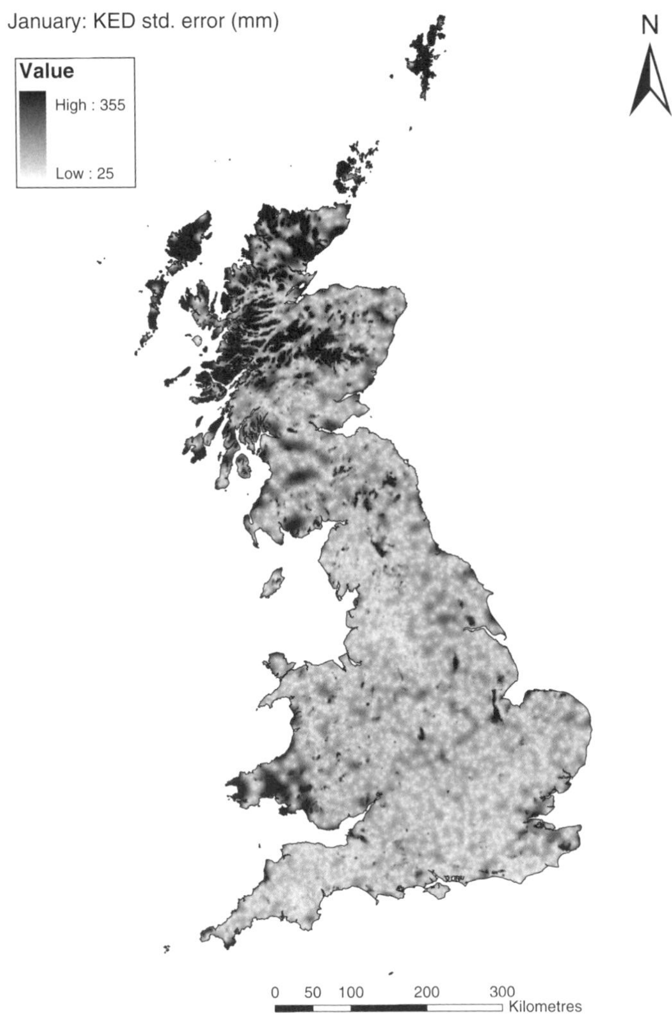

Figure 15.16 Map of KED standard error, January 1999

In months where large areas have similar patterns of precipitation, predictions are likely to be more accurate than they are in months where there is much variation in precipitation over quite small areas. In addition, the observed relations between elevation and precipitation vary with season, so the use of elevation as a covariate in prediction will have greater benefits in some seasons than in others.

Future work will explore changes in the accuracy of predictions for each month over a year. In addition, other techniques will be used and other secondary variables, such as precipitation predicted by radar imagery, will be introduced. Cross-validation is a limited way of comparing the results of different methods, therefore external validation through acquisition of additional data or jackknifing (prediction of values in one subset of the data using the remaining data) might be worthwhile.

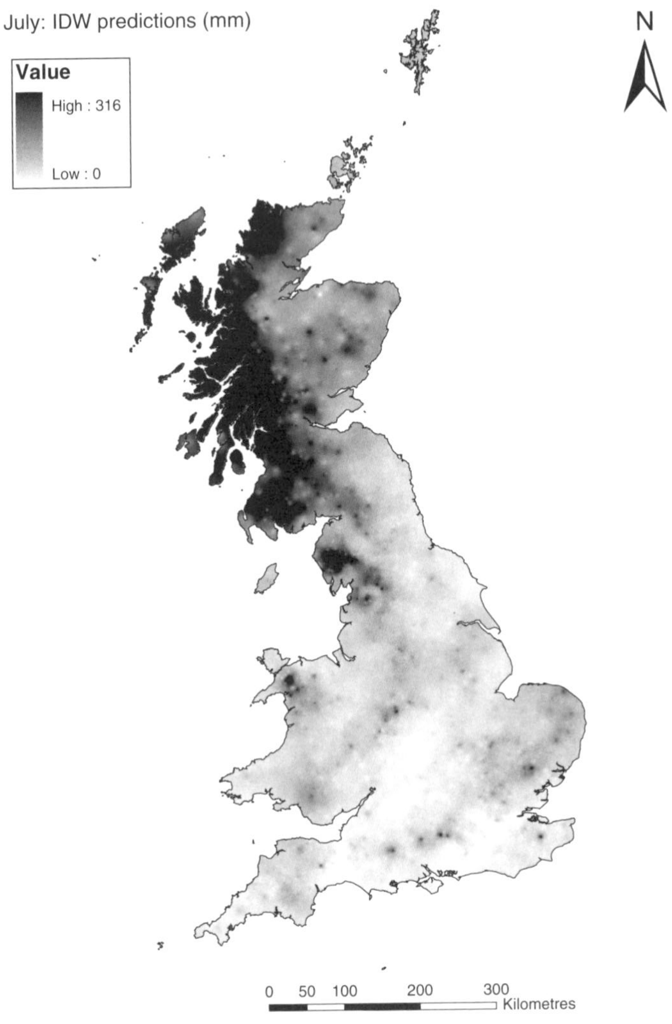

Figure 15.17 Map of precipitation derived using IDW, July 1999

More specifically, exploration of other ways of estimating the trend-free variogram for KED (for example, by estimating the variogram of residuals from a polynomial trend) may provide benefits.

Acknowledgements

The British Atmospheric Data Centre is thanked for making the precipitation data available. The comments on an earlier draft of this chapter by the three referees improved the chapter greatly and they are all acknowledged gratefully.

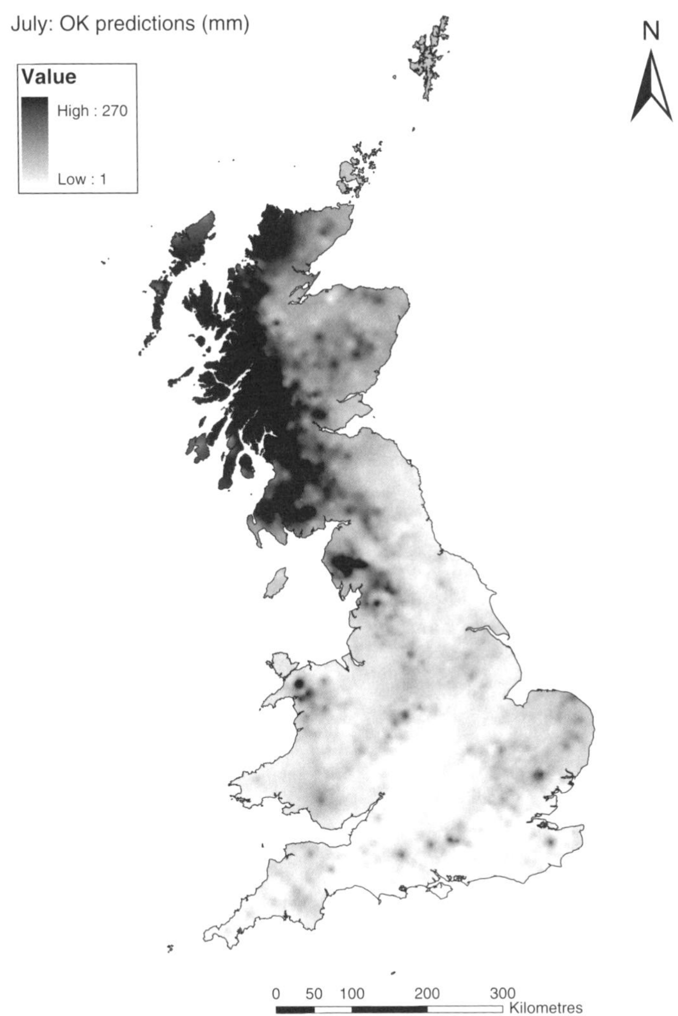

Figure 15.18 *Map of precipitation derived using OK, July 1999*

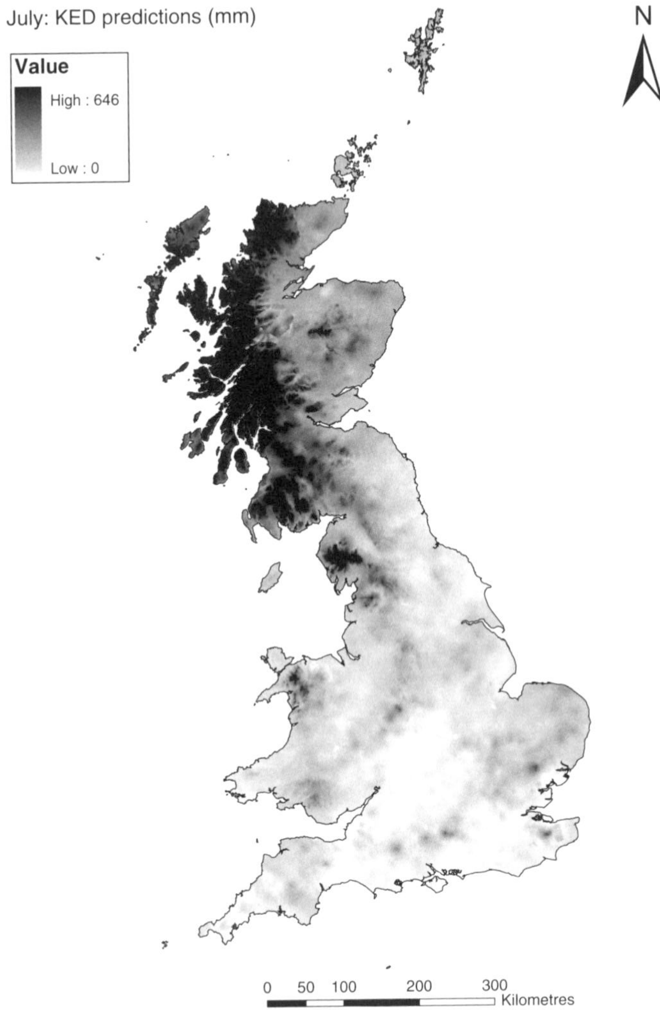

Figure 15.19 *Map of precipitation derived using KED, July 1999*

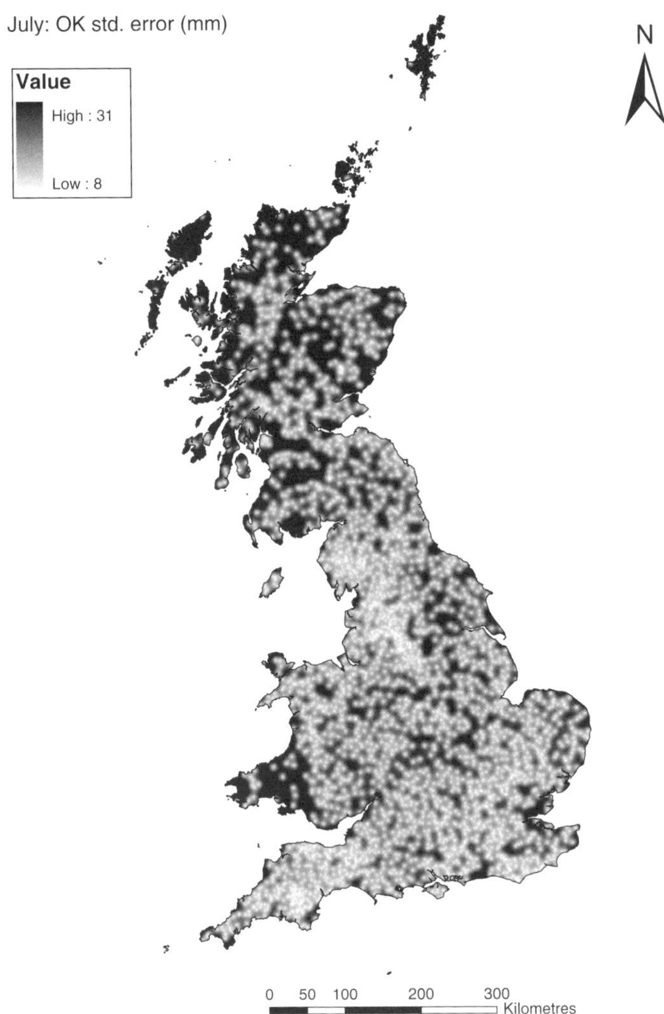

Figure 15.20 *Map of OK standard error, July 1999*

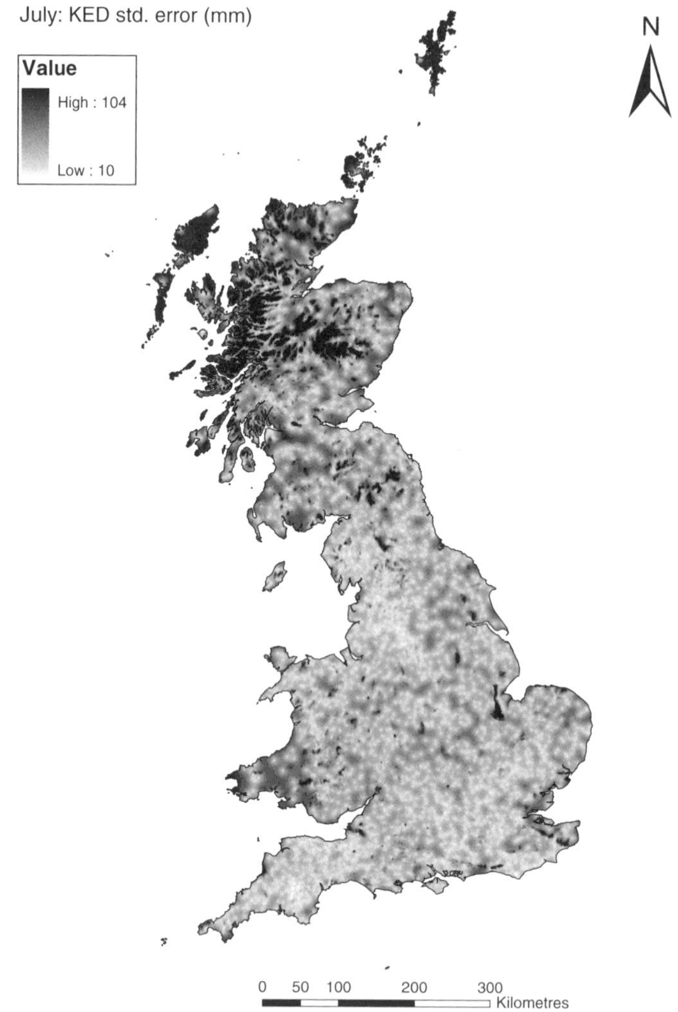

Figure 15.21 *Map of KED standard error, July 1999*

References

Armstrong, M., 1998, *Basic Linear Geostatistics* (Berlin: Springer).

Atkinson, B. W. and Smithson, P. A., 1976, Precipitation, in T. J. Chandler and S. Gregory (eds), *The Climate of the British Isles* (London: Longman), pp. 129–82.

Brunsdon, C., McClatchey, J. and Unwin, D. J., 2001, Spatial variations in the average rainfall–altitude relationship in Great Britain: an approach using geographically weighted regression, *International Journal of Climatology*, **21**, 455–66.

Burrough, P. A. and McDonnell, R. A., 1998, *Principles of Geographical Information Systems* (Oxford: Oxford University Press).

Chilès, J. P. and Delfiner, P., 1999, *Geostatistics: Modeling Uncertainty* (New York: Wiley).

Cressie, N. A. C., 1991, *Statistics for Spatial Data* (New York: Wiley).

Deraisme, J., Humbert, J., Drogue, G. and Freslon, N., 2001, Geostatistical interpolation of rainfall in mountainous areas, in P. Monestiez, D. Allard and R. Froidevaux (eds) *GeoENV III: Geostatistics for Environmental Applications* (Dordrecht: Kluwer Academic), pp. 57–66.

Deutsch, C. V. and Journel, A. G., 1998, *GSLIB: Geostatistical Software and User's Guide*, second edition (New York: Oxford University Press).

Gòmez-Hernàndez, J., Cassiraga, E., Guardiola-Albert, C. and Álvarez Rodríguez, J., 2001, Incorporating information from a digital elevation model for improving the areal estimation of rainfall, in P. Monestiez, D. Allard and R. Froidevaux (eds) *GeoENV III: Geostatistics for Environmental Applications* (Dordrecht: Kluwer Academic), pp. 67–78.

Goovaerts, P., 1997, *Geostatistics for Natural Resources Evaluation* (New York: Oxford University Press).

Goovaerts, P., 2000, Geostatistical approaches for incorporating elevation into the spatial interpolation of rainfall, *Journal of Hydrology*, **228**, 113–29.

Hevesi, J. A., Istok, J. D. and Flint, A. L., 1992a, Precipitation estimation in mountainous terrain using multivariate geostatistics. Part I: structural analysis, *Journal of Applied Meteorology*, **31**, 661–76.

Hevesi, J. A., Flint, A. L. and Istok, J. D., 1992b, Precipitation estimation in mountainous terrain using multivariate geostatistics. Part II: isohyetal maps, *Journal of Applied Meteorology*, **31**, 677–88.

Hudson, G. and Wackernagel, H., 1994, Mapping temperature using kriging with external drift: theory and an example from Scotland, *International Journal of Climatology*, **14**, 77–91.

Isaaks, E. H. and Srivastava, R. M., 1989, *An Introduction to Applied Geostatistics* (New York: Oxford University Press).

Matheron, G., 1971, *The Theory of Regionalised Variables and its Applications* (Fontainebleau: Centre de Morphologie Mathématique de Fontainebleau).

McBratney, A. B. and Webster, R., 1986, Choosing functions for semi-variograms of soil properties and fitting them to sampling estimates, *Journal of Soil Science*, **37**, 617–39.

Myers, D. E., 1989, To be or not to be... stationary? That is the question, *Mathematical Geology*, **21**, 347–62.

Pebesma, E. J. and Wesseling, C. G., 1998, Gstat, a program for geostatistical modelling, prediction and simulation, *Computers and Geosciences*, **24**, 17–31.

Wackernagel, H., 1998, *Multivariate Geostatistics. An Introduction with Applications*, second edition (Berlin: Springer).

Webster, R. and Oliver, M. A., 2000, *Geostatistics for Environmental Scientists* (Chichester: Wiley).

16

Conditional Simulation Applied to Uncertainty Assessment in DTMs

J. Sénégas, M. Schmitt and P. Nonin

1 Introduction

The computation of a digital terrain model (DTM) can be achieved using a stereoscopic pair of satellite sensor images. When an object is observed from two different view points, its projections on each image, which form a pair of homologous points, appear shifted and this shift, the disparity, is a known function of the height (see Figure 16.1). If I_l and I_r denote the left and right images of the stereoscopic pair, and d the disparity, we have in absence of radiometric noise the following relation:

$$I_l(\mathbf{x} + d(\mathbf{x})) = I_r(\mathbf{x})$$

Amongst the available techniques (see Dhond and Aggarwal, 1989, for a review), stereoscopy by correlation is a widely used method to compute a dense disparity map. This chapter addresses the problem of assessing the uncertainty linked to the computation of disparity, which is the most important source of error in producing DTMs.

Bayesian inference has been used widely in image analysis (e.g. Besag, 1986; Besag *et al.*, 1995), and efficient algorithms have been proposed to solve the prediction problem, as in Geman and Geman (1984). More recently, this formalism has been applied to the stereo-reconstitution problem (e.g. Szeliski, 1989; Belhumeur, 1996), where not a characteristic of the image itself has to be predicted, but an underlying variable deduced from the stereoscopic pair: the disparity. The main difficulty lies in the nature of the information contained in the initial images. Indeed, the signal is often perturbed by noise, but the correspondence problem is also altered by the presence of repetitive features, the lack of texture and the existence of occlusions

Uncertainty in Remote Sensing and GIS. Edited by G.M. Foody and P.M. Atkinson.
 ISBN: 0–470–84408–6

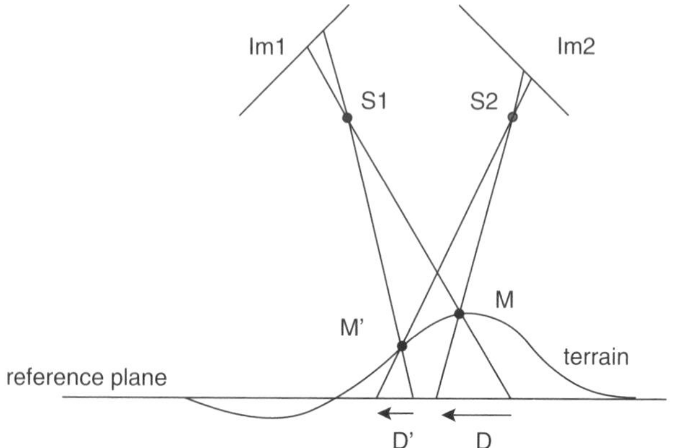

Figure 16.1 When viewed with two different sensors (or the same sensor at two different times), points appear shifted on the image reference plane. This shift, the disparity, is a mathematical function of the sensor geometry and the terrain elevation

or diachronism. In other words, the information is often ambiguous and there is no one-to-one mapping between the stereoscopic pair and the underlying disparity. As a consequence, the errors which occur in the disparity maps computed by maximum likelihood or correlation, which are often auto-correlated and dependent on the initial information, cannot be detected automatically. From a statistical point of view, a solution to this problem can be given in terms of posterior probability, which enables assessment of the uncertainty associated with a given prediction. Typically, one would like to compute the probability of exceeding a given error threshold s:

$$P(Z - \hat{Z} \geq s)$$

where $\hat{Z}$ is a prediction of the true elevation Z. Other useful information is the specification of confidence intervals $[z_{\text{inf}},\ z_{\text{sup}}]$, given a confidence probability α:

$$P(Z \in [z_{\text{inf}},\ z_{\text{sup}}]) \geq \alpha$$

This task can be led under a Bayesian framework, according to which the posterior distribution π is split into a likelihood term L and a prior distribution p:

$$\pi(z|y) \propto L(y|z)p(z) \tag{1}$$

where z denotes the variable of interest (the disparity) and y the observations (the stereoscopic pair). In this perspective, the stereoscopic pair is rather considered as a constraint to be fulfilled than as hard information. This constraint is expressed in the likelihood function, which attributes a cost to a given disparity map. Typically, if Gaussian noise is assumed, this cost is an exponential function of a quadratic form of the intensity difference between the two images. The prior distribution p is a subjective term, which describes the nature of the disparity, and especially its regularity. For example, a Gaussian distribution with a continuous covariance at the origin will refer to a continuous disparity map. Belhumeur (1993) lists several types of prior distributions in association with different features likely to occur in stereovision.

From a practical point of view, a solution is to generate a set of simulations according to the distribution π and to estimate the previous probabilities using a Monte Carlo algorithm, since analytical evaluation of π is not feasible. However, since the observations y are not linearly linked to the variable of interest, the disparity, it is not possible to simulate directly from π. Markov chain Monte Carlo (MCMC) algorithms provide a convenient framework to sample from such conditional distributions. For an introduction to MCMC simulations, with a wide range of applications, the interested reader is referred to Gilks *et al.* (1996); a specific chapter (Green, 1996) is dedicated to image analysis. The principle is to generate a chain of consecutive states which converges to the target distribution. The evolution of the chain is driven by a transition kernel, which describes the transition between two consecutive states.

The need for a robust algorithm which converges rapidly enough has led us to propose a new sampling algorithm: the Multiple Metropolis–Hastings algorithm. It is based on a Gaussian proposal transition kernel and on the construction of a set of possible transitions. We show that with this construction the chain has the appropriate stationary distribution. A comparison with usual Markov chain sampling algorithms, such as Metropolis–Hastings and Langevin diffusions, is presented in a monovariate case. The Multiple Metropolis–Hastings algorithm is shown to perform well in terms of convergence and mixing.

This chapter is organized as follows. In section 2, a rapid introduction to MCMC simulations is presented. The details of the simulation algorithm are described in section 3, and a convergence theorem is stated. Section 4 gives an application taken from the stereo-vision problem, which allows one to compare in a monovariate case the behaviour and the efficiency of this algorithm. Lastly, in section 5, a real case study is presented: our sampling algorithm is applied to the study of a stereoscopic pair of SPOT (Système Pour l'Observation de la Terre) HRV Panchromatic images.

2 Some Reminders on Markov Chains and Sampling Algorithms

2.1 Markov chains

Simulation methods which rely on Markov chain theory are iterative: the principle is to build a succession of states, and once convergence is reached, the consecutive states are assumed to be drawn from the target probability distribution. With these methods, it is possible to sample from general conditional distributions, whereas algorithms such as conditioning by kriging (Chilès and Delfiner, 1999) only apply to the Gaussian case. In this section, we give some reminders on the definition and basic properties of Markov chains defined on countable state spaces. For an extension to general state spaces, the interested reader is referred to Meyn and Tweedie (1993) and Tierney (1996).

A Markov chain $Z = (Z_k)_{k\geq 0}$ is a discrete time stochastic process with the property that the distribution of Z_k given all previous values of the process $Z_0, \ldots, Z_{k-1}$ only depends upon Z_{k-1}. Typically, the Markov chain Z takes values in R^d. However, to illustrate the main ideas, we restrict our attention to discrete state spaces (i.e. Z_k can take only a finite or countable number of different values).

The transitions between two states z and z' are defined by the transition kernel K:

$$P(Z_k = z'|Z_{k-1} = z) = K(z, z') \tag{2}$$

A central point in the study of Markov chains is the behaviour of the chain as k grows, i.e. for large values of k can we consider Z_k to be drawn from a probability distribution π, whatever the starting value z_0? If it is so, we say that the chain has π as stationary distribution. A necessary condition for the chain to have π as stationary distribution is the invariance of π under the transition kernel K:

$$\sum_y \pi(y)K(y, z) = \pi(z) \ \forall z \tag{3}$$

A sufficient condition for the invariance of π under the transition kernel K is the following reversibility condition:

$$\pi(z)K(z, z') = \pi(z')K(z', z) \ \forall z, z' \tag{4}$$

Two notions define the behaviour of the chain: the irreducibility and the aperiodicity. Is is beyond the scope of this chapter to describe in detail these properties, but we will review the ideas behind them. A chain is said to be irreducible if each state can communicate with every other one, i.e. for every z and y, there exists $k > 0$ such that $P(Z_k = y|Z_0 = z) > 0$. An irreducible chain is said to be (strongly) aperiodic if for some state z the probability of remaining in z is strictly positive: $K(z, z) > 0$. This prevents the chain from having a cyclic behaviour.

We finally have the following convergence theorem (Meyn and Tweedie, 1993):

Theorem 1 *Let $Z = (Z_k)_{k \geq 0}$ be a Markov chain with transition kernel K. If Z is irreducible, aperiodic and has π as invariant distribution, then:*

$$P(Z_k = z|Z_0 = z_0) \to_{k \to \infty} \pi(z) \ \forall z \tag{5}$$

for every initial state z_0.

The extensive use of Markov chains in statistical applications relies on the possibility of sampling from the stationary distribution π through Monte Carlo simulations. The principle is to generate a Markov chain with transition kernel K which admits π as invariant distribution. Then, every characteristic of π, writes as follows:

$$\mathrm{E}_\pi(f) = \int_z f(z)\pi(z)dz$$

where f is a π-measurable application, and can be approached by the sum:

$$\bar{f}_m = \frac{1}{m}\sum_{k=1}^{m} f(z_k)$$

where z_k denotes a state of the chain, and m is the length of the chain.

Especially, the probability under π of a particular event can be computed:

$$\pi(A) = \mathrm{E}_{\pi}(1_A) \approx \frac{1}{m}\sum_{k=1}^{m} 1_{z_k \in A}$$

Therefore, the problem of simulating from π amounts to determining a transition kernel K which verifies the invariance relation (3).

In the following, we give two examples of sampling algorithms which use different types of transition kernel with invariant distribution π.

2.2 Two sampling algorithms

The reference sampling algorithm in Bayesian statistics is the Metropolis–Hastings algorithm. It allows to construct from a proposal transition probability q a transition kernel which has π as invariant distribution.

Let us consider a transition kernel with density $q(z, z')$, which verifies the following property: $q(z, z') > 0 \leftrightarrow q(z', z) > 0$, and guarantees the aperiodicity and the irreducibility of the chain. A transition to the state z' is generated from q, and accepted with probability:

$$a(z, z') = \min\left(1, \frac{\pi(z')q(z', z)}{\pi(z)q(z, z')}\right) \tag{6}$$

It can be easily shown that the resulting transition kernel verifies the reversibility condition (4), which is enough to ensure that the Markov chain admits π as stationary distribution. More details on the use of Metropolis–Hastings algorithms can be found in Besag *et al.* (1995) or in Gilks *et al.* (1996).

In a Bayesian setting and in the absence of any information about the form of the posterior distribution π, we can take the proposal transition distribution equal to the prior p, i.e.

$$q(z, z') = p(z')$$

The acceptance probability a becomes then (see Hastings, 1970, for details):

$$a(z, z') = \min\left(1, \frac{L(y|z')}{L(y|z)}\right) \tag{7}$$

In other words, every transition is accepted if the proposed state is more likely than the current state, and if not is rejected with a probability which depends only upon the likelihood function.

An interesting sampling algorithm for differentiable distributions, the Langevin diffusions, has been proposed in Roberts and Tweedie (1996). It relies on the continuous-time stochastic processes, whose evolution is driven by a diffusion equation.

Let us assume that the distribution π is non-zero and differentiable, so that $\nabla \log \pi(z)$ is well defined. We consider the following diffusion equation:

$$dZ_t = dW_t + \frac{1}{2}\nabla \log \pi(Z_t)dt \tag{8}$$

where W_t is a standard Brownian motion. It can be established that, under some regularity conditions on $\nabla \log \pi$, the diffusion $Z = (Z_t)_{t \geq 0}$ has π as stationary distribution (Roberts and Tweedie, 1996).

In practice, only discretized versions of equation (8) can be implemented to generate simulations. However, crude implementations may fail to converge to the target distribution, as shown in Roberts and Tweedie (1996). To avoid this feature, Roberts and Tweedie propose a metropolized version of the Langevin diffusion, which guarantees convergence.

The proposal transition is generated according to equation (8), i.e.

$$q(z,\ z') = \phi\left(\frac{z' - (z + \frac{1}{2}h\nabla \log \pi(z))}{\sqrt{h}}\right) \tag{9}$$

where ϕ is the density of the standard Gaussian distribution and h is the discretization lag. The state z' is then accepted with probability:

$$a(z, z') = \min\left(1,\ \frac{\pi(z')\phi\left(\frac{z - (z' + \frac{1}{2}h\nabla \log \pi(z'))}{\sqrt{h}}\right)}{\pi(z)\phi\left(\frac{z' - (z + \frac{1}{2}h\nabla \log \pi(z))}{\sqrt{h}}\right)}\right) \tag{10}$$

This algorithm can be very useful when the gradient of the target distribution is meaningful, i.e. when the distribution itself is relatively smooth. Otherwise, the drift term $h\nabla \log \pi$ can be misleading.

3 Multiple Metropolis–Hastings

In Metropolis–Hastings, the transitions rely on a proposal transition kernel. Although this algorithm may be applied to generate samples from every probability distribution, the rejection rate may be too large if the proposal distribution q and the target distribution π are not very close. There is therefore a need for algorithms which take into account the form of the target distribution to generate the transitions, but are not too sensitive to local characteristics of π for the sake of robustness. The idea behind the following algorithm is to construct a set of proposal transitions based on a Gaussian distribution and to choose the transition according to the target distribution.

Let ϕ be the density of the Gaussian distribution, with zero mean and covariance C:

$$\phi(z) = \frac{1}{\sqrt{2\pi \det(C)}} \exp\left(-\frac{1}{2} z^{\top} C^{-1} z\right)$$

We denote r the ratio between the target distribution π and the Gaussian density ϕ:

$$r(z) = \frac{\pi(z)}{\phi(z)} \tag{11}$$

Let us consider the combinations of the form:

$$Z'(\theta) = Z \cos \theta + W \sin \theta \tag{12}$$

where Z and W are independent Gaussian random functions with covariance C and zero-mean. Since $Z'(\theta)$ is a linear combination of two independent Gaussian functions, it is also a Gaussian function. It is moreover straightforward to show that $Z'(\theta)$ has zero-mean and covariance C (Matheron, 1982). Let us now consider a Markov chain with stationary distribution π. It is clear that for every θ in $[0, 2\pi)$, $z'(\theta)$, as defined in equation (12), represents a possible transition from the current state z. Instead of fixing θ to a given value and to add a Metropolis–Hastings-like rejection step, as in Barone and Frigessi (1990), we can build a set of possible transitions z'_i for different values θ_i and choose z'_i with an appropriate transition probability, to guarantee the invariance condition (3). As in the Metropolis–Hastings algorithm, it is sufficient to choose a transition probability $a(z, z'_i)$ which verifies the reversibility condition (4). Then, we set $z' = z'_i$ with probability $a(z, z'_i)$, i.e. we choose z' amongst the set $(z'_i)_{i=0,\ l}$. An intuitive choice is to take $a(z, z'_i)$ proportional to the ratio $r(z'_i) = \frac{\pi(z'_i)}{\phi(z'_i)}$.

The following theorem defines the sampling algorithm and guarantees the convergence to the target distribution π.

Theorem 2 *Let π be a probability with density $\pi(z) = r(z)\phi(z)$ where ϕ denotes a multi-Gaussian centred distribution, and mean m.*

Let $Z = (Z_k)_{k\geq 0}$ be the Markov chain, whose transitions are constructed according to the following algorithm:

$$w \sim \phi, \tag{13}$$

$$z'_i = m + (z - m)\cos\left(i\frac{2\pi}{l+1}\right) + (w - m)\sin\left(i\frac{2\pi}{l+1}\right), \ \forall i = 0,\ l, \tag{14}$$

$$a(z, z'_i) = \frac{r(z'_i)}{\sum_{j=0}^{l} r(z'_j)}, \ \forall i = 0,\ l, \tag{15}$$

$$z' = z'_i \text{ with probability } a(z, z'_i) \tag{16}$$

for some fixed integer l. Then, Z has π as stationary distribution.

Intuitively, the chain constructed according to this algorithm is irreducible and aperiodic, since all the states can communicate and the probability to remain at the same state (i.e. $a(z, z'_o)$) is non-zero. A proof of this theorem is given at the end of the chapter.

In a Bayesian setting where the prior p is Gaussian, it seems natural to choose $\phi(z) = p(z)$. Hence, the transition probability $a(z, z'_i)$ reduces to $\frac{L(y|z'_i)}{\sum_{j=0}^{l} L(y|z'_j)}$.

Figure 16.2 illustrates the principle of this algorithm. First of all, a new point in the state-space is simulated according to the Gaussian prior distribution. Then, a set of possible transitions is generated. Within this set, the transition probability is proportional to the likelihood function. It is clear from Figure 16.2 that a much wider set of states can be reached; especially, areas of high probability, although both current state and independent simulation may lie in low probability areas.

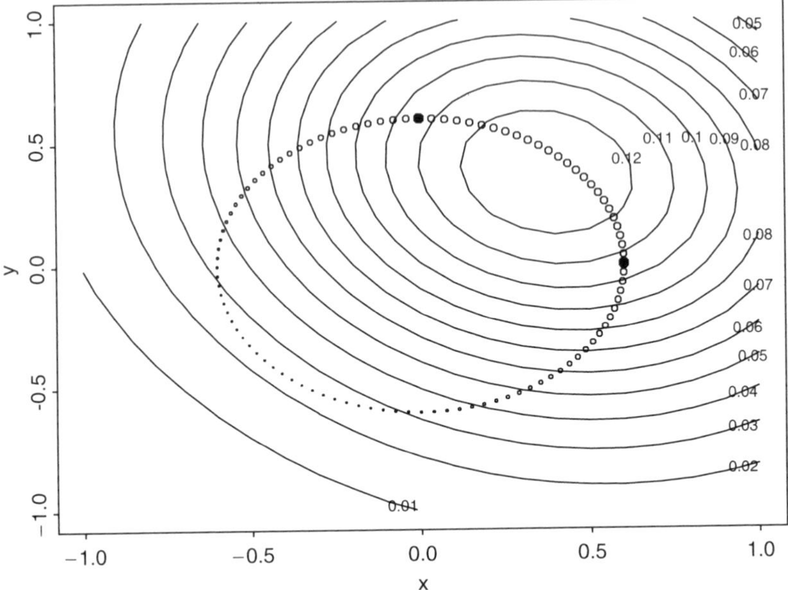

Figure 16.2 Transition set generated from the current point and an independent simulation (black dots) in a two-dimensional state space. The circles along the path are proportional to the transition probability. In the background, the surface corresponding to the target distribution has been plotted

4 Illustration

4.1 Model

To illustrate the principal characteristics of the algorithm introduced in the previous section, we propose to consider an illustration taken from the stereo-reconstitution problem. To simplify, we assume that we are provided with two known deterministic signals I_l and I_r; one being translated globally by a given shift and being corrupted with white noise. The following model holds:

$$I_l(x) = \cos(x)$$
$$I_r(x) = \cos(x + d) + \varepsilon$$
$$\varepsilon \sim N(0, \sigma_\varepsilon^2)$$

The situation is similar to a stereovision problem reduced to one dimension, except that the translation to be determined is not local, but global. We assume that we observe I_l and I_r at fixed positions.

We first need to express the likelihood function of I_l and I_r given d. Since ε is a non-correlated Gaussian noise, we can derive the following expression:

$$L(I_l, I_r|d) \propto \exp\left(-\sum_{i=0}^{l} \frac{(I_l(x_i + d) - I_r(x_i))^2}{2\sigma_\varepsilon^2}\right)$$

where x_i, $i = 0$, l are the observation locations.

We assume moreover a Gaussian distribution for the prior, i.e.

$$d \propto \exp\left(-\frac{(d - m_d)^2}{2\sigma_d^2}\right)$$

For the simulation experiments, we have used the parameter values of Table 16.1.

4.2 Simulations

We first compared the experimental histograms obtained with Metropolis–Hastings, a Langevin diffusion and the Multiple Metropolis–Hastings algorithm. In each case, 5000 simulations were generated, and the value after 1000 iterations was retained, to ensure convergence. The three experimental histograms turn out to be in very close agreement with the theoretical distribution.

Then, we compared the efficiency of the three algorithms for computing Monte Carlo integrals. Indeed, the interest of MCMC lies in the estimation of expectations of a given event under the probability π of interest. For our application, probabilities of exceeding a threshold were computed, i.e. $P(d \geq \alpha) = E_\pi(1_{d \geq \alpha})$ where α is the threshold. These results are shown in Figure 16.3, which displays the evolution of the estimated probabilities with the integration time, for four different thresholds. The simulations were stopped after 10 000 iterations. We observe convergence in the three cases, but the speeds of convergence differ strongly. The relatively high rejection rate in Metropolis–Hastings ends in a lower rate of convergence, as expected. Surprisingly, the convergence rate for the Langevin diffusion is relatively low also. This may be due to high correlations within the chain, so that the mixing is not sufficient, even though transitions occur frequently. Note that this behaviour depends directly on the discretization parameter h, which governs the variance of the proposal transition. Increasing h to increase mixing increases the instability and the rejection rate, so that a reasonable balance has to be found. The fastest convergence is observed for the Multiple Metropolis–Hastings algorithm, for which convergence is achieved after 2000 iterations approximately. The construction of a subspace of proposal transitions allows one to increase greatly mixing within the chain while keeping the rate of effective transitions high. From this point of view, the Multiple Metropolis–Hastings algorithm represents a nice balance between Metropolis–Hastings and the discretized version of the Langevin diffusion.

Table 16.1 *Simulation parameters*

m_d	σ_d	σ_ϵ	l	d
0	1	0.05	10	1

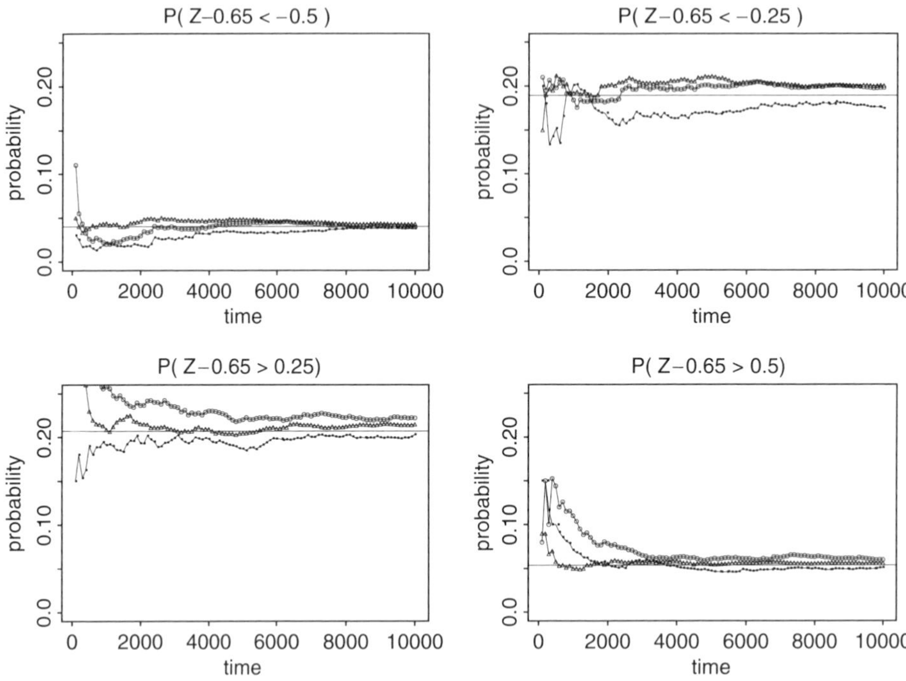

Figure 16.3 Probabilities of exceeding a given threshold as a function of the iteration number. The solid line represents the theoretical value. Dots: Metropolis–Hastings; circles: Langevin; triangles: Multiple Metropolis–Hastings

5 Application to SPOT HRV Panchromatic Images

In this section, we propose to apply the Multiple Metropolis–Hastings algorithm to the study of a stereoscopic pair of rectified SPOT HRV Panchromatic images. Such images are used routinely to produce mid-scale DTMs, with a spatial resolution around 10 m (Girard and Girard, 1999).

In rectified images, homologous points belong to the same line. In this system, the disparity becomes then a scalar value, and the search for pairs of homologous points can be performed along a single line. An introduction to stereo-reconstitution can be found in Faugeras (1993), a review of the most common algorithms in use, and amongst them correlation-based algorithms, is presented in Dhond and Aggarwal (1989).

5.1 Data and model

The stereoscopic pair is represented in Figure 16.4. The scene represents an approximately 10 km × 10 km part of Marseille, south of France. Although this is mostly an urban scene, noticeable relief features can be noted, especially on the upper right part of the images. This stereoscopic image pair suffers from some common problems

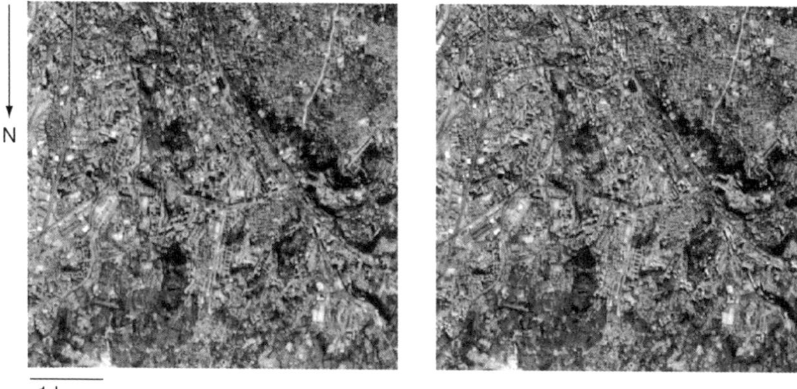

Figure 16.4 Rectified stereoscopic pair (SPOT HRV Panchromatic images). Orientation and scale bars are only indicative since the rectification distorts the images

encountered in stereovision, such as radiometric discrepancies (e.g. difference in image tone) between the same items of the two images, or the presence of shadow areas where little texture exists.

We computed the disparity map using a correlation-based algorithm (see Faugeras, 1993, and Dhond and Aggarwal, 1989, for more details). In order to assess the accuracy of the results, we also computed a reference disparity map from high spatial resolution images. This reference map was then degraded to the spatial resolution of the SPOT HRV Panchromatic images, to simplify the comparison. We consider therefore this reference map as a kind of 'ground data'. Both disparity maps are represented in Figure 16.5. The black pixels represent non-predicted values. Note that a crude comparison of these two maps is not appropriate: despite the spatial resolution degradation, small features like buildings can be clearly recognized on the reference map. These details do not appear on the disparity map computed from the SPOT HRV Panchromatic images. We are clearly not interested in these details, but will in the sequel only focus on the mid-sized features.

To give an order of magnitude, the mean value of the estimated disparity is -0.46, and maximum and minimum values are respectively 7.4 and -7.3 pixels.

It is beyond the scope of this chapter to propose a detailed Bayesian modelling of the stereovision problem. Some examples can be found in Belhumeur (1996). The sensor model we use assumes uncorrelated Gaussian noise, which is certainly not correct, but is a first approximation. In order to take radiometric discrepancies into account, it may be preferable to first normalize the two images. We have therefore the following model:

$$L(I_l,\ I_r|d) \propto \exp\left(-\sum_i \frac{(\tilde{I}_l(\mathbf{x}_i + d_i) - \tilde{I}_r(\mathbf{x}_i))^2}{2\sigma^2} \right)$$

where $\tilde{I}_r$ and $\tilde{I}_l$ are the right and left normalized images of the pair, d the disparity and σ is the standard deviation of the noise. It is possible to use the disparity

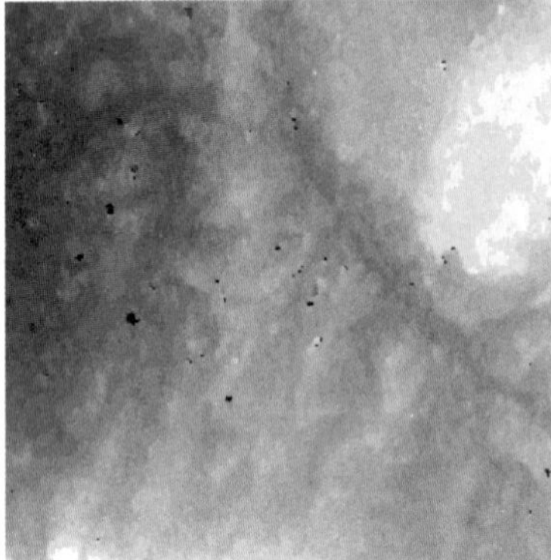
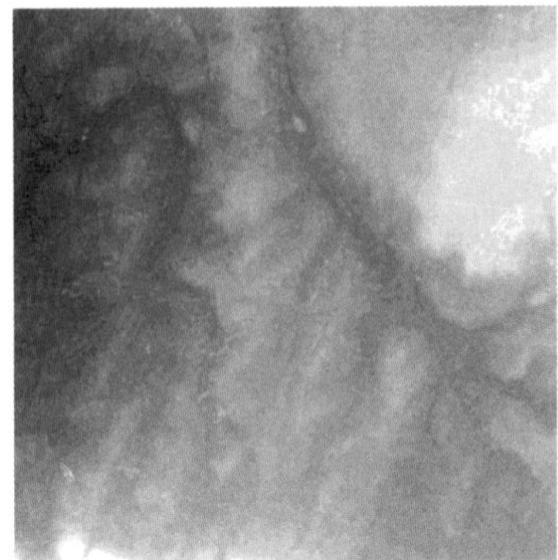

Figure 16.5 Disparity map (left) and reference disparity map obtained from high spatial resolution images (right)

map obtained from correlation to estimate the standard deviation σ of the residual noise.

As stated previously, we model the disparity as a Gaussian random function. The use of Gaussian models is rather common (see Belhumeur, 1996, for examples), but very often the problem of estimating the spatial structure is not considered. We prefer to assume spatial stationarity and to estimate the spatial model from the disparity map obtained with correlation. However, this approach is not totally satisfactory, especially since the behaviour at the origin of the covariance will reflect merely the artefacts of the correlation algorithm, and is not representative of the underlying structure. Therefore, this covariance estimation should be used only as a rough estimation, which provides an idea of the spatial model to use and an estimation of the range. It is then necessary to perform first a batch of training simulations, with different parameter values, and to fit them. Note that a Bayesian treatment of this problem may also be feasible.

Finally, we can write the prior model as follows:

$$p(d) \propto \exp\left(-\frac{1}{2}(d-m)^{\top} C^{-1}(d-m)\right)$$

where m and C are respectively the mean and the spatial covariance of the disparity map. In our application, the covariance model C is a cubic model, with range of 128 pixels, and a variance of 7.

Lastly, the Multiple Metropolis–Hastings algorithm requires the simulation of independent multi-Gaussian random functions with known covariance. A wide panel of algorithms can be used, see Chilès and Delfiner (1999) for a presentation. We have used a Fourier-based algorithm, which is well adapted to the simulation of finite range covariance models on regular grids, see Pardo-Igúzquiza and Chica-Olmo (1993) or Chilès and Delfiner (1997) for details.

5.2 Simulation results

A total of 2000 simulations were generated, which were used to compute the conditional mean, the standard deviation map and maps of probability of large errors, for different thresholds.

The standard deviation map of Figure 16.6 gives an idea of the local uncertainty associated with the computation of disparity. The dark grey areas correspond to low standard deviation values (less than 0.1 pixel) and indicate therefore that there is little variability in these zones. Opposite, there are a number of regions for which the standard deviation reaches larger values (up to 1.2). Note also the existence of border effects, with large standard deviation values, which are explained by the fact that at the borders of the images, a feature in one image may not appear at all in the other image. This standard deviation map gives a very useful representation of the amount of information in the stereoscopic pair which is useful for disparity prediction: the low variance areas correspond systematically to features which are clearly identifiable on the two images, such as white buildings in a darker toned background, whereas the higher variance regions correspond exactly to the shadow or poor textured areas of the stereoscopic image pair. The influence of the spatial structure of the prior model is also clearly to be seen in this map.

To conclude this study, we computed probability maps of errors, which are shown in Figure 16.7. The left and the right maps represent probability of respectively positive and negative errors larger than 2 pixels. Using this kind of information, it is therefore possible to locate the points where the estimation is likely to be erroneous. Moreover, the use of the simulation allows us to quantify the errors. To assess the accuracy of the method, we first computed an error map as a difference between the reference disparity and the estimated disparity. We were therefore able to locate and quantify the disparity errors. However, it is not possible to compare directly this error map with the probability maps of Figure 16.7. To do so, we therefore thresholded the probability maps to a risk value equal to 5%. Finally, we compared the number of pixels detected as erroneous in both maps. Using this method, 84% of the negative errors could be detected, and 46% of the positive errors. However, most

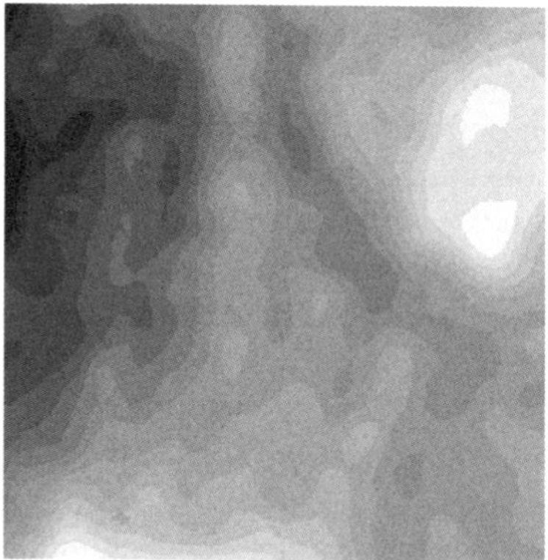

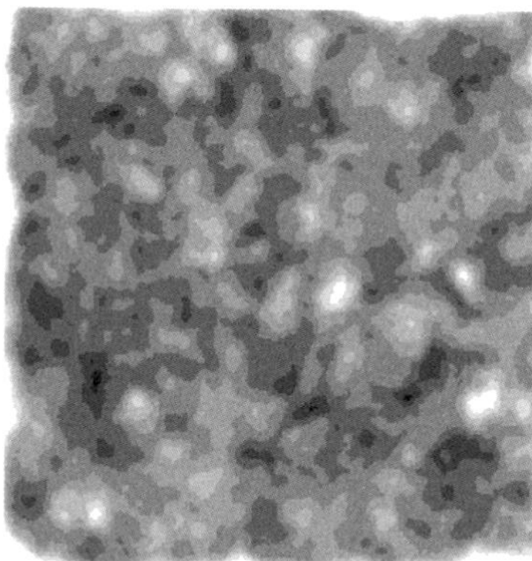

Figure 16.6 *Conditional mean (left) and standard deviation map (right)*

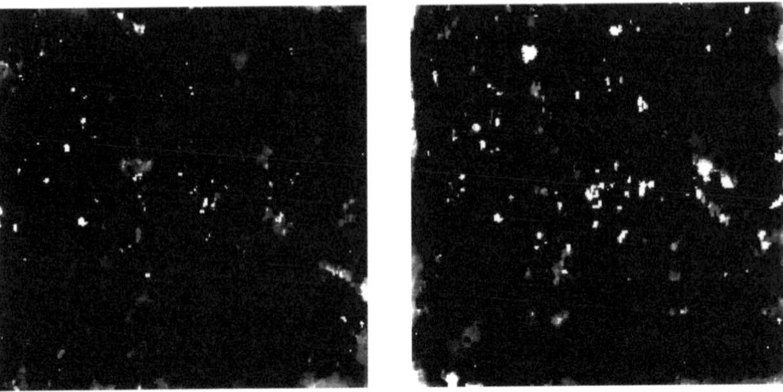

Figure 16.7 Probability map of positive (left) and negative (right) errors larger than 2 pixels

of the non-detected positive errors turn out to correspond to buildings, some of which were not present at the time of the SPOT HRV Panchromatic data acquisition.

6 Remarks and Conclusion

Although Bayesian inference and MCMC methods are used commonly in stereo-vision and image analysis in general, the use of simulations to assess the uncertainty of a disparity map seems to be a new application. Considering the increasing use of digital terrain models in geographical information systems or in the aerospace industry, it is an important application.

In a Bayesian framework, assessing the uncertainty can be interpreted as the sampling of the posterior probability of the disparity map given the stereoscopic pair. For this type of problem, MCMC simulations are efficient tools to estimate any functional of the posterior distribution. However, their application requires suitable algorithms, especially to reduce the convergence time.

In this chapter, a new Markov chain sampling algorithm has been introduced, whose principle is to generate a whole set of possible transitions based on the Gaussian distribution. With this algorithm, it is possible to generate chains with fast convergence rate and increased mixing. Therefore, it seems to be well adapted to applications in image analysis, where the computational burden is often important.

Finally, we have applied this formalism to the study of a stereoscopic pair of SPOT HRV Panchromatic images and have shown that it provides tools to identify areas of high variability and locate potential errors.

Beyond the practical aspects of MCMC simulations, particular attention must be paid to the interpretation of the results. Indeed, the significance of a probability is only valid within a given statistical model and has no general meaning. For example, the results are largely dependent on the form of the likelihood function or the choice of the prior distribution. This means in practice that a large part of the work must be dedicated to a sensitivity study to the different parameters. Especially, suitable tools still need to be developed to assess the spatial structure for the prior model. Moreover, it would be worth analysing the advantages of taking into account the spatial structure of the radiometric discrepancies in the stereoscopic pair. These remarks provide suitable directions for future work.

Proof of theorem 2

We consider the transition driven by equations (13)–(16), where, without loss of generality, $m = 0$. The goal is to verify that the corresponding transition kernel verifies the reversibility condition (4).

Let us denote $\theta_i = i\frac{2\pi}{l+1}$, $i = 0,\ l$, remarking that $z = z'_0$. We introduce the multiple transition kernel:

$$q(z'_1,\ \ldots,\ z'_l|z) = \int_w \phi(w) \prod_{j=1}^{l} 1_{z\cos\theta_j + w\sin\theta_j}(z'_j)dw$$

The transition kernel corresponding to equations (13)–(16) is:

$$K(z,\ dz') = \sum_{i=0}^{l} \int_{z'_1,\ldots,z'_l} a(z,\ z'_i)\delta_{z'_i}(dz')q(z'_1,\ \ldots, z'_l|z)dz'_1 \ldots dz'_l$$

Let us first show that the multiple transition kernel q verifies a multiple reversibility condition for ϕ, i.e. for each i there exists a permutation σ_i of $\{1, \ldots, l\}\backslash\{\text{i}\}$ such that:

$$\phi(z)q(z'_1,\ \ldots, z'_l|z) = \phi(z'_i)q(z'_{\sigma_i(1)}, \ldots, z'_{\sigma_i(i-1)},\ z, z'_{\sigma_i(i+1)}, \ldots, z'_{\sigma_i(l)}|z'_i)$$

First of all, we have:

$$\phi(z)q(z'_1,\ \ldots, z'_l|z) = \int_w \phi(z)\phi(w) \prod_{j\leq l,\ j\neq 0} 1_{z\cos\theta_j + w\sin\theta_j}(z'_j)dw$$

$$= \int_w \int_y \phi(y)\phi(w) \prod_{j=0}^{l} 1_{y\cos\theta_j + w\sin\theta_j}(z'_j)dwdy$$

For i fixed, we consider the following change of variable:

$$y' = y\cos\theta_i + w\sin\theta_i$$
$$w' = y\sin\theta_i - w\cos\theta_i$$

It is straightforward to show that:

$$\phi(y)\phi(w) = \phi(y'\cos\theta_i + w'\sin\theta_i)\phi(y'\sin\theta_i - w'\cos\theta_i) = \phi(y')\phi(w')$$

and

$$y\cos\theta_j + w\sin\theta_j = y'\cos(\theta_i - \theta_j) + w'\sin(\theta_i - \theta_j)$$

After the change of variable, we therefore obtain:

$$\phi(z)q(z'_1, \ldots, z'_l|z) = \int_{w'}\int_{y'} \phi(y')\phi(w') \prod_{j=0}^{l} \delta_{y'\cos(\theta_i-\theta_j)+w'\sin(\theta_i-\theta_j)}(z'_j)dw'dy'$$

thus:

$$\phi(z)q(z'_1, \ldots, z'_l|z) = \int_{w'} \phi(z'_i)\phi(w') \prod_{j\leq l,\ j\neq i} \delta_{z'_i\cos(\theta_i-\theta_j)+w'\sin(\theta_i-\theta_j)}(z'_j)dw'$$

We now remark that $\theta_i - \theta_j = (i-j)\frac{2\pi}{l+1}$. Hence by defining the permutation σ_i as follows:

$$\sigma_i(j) = i - j \text{ if } j \leq i$$
$$\sigma_i(j) = i - j + l + 1 \text{ if } j > i$$

the multiple reversibility condition holds for q:

$$\phi(z)q(z'_1, \ldots, z'_l|z) = \phi(z'_i)q(z'_{\sigma_i(1)}, \ldots, z'_{\sigma_i(i-1)}, z, z'_{\sigma_i(i+1)}, \ldots, z'_{\sigma_i(l)}|z'_i)$$

The reversibility of the transition kernel K for π is then proved by remarking that for each i:

$$\pi(z)\int_{z'_1,\ldots,z'_l} a(z,z'_i)\delta_{z'_i}(dz')\,q\,(z'_1,\ldots,z'_l|z)dz'_1\ldots dz'_l$$
$$= \int_{z'^i} \phi(z)\,r\,(z)\,a\,(z,z')\;q\;(z'_1,\ldots,z'_{i-1},\;z',\;z'_{i+1},\ldots,z'_l|z)dz'^i$$
$$= \int_{z'^i} \phi(z')\,r\,(z')\,a\,(z',z)\;q\;(z'_{\sigma_i(1)},\ldots,z'_{\sigma_i(i-1)},z,z'_{\sigma_i(i+1)},\ldots,z'_{\sigma_i(l)}|z')dz'^i$$
$$= \pi(z')\int_{z'_1,\ldots,z'_l} a\,(z',z'_i)\delta_{z'_i}(dz)\,q\,(z'_1,\ldots,z'_l|z)dz'_1\ldots dz'_l$$

where $z'^j = \left\{z'_1, \ldots, z'_{j-1}, z'_{j+1}, \ldots, z'_l\right\}$.

References

Barone, P. and Frigessi, A., 1990, Improving stochastic relaxation for Gaussian random fields, *Probability in the Engineering and Informational Sciences*, **4**, 369–89.

Belhumeur, P. N., 1993, *A Bayesian Approach to the Stereo Correspondence Problem*, PhD thesis, Harvard University.

Belhumeur, P. N., 1996, A Bayesian approach to binocular stereopsis, *International Journal of Computer Vision*, **19**, 237–62.

Besag, J., 1986, On the statistical analysis of dirty pictures (with discussion), *Journal of the Royal Statistical Society B*, **48**, 259–302.

Besag, J., Green, P., Higdon, D. and Mengersen, K., 1995, Bayesian computation and stochastic systems, *Statistical Science*, **10**, 3–66.

Chilès, J. P. and Delfiner, P., 1997, Discrete exact simulation by the Fourier method, in E. Y. Baafi and N. A. Schofield (eds), *Geostatistics Wallagong '96* (Dordrecht: Kluwer Academic).

Chilès, J. P. and Delfiner, P., 1999, *Geostatistics: Modeling Spatial Uncertainty* (New York: Wiley).

Dhond, U. R. and Aggarwal, J. K, 1989, Structure from stereo – a review, *IEEE Transactions on Systems, Man and Cybernetics*, **19**, 1489–510.

Faugeras, O., 1993, *Three-Dimensional Computer Vision* (Boston: Massachusetts Institute of Technology).

Geman, S. and Geman, D., 1984, Stochastic relaxation, Gibbs distribution, and the Bayesian restoration of images, *IEEE Transactions on Pattern Analysis and Machine Intelligence*, **6**, 721–41.

Gilks, W. R., Richardson, S. and Spiegelhalter, D. J., 1996, *Markov Chain Monte Carlo in Practice* (London: Chapman and Hall).

Girard, M. C. and Girard, C., 1999, *Traitement des données de télédétection* (Paris: Dunod).

Green, P. J., 1996, MCMC in image analysis, in W. R. Gilks, S. Richardson and D. J. Spiegelhalter (eds), *Markov Chain Monte Carlo in Practice* (London: Chapman and Hall), pp. 381–400.

Hastings, W. K., 1970, Monte Carlo sampling methods using Markov chains and their applications, *Biometrika*, **57**, 97–109.

Matheron, G., 1982, La destructuration des hautes teneurs et le krigeage des indicatrices, Technical report, CGMM N-761 (Paris: Centre de Géostatistique, Ecole des Mines de Paris).

Meyn, S. P. and Tweedie, R. L., 1993, *Markov Chains and Stochastic Stability* (London: Springer-Verlag).

Pardo-Igúzquiza, E. and Chica-Olmo, M., 1993, The Fourier integral method: an efficient spectral method for simulation of random fields, *Mathematical Geology*, **25**, 177–217.

Roberts, G. O. and Tweedie, R. L., 1996, Exponential convergence of Langevin diffusions and their discrete approximations, *Bernoulli*, **2**, 341–64.

Szeliski, R., 1989, *Bayesian Modeling of Uncertainty in Low-level Vision* (Dordrecht: Kluwer Academic).

Tierney, L., 1996, Introduction to general state-space Markov chain theory, in W. R. Gilks, S. Richardson and D. J. Spiegelhalter (eds), *Markov Chain Monte Carlo in Practice* (London: Chapman and Hall), pp. 59–74.

17

Current Status of Uncertainty Issues in Remote Sensing and GIS

Giles M. Foody and Peter M. Atkinson

1 Introduction

Uncertainty is a characteristic feature of research undertaken at the frontiers of science and technology (May, 2001). Although uncertainty has been recognized for some time as being important within the realm of geographical information science it is, however, a very problematical issue to accommodate (Zhang and Goodchild, 2002). Problems associated with uncertainty are particularly evident in research using remote sensing and geographical information systems. Throughout this book, a range of issues connected with uncertainty has been raised. The chapters have discussed, for example, the existence of different types and sources of uncertainty, highlighted the problems of measuring and predicting uncertainty and evaluating the implications of uncertainty on geographical investigations. It is evident from the material in the previous chapters, and the literature in general, that uncertainty limits our ability to study and understand the geographical world. As awareness of uncertainty and its implications has grown, studies have focused increasingly on the topic explicitly (Figure 17.1) and uncertainty is now recognized as one of the top 10 research priorities in geographical information science research (Cobb *et al.*, 2000).

Uncertainty has increasingly been noted as a major issue across a broad range of studies including those on conservation planning (Lister, 1998; Guikema and Milke, 1999), stream modification (Johnson and Brown, 2001), water quality (Gurian *et al.*, 2001), urban precipitation (Lowry, 1998), soil carbon dynamics (Torn *et al.*, 1997; Richter *et al.*, 1999), forestry modelling (Eid, 2001; Eid and Tuhus, 2001), boundary location (Edwards and Lowell, 1996; Cheng and Molenaar, 1999), population

Uncertainty in Remote Sensing and GIS. Edited by G.M. Foody and P.M. Atkinson.
 ISBN: 0–470–84408–6

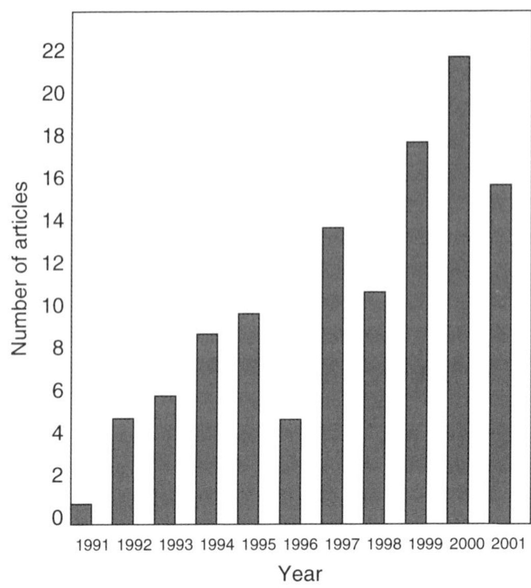

Figure 17.1 *The recent growth in research articles addressing uncertainty in geographical information science. The chart shows the number of articles published in each year over the period 1991–2001 that are recorded in a major bibliographic database and contained the terms 'uncertainty' and 'GIS' somewhere within the article's title, abstract and keywords. Note that this chart greatly under-estimates the work on the subject area (had terms such as accuracy, error, fuzziness etc. been used the estimated number of articles would be orders of magnitude greater than that shown for the tightly constrained search used) and is used simply to illustrate a general trend in the literature*

projections (Lutz *et al.*, 1997, 2001), scaling ecological knowledge (Loreau *et al.*, 2001), estimating ice sheet thickness (Kleman and Hattestrand, 1999), assessing the impacts of greenhouse warming on the strength and magnitude of ENSO events (Tudhope *et al.*, 2001), deforestation and climate (O'Brien, 1998) and in the analysis of remotely sensed data (Thome *et al.*, 1997; Miura *et al.*, 2000; Carmel *et al.*, 2001; de Groeve and Lowell, 2001). Focusing on climate change there are, for example, many sources of uncertainty associated with the links between the terrestrial biosphere and the atmosphere which add to the inherent uncertainty of predictions of future climate (Conway, 1998; Hurtt *et al.*, 1998; Lindroth *et al.*, 1998; Houghton and Hackler, 1999; Mitchell and Hulme, 1999; Allen *et al.*, 2000; Pittock and Jones, 2000) and estimating the potential of carbon mitigation strategies (de Jong, 2001; Royal Society, 2001). For example, the magnitude of the global methane sink has been markedly reassessed (Smith *et al.*, 2000), biomass estimates of Amazonian forests differ by a factor of more than two (Houghton *et al.*, 2001), poorly understood processes are often associated with large uncertainties (Clein *et al.*, 2000; Winiwarter and Rypdal, 2001) and the outputs of environmental models can be highly sensitive to their parameter settings (Hallgren and Pitman, 2000). Consequently, it is not surprising that there is considerable uncertainty over the extent, rate and magnitude of climate change (Seki and Christ, 1995). These problems all limit knowledge which, in turn, can lead to a resistance against calls for change because of the uncertainties involved (Matthews *et al.*, 2001), thus providing a rationale for inaction (Bradshaw and Borchers, 2000). Consequently, managing uncertainty is a major challenge (Rotmans and Asselt, 2001).

The growing awareness of, and interest in, uncertainty and its impacts upon geographical research, particularly that using remote sensing and geographical information systems, formed the background context to this book. The previous chapters have given a snapshot of some recent work from which it is possible to appreciate the current status of work on the topic and identify key issues for future research. Before addressing the material raised in the preceding chapters, it may be helpful to briefly review uncertainty.

2 Uncertainty

Chapter 1 (Atkinson and Foody) provided an overview of some important issues connected with uncertainty. This helped in setting the scene for the rest of the book but also in clarifying problematical issues. This is important as the remote sensing and GIS literature is often unclear in relation to uncertainty, which only acts to further complicate a difficult topic. For example, the literature uses a range of terms in the discussion of issues connected with uncertainty (Crosetto and Tarantola, 2001; Duckham *et al.*, 2001; Zhang and Goodchild, 2002), notably accuracy, reliability, error, ignorance, precision, inexactness, doubt, indeterminacy and distinctiveness. Each of these terms may be used to convey the message that in some way our knowledge or data set on the property under discussion is imperfect. The meaning of the various terms used also differs considerably between disciplines. The use of various terms, compounded by inconsistency in their use, often leads to further confusion that complicates an already difficult topic. This can be especially unhelpful when emotionally coloured language is also used and/or a large degree of risk is involved (ESRC Global Environmental Change Programme, 2000; May, 2001).

For the purposes of this book, uncertainty, and some of the key measures related to it, were defined in the first chapter. In general terms, however, two types of uncertainty can be recognized; ambiguity and vagueness (Klir and Folger, 1988; Longley *et al.*, 2001). Ambiguity is, essentially, the problem of making a choice between two or more alternatives and is probably the more commonly studied form of uncertainty. Vagueness is associated with making sharp or precise distinctions. Each type of uncertainty can result in major problems for research. Often the existence, let alone the magnitude, of the problems connected with uncertainty are under-estimated. A further related, but separate, issue discussed in the foreword by Curran is that of ignorance, which is also a major limitation to science and the public's confidence in science (Hoffmann-Riem and Wynne, 2002).

3 Uncertainty in Understanding and Managing the Geographical World

Change is a key component of human and physical environments. Understanding change and making appropriate decisions in relation to it may require the analysis of vast amounts of geographical data. These data may sometimes be derived by remote sensing and are often analysed most appropriately within a geographical information system. Consequently, geographical information science has an important role to play in understanding the environment and informing end users and policy makers. While geographical information scientists may be aware of uncertainty, end users and decision makers typically under-estimate it (ESRC Global Environmental Change Programme, 2000) and disregard ignorance (Hoffmann-Riem and Wynne, 2002). Moreover, end users and decision makers want advice from the scientific community that is certain (Bradshaw and Borchers, 2000; May, 2001). There is, therefore, a general desire to reduce uncertainty if possible (e.g. Varis *et al.*, 1993; Pan *et al.*, 1996; Lunetta, 1998). However, uncertainty is a difficult issue. It is connected with the closely interrelated issues of complexity and information (Klir and Folger, 1988). These are context-dependent issues that are greatly influenced by how much is known and unknown. Unfortunately, the amount we do not know about some issue increases with the amount that we do know. This is reflected in a quote attributed to John A. Wheeler:

> We live on an island surrounded by a sea of ignorance. As our island of knowledge grows, so does the shore of our ignorance.

What we can be sure about, however, is that if uncertainty cannot be removed or reduced substantially, end users of research results should at least be provided with some information on uncertainty and its effects (Stoms *et al.*, 1992).

Traditionally, one major source of uncertainty has been a lack of data or knowledge. The last four or five decades have seen tremendous technological advances in remote sensing and geographical information systems such that many topics of research have moved from being handicapped by a lack of data to now having, and often being unable to cope with, large data sets. Consequently, some argue that we now live in an information age or society. In some instances, the sizes of data sets are now so overwhelmingly large that intelligent assistance, perhaps through knowledge discovery and data mining techniques are required in some studies (Clementini *et al.*, 2000; Gluck, 2001). The availability of increasing amounts of data and growing knowledge is, within the confines of a traditional scientific paradigm, a desirable situation as knowledge and information are believed to increase precision and reduce uncertainty. However, the principle of incompatibility (Zadeh, 1973) argues that as complexity increases the meaningfulness of precise statements declines. Means to cope with complexity are required to facilitate scientific and technological advancement. One commonly used approach to coping with complexity is to simplify the situation by trading off the amount of information available with the amount of uncertainty allowed. The simplification of a complex system in a way that minimizes the loss of information relevant to the problem in-hand can provide a robust, but uncertain, summary (Klir and Folger, 1988). For this reason, many researchers have moved away from the precision associated with crisp sets and adopted fuzzy sets as one step in accommodating uncertainty issues in their work.

There is a great need to recognize and accommodate uncertainty as it appears to be ubiquitous. Uncertainty is an inherent feature of geographical data and concepts

(Fisher, 2000; Bennett, 2001; Phillis and Andriantiatsaholiniaina, 2001; Zhang and Goodchild, 2002). Fundamental spatial relations used in geographical information science, such as nearness, are, for example, vague (Worboys, 2001). A further problem is that uncertainty also arises in analyses with frightening ease. Take, for example, the very simple representation depicted in Figure 17.1. The data plotted in this graph were derived by a basic search of a bibliographic database. Each bar plotted represents the number of journal articles stored in the database that were published in a specified calendar year and which contain the terms GIS and uncertainty somewhere within their title, abstract and listed keywords. There seems to be no reason for there to be any uncertainty in the results of the search and in the interpretation of Figure 17.1. Unfortunately, this is not the case. If one inspects the articles selected for any one year, it is common to find that articles published in different years are included. No doubt this arises from simple processes. For example, the first issue of a new volume of a journal may actually be published slightly ahead of the publication date shown on the journal. Perhaps more commonly (at least from the authors' experience!), a journal may be delayed in publishing and an issue dated from the previous year may be found to contribute articles to the estimate derived for the year after its stated publication date. Manually correcting for this effect yields the data shown in Figure 17.2. Although the difference between the estimates derived for any year, and in the overall trend, are small this example does show how, in even very simple analyses, uncertainty can arise. Manslow and Nixon provide a more useful and relevant example in Chapter 4 that demonstrates clearly that information

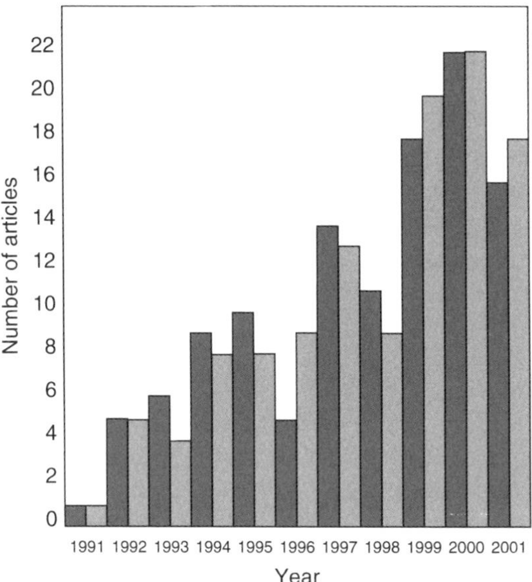

Figure 17.2 *The original and revised estimates of articles on uncertainty and GIS. The original values, shown in dark tone, correspond to the values depicted in Figure 17.1 while the values revised after manual correction are shown in light tone. While the difference between the two estimates is an error this simple example highlights how uncertainty in interpreting data can arise*

extracted from remotely sensed data can be ambiguous because of the point spread function of a sensor. Uncertainty is, therefore, commonplace, as evident from recent research across a broad spectrum of geographical investigation.

With increasing recognition of the importance of uncertainty, recent research has focused on key topics such as the measurement, representation and reduction of uncertainty as well as the identification of sources of uncertainty and their propagation through analyses, especially those undertaken within a geographical information system (e.g. Jager *et al.*, 2000; Goovaerts, 2001; Kiiveri *et al.*, 2001). Chapters in this book have addressed such issues. For example, Dungan (Chapter 3) identifies a range of uncertainty sources, Manslow and Nixon (Chapter 4) quantify the impacts of a major source of ambiguity in remote sensing studies, Smith and Fuller (Chapter 9) provide an example of how uncertainty information may be provided to help end users and Heuvelink (Chapter 10) discusses important issues in error propagation. Before looking at issues covered in this book more closely it is worth looking briefly at how the geographical information science community has dealt with uncertainty in the past.

4 Dealing with Uncertainty

Although uncertainty has been recognized for some considerable time, many early studies using geographical information systems were relatively deterministic and so tended to under-estimate if not disregard uncertainty (Crosetto and Tarantola, 2001; Greenland, 2001; Halls, 2001). The imperfections that are intrinsic in geographical data (Duckham *et al.*, 2001) and the dangers inherent in spatial analyses based on such data (Veregin, 1994) were disregarded in the way reality was represented in the system. Furthermore, conventional statistical analyses that may be inappropriate in the presence of uncertainty were generally used, yielding erroneous and misleading outputs (ESRC Global Environmental Change Programme, 2000). It is, therefore, unfortunate that uncertainty is ubiquitous and is inherent in geographical data (Fisher, 2000; Bennett, 2001). Moreover, the inherent uncertainty may be compounded in data processing and analysis. Actions such as rescaling or rezoning data sets may introduce uncertainty. Changes in administrative boundaries, for example, greatly handicap studies as it results in the use of incompatible zoning systems over time (Martin, 2001). This adds to problems in analysing data sets in which locational uncertainty may be large (Carmel *et al.*, 2001; Dearwent *et al.*, 2001). This is common in remote sensing and GIS. For example, the inability to perfectly co-register remotely sensed data sets may substantially limit the ability to accurately monitor major land cover changes (e.g. deforestation) and so limit scientific understanding of environmental processes and change. Roy (2000), for example, illustrates how misregistration errors can result in land cover changes being missed or exaggerated.

The failure to recognize uncertainty, whatever its source, may lead to erroneous and misleading interpretations. Yet to deal with uncertainty a high level of awareness of uncertainty and methods to accommodate it is required (Goodchild, 2000). Thus, although conventional scientific studies aim commonly to reduce uncertainty and recent studies have tried increasingly to accommodate it explicitly, uncertainty has often been ignored. Depending on the nature of the uncertainty, a variety of approaches such as those based upon the use of fuzzy sets, rough sets, multivalued logic and supervaluation as well as probabilistic and possibilistic assessments have been suggested as means to accommodate it (Goodchild, 2000; Martin-Clouaire *et al.*, 2000; Wang, 2000, Bennett, 2001; Duckham *et al.*, 2001). Although this is an area of active research, some techniques are now used widely in relation to uncertainty. Monte Carlo simulation has, for example, become a popular technique to study uncertainty propagation (Ellis *et al.*, 2000; Dunn and Lilly, 2001; Damian *et al.*, 2001). In the assessment of uncertainty impacts, uncertainty analysis and sensitivity analysis have also been used widely to evaluate the uncertainties associated with model outputs that result from errors propagated from the input data and the model uncertainties (e.g. the model's parameter settings, assumptions etc.) and the relative importance of sources of uncertainty respectively (Crosetto and Tarantola, 2001; Crosetto *et al.*, 2000). Such analyses yield information on uncertainty that may be used in a variety of ways. The information may, for example, help in selecting data sets on the basis of their fitness for a particular application (de Bruin *et al.*, 2001; Crosetto and Tarantola, 2001). Knowledge of uncertainty will also provide an overall assessment of the quality of model outputs with the propagation of uncertainty helping to qualify predicted outcomes by providing information on the risk of assuming a particular scenario (Crosetto *et al.*, 2000; Felguiras, 2001).

The need for a high level of awareness of uncertainty when using geographical data (Goodchild, 2000) is unfortunate as the geographical science community appears to be becoming increasingly diversified. Early geographical information systems, for example, were the domain of a relatively specialist group of users (Sui and Goodchild, 2001). Because of considerable recent technological advancement, modern geographical information systems are, however, both substantially more sophisticated and easier to use than early systems. Specialists are, therefore, no longer always required to operate a geographical information system. Indeed within the realm of information technology the variety of users is probably greatest with geographical information systems (Wang, 2000). When GIS is thought of as primarily a means of communicating information or media (Sui and Goodchild, 2001) this can be a relatively positive situation. However, when uncertainty is present it means that many system users will have an inappropriate background that may be a source of important problems. As it is unlikely that the average user of a geographical information system will be inclined or able to develop relevant skills in handling uncertainty, the problems of dealing with uncertainty will remain. This is particularly the case if the geographical science community is segregating into two distinct groups, those that use standard tools relatively unquestioningly and those involved in research and development of techniques (Martin and Atkinson, 2000; Halls, 2001). In this respect, it is, therefore, most unfortunate that the ease of use of many geographical information systems enables users with little knowledge or appreciation of uncertainty to derive outputs that appear very impressive but in reality could be fundamentally flawed (Goodchild, 2000). The development of error-aware geographical information systems may help reduce the problem by maintaining the required information for evaluation (Heuvelink, Chapter 10). Similarly, if a broad range of users are to be able to use geographical information systems effectively and sensibly, a natural language interface to facilitate meaningful analysis may be required (Wang, 2000). These are important developments, as without correct allowance for uncertainty the use of sophisticated methods may ultimately obfuscate rather than clarify (Gupta, 2001).

5 Uncertainty in Remote Sensing and GIS

Against the backdrop discussed above, it is apparent that the contributions to this book have provided a snapshot of the topic that both reiterates important general trends noted in the literature as well as highlights potential future directions. Curran's Foreword highlighted some of the basic issues and concerns of uncertainty in the broad area of geographical information science. In particular, the distinction between uncertainty in measurement and uncertainty in understanding is both fundamental and important. This book has focused mostly on issues of uncertainty in measurement and prediction that are currently key research foci in remote sensing and GIS. The book has also addressed issues of understanding uncertainty itself and the techniques that may be used because of it. It is hoped that an increased understanding of and ability to accommodate measurement uncertainty will help ultimately to reduce the uncertainty in understanding that Curran refers to.

The chapters by Woodcock (Chapter 2) and Dungan (Chapter 3) provided clear overviews of uncertainty in remote sensing. Both Woodcock and Dungan make use of Schott's (1997) image chain as a means to describe the overall remote sensing process. This is particularly valuable not only in highlighting the sources of uncertainty that may occur in the various steps (or chain links) of a remote sensing process such as land cover mapping but also in identifying the weakest links and so future research priorities.

Dungan (Chapter 3) outlines the key sources of uncertainty in remote sensing. This discussion included important issues such as uncertainty in location and parameters that have been the focus of considerable attention and are issues raised in later chapters. However, it also stressed topics that have been relatively rarely studied, especially structural uncertainty and uncertainty about the support. The latter is of particular importance within geographical information science as the size of the support is a fundamental property of the data and commonly analyses are undertaken with data acquired at differing and possibly inappropriate supports.

Although uncertainty is an issue along the complete length of the image processing chain, most studies have focused on the final stage. Thus most research on uncertainty in land cover mapping projects have focused on the classification stage. As Woodcock (Chapter 2) notes, this is because it is the stage that the researcher has most control over. This may help in understanding uncertainty and its impacts as well as lead to its reduction but may not be the most profitable part of the chain to study; a weak link prior to the classification may limit the analysis substantially. For this reason, it is important to study uncertainty along the entire length of the image chain or perhaps more fully the entire chain of analysis. Several chapters in this book address uncertainty at points along the image processing chain. While many address issues associated with the classification stage of a land cover mapping application this is, of course, merely a link in a longer chain when the mapped data are used within a geographical information system. As such, this research has an important role to play in understanding and ultimately reducing uncertainty in remote sensing and geographical information science.

Chapter 4 by Manslow and Nixon addresses uncertainty issues prior to the classification stage of a typical image processing chain. It is an excellent example of why it is important to consider uncertainty in the early links of the image processing chain. This chapter addresses uncertainty arising from the point spread function (PSF) of a sensor. Since the PSF's impacts on the measured remotely sensed response are shown to be considerable all later analyses of the data are impacted. Manslow and Nixon demonstrated with great clarity that, as a consequence of the PSF, estimates of properties derived from remote sensing are ambiguous, with the degree of uncertainty a function of the composition and distribution of sub-pixel thematic components. Fortunately, real sensors have a less extreme PSF than that assumed in the chapter and Manslow and Nixon present a means of quantifying the bounds on the uncertainty induced by a sensor's PSF. They also illustrate the benefits that arise through using probability distributions in the representation of quantities estimated from the remotely sensed data.

Several chapters focus on image classification issues. These are discussed in relation to conventional hard classifications and the soft or fuzzy classifications that may sometimes be more appropriate (Foody, 1996). Although a soft classification provides a richer and more informative representation than a conventional hard classification it is far from perfect. Even if the class composition of each pixel is represented accurately in the soft classification, the spatial representation of land cover is limited. The chapters by Zhan *et al.* (Chapter 5) and Tatem *et al.* (Chapter 6) report on issues connected with refining soft classification outputs. In particular, both address super-resolution mapping. Zhan *et al.* use an inverse distance weighted average of *a posteriori* probabilities in neighbouring pixels in their mapping method. In this method, the sub-pixel to be predicted (hard class) is thus a distance weighted function of neighbouring pixels (*a posteriori* probabilities). A problem with this approach is that the sub-pixels are compared to neighbouring pixels, thereby mixing scales of measurement. However, the advantage of the approach is that the model is linear and can be programmed readily within a mixture model (i.e. with linear algebra), and the predictions obtained may be useful for a range of applications. Alternatively, Tatem *et al.* present a very different approach to usefully extend soft classification analyses for thematic mapping applications. They use a Hopfield neural network and show how it is possible to model the distribution of a mixed pixel's component land cover over the area that it represents. Although the methods proposed by Zhan *et al.* and Tatem *et al.* do not allow for problems such as the PSF discussed by Manslow and Nixon (Chapter 4), they are important in allowing the production of more spatially detailed thematic maps and can be modified to accommodate the effects of the PSF.

In many instances, conventional techniques will be appropriate for the analysis of remotely sensed data. Thus, for example, standard hard classification techniques are generally appropriate for mapping land cover from imagery with a spatial resolution that is much finer than the typical size of features to be mapped. There are, however, problems with the use of conventional methods. These relate especially to the aim of extracting fully the desired information from the remotely sensed data. The chapters by Jakomulska and Radomski (Chapter 7) and Boyd *et al.* (Chapter 8) address two different issues connected with the extraction of information from remotely sensed data using established methods. In each case, uncertainties remain about how to use the techniques most appropriately to extract the desired information accurately from the imagery. Jakomulska and Radomski, for example, investigate the problem of specifying the kernel size for the quantification of image texture. This fits well with the image chain representation of the analysis of remotely sensed data with the use of an incorrectly sized kernel clearly resulting in an inappropriate textural description that may ultimately limit classification accuracy.

Boyd *et al.* (Chapter 8) focus on a very different aspect of uncertainty in the extraction of information from remotely sensed data. Here, the concern is with uncertainty associated with the use of NOAA AVHRR middle infrared reflectance data, and in particular the emissivity value to use in extracting the reflected signal from the total spectral response measured by the sensor. Boyd *et al.* show that the magnitude of the emissivity value used in the methodology to estimate middle infrared reflectance impacts upon the ability to extract useful environmental information from the remotely sensed data. Moreover, they show that in tackling one problem another became apparent and, essentially, pose a general question 'when is a known uncertainty a problem?'. The answer to this question is no doubt application-dependent.

The final chapter focusing predominantly on remote sensing was that by Smith and Fuller (Chapter 9). This chapter returns to land cover mapping applications but provides a different perspective on the problems of uncertainty associated with the classification of remotely sensed data. Their perspective was essentially that of a map producer sympathetic to the various needs of map users and so relates to a late link in the image chain. Smith and Fuller review the general lack of uncertainty information provided with land cover maps. This is illustrated with the Land Cover map of Great Britain derived from Landsat sensor data in the 1990s. They explain that the production of a meaningful and consistent index of uncertainty is difficult but make an important step in the provision of uncertainty meta-data. This is through the provision of process history descriptors (PHD) to accompany the Land Cover Map 2000 that replaces the earlier Land Cover Map of Great Britain. For each classified unit, the PHD provides information on the image data sets used in its classification, a summary of the probability of class membership and a range of other information relating to the allocation made that may be of use to a range of users later in the chain.

The remaining chapters focused mainly on issues connected with geographical information systems and science. Heuvelink (Chapter 10) uses his chapter to ask some fundamental questions of the research and user communities. The key issue raised is that as the techniques for handling uncertainty become ever more sophisticated, so the ability of users and operators of geographical information systems (who are not experts in spatial statistics and uncertainty analysis) to appropriately implement such techniques becomes increasingly reduced. Perhaps what is needed is an 'uncertainty button' in geographical information systems. However, this is beyond the current state of the art. Heuvelink also draws attention to the computational limitations on uncertainty analysis, with particular reference to Monte Carlo analysis. Such methods provide a solution for only limited error propagation problems, and are not applicable to large multi-dimensional data sets because of computational cost. He also uses the chain of operation analogy and discusses how meta-data on data quality, such as that derived by Smith and Fuller (Chapter 9), may usefully be used to help construct an error-aware geographical information system.

The following two chapters provide interesting case studies on issues connected to uncertainty and its propagation in geographical information system based analyses. Warren *et al.* (Chapter 11) report on uncertainties in a system designed to estimate the biodiversity of boreal forest at the scale of the landscape. In particular, they focus on issues connected with the estimation of uncertainties in landscape variables that influence biodiversity and the propagation of uncertainty through a model into a final map of biodiversity. Similarly, Heywood *et al.* (Chapter 12) focus on how uncertainties in sulphur and nitrogen deposition impact upon the exceedances of acidity critical loads in the UK. Using simplified means of simulating uncertainty, notably in the use of fixed value uncertainties and deposition distributions with simple shapes, they show that uncertainties in the deposition estimates used have a large impact upon exceedance calculations. Although there are concerns, including those raised by the authors, about the use of both critical loads (as the definition of thresholds is problematical (e.g. O'Riordan, 2000)) and the shape of the distributions used, the results nonetheless demonstrate the value of studying uncertainty impacts and may have important implications for policy development.

The remaining four chapters have a strong geostatistical flavour. Wameling (Chapter 13) considers uncertainty within a geostatistical modelling framework. However, the interest is not in the kriging (i.e. minimum prediction) variance, but in the extension of this measure to the spatial rather than attribute dimension. The particular focus of Wameling's chapter is uncertainty in contours, where that uncertainty must be in location (spatial), rather than actual elevation (attribute). The problem here is clearly that a linear model is being used to predict a non-linearly related variable. However, the model promoted by Wameling allows prediction of precision in contours under a restricted set of cirucmstances. With a similar objective to Wameling, in that a non-linearly related variable (location) is to be predicted, Warr *et al.* (Chapter 14) use a different approach in their study of soil horizon data. In particular, they use the indicator formalism along with conventional sequential conditional simulation algorithms to simulate lateral soil horizon thickness. Such algorithms are suitable for the prediction of non-linearly related variables.

Lloyd (Chapter 15) presents a geostatistical analysis of rainfall data in Great Britain. The particular focus of this chapter is the prediction of monthly rainfall given sparsely distributed Meteorological Office data on rainfall, and 'complete' coverage of elevation. Since the utility of the secondary variable (elevation) depends on the correlation between rainfall and elevation locally, Lloyd uses several local techniques for geostatistical prediction of rainfall based on both data sets. The results add to the body of empirical knowledge on the utility of these techniques.

In the penultimate chapter Sénégas *et al.* (Chapter 16) tackle the problem of blunders and their replacement in digital photogrammetry. A Markov chain Monte Carlo (MCMC) approach to uncertainty evaluation was chosen to provide a solution. The framework was heavily model-based, and the case study illustrates well what can be achieved when a mathematical model is used to represent and handle uncertainty in geographical data. In particular, the use of computational (Monte Carlo) (cf. analytical) solutions for statistical inference has much to offer.

A variety of issues connected with uncertainty have been covered, and from a range of viewpoints. A range of sources of uncertainty and issues connected with the propagation of uncertainty through analyses and along a chain of processing have, in particular, been presented. While the general desire has been to reduce the deleterious effects of uncertainty it has also been recognized that uncertainty need not always be viewed negatively (Trudgill, 2001; Dungan, Chapter 3) and that people differ greatly in their interest in, knowledge of, and ability to accommodate uncertainty.

6 Closing Comments

The previous chapters have shown that some major advances have been made in addressing issues of uncertainty in remote sensing and geographical information systems research. There is still, however, considerable scope for further work. We are a long way from having an 'uncertainty button' in our geographical information systems. Indeed, we are a long way from appreciating fully uncertainty and its impacts. Not all aspects of uncertainty are, for example, sufficiently recognized today. Commonly, researchers may overlook the intrinsic uncertainty of the issues or concepts under study (Bennett, 2001; Phillis and Andriantiatsaholiniaina, 2001) and can never fully avoid the problem of ignorance (Hoffmann-Riem and Wynne, 2002). Thus while much work has addressed issues such as the quality and accuracy of data (ESRC Global Environmental Change Programme, 2000; Foody, 2002; Remmel and Perera, 2002) little has addressed directly the problem that objects used in geographical information systems are often inadequately defined (Fisher, 2000; Bennett, 2001). Fundamental spatial relations, such as nearness, are also vague concepts that are often not treated as such yet are a component of any comprehensive ontology of space (Wang, 2000; Worboys, 2001). Despite the problems, however, we anticipate that the growing awareness of uncertainty and its impacts in remote sensing and geographical information systems and science will ultimately lead to an increase in our ability to study and understand the geographical world.

References

Allen, M. R., Sott, P. A., Mitchell, J. F. B., Schnur, R. and Delworth, T. L. 2000, Quantifying the uncertainty in forecasts of anthropogenic climate change, *Nature*, **407**, 617–20.

Bennett, B., 2001, What is a forest? On the vagueness of certain geographic concepts, *Topoi*, **20**, 189–201.

Bradshaw, G. A. and Borchers, J. G., 2000, Uncertainty as information: narrowing the science-policy gap, *Conservation Ecology*, **4**, 120–34.

Carmel, Y., Dean, D. J. and Flather, C. H., 2001, Combining location and classification error sources for estimating multi-temporal database accuracy, *Photogrammetric Engineering and Remote Sensing*, **67**, 865–72.

Cheng, T. and Molenaar, M., 1999, Objects with fuzzy spatial extent, *Photogrammetric Engineering and Remote Sensing*, **65**, 797–801.

Clein, J. A., Kwiatkowski, B. L., McGuire, A. D., Hobbie, J. E., Rastetter, E. B., Melillo, J. M. and Kicklighter, D. W., 2000, Modeling carbon responses of tundra ecosystems to historical and projected climate: a comparison of a plot and global-scale ecosystem model to identify process-based uncertainties, *Global Change Biology*, **6**, 127–40.

Clementini, E., Di Felice, P. and Koperski, K., 2000, Mining multiple-level spatial association rules for objects with broad boundary, *Data and Knowledge Engineering*, **34**, 251–70.

Cheng, T. and Molenaar, M., 1999, Objects with fuzzy spatial extent, *Photogrammetric Engineering and Remote Sensing*, **65**, 797–801.

Cobb, M., Petry, F. and Robinson, V., 2000, Uncertainty in geographic information systems and spatial data, *Fuzzy Sets and Systems*, **113**, 1.

Conway, D., 1998, Recent climate variability and future climate change scenarios for Great Britain, *Progress in Physical Geography*, **22**, 350–74.

Crosetto, M. and Tarantola, S., 2001, Uncertainty and sensitivity analysis: tools for GIS-based model implementation, *International Journal of Geographical Information Science*, **15**, 415–37.

Crosetto, M., Tarantola, S. and Saltelli, A. 2000, Sensitivity and uncertainty analysis in spatial modelling based on GIS, *Agriculture Ecosystems and Environment*, **81**, 71–9.

Damian, D., Sampson, P. D. and Guttorp, P., 2001, Bayesian estimation of semi-parametric non-stationary spatial covariance structures, *Environmetrics*, **12**, 161–78.

Dearwent, S. M., Jacobs, R. R. and Halbert, J. B., 2001, Locational uncertainty in georeferencing public health datasets, *Journal of Exposure Analysis and Environmental Epidemiology*, **11**, 329–34.

de Bruin, S., Bregt, A. and van de Ven, M., 2001, Assessing fitness for use: the expected value of spatial data sets, *International Journal of Geographical Information Science*, **15**, 457–71.

de Groeve, T. and Lowell, K., 2001, Boundary uncertainty assessment from a single forest-type map, *Photogrammetric Engineering and Remote Sensing*, **67**, 717–26.

de Jong, B. H. J., 2001, Uncertainties in estimating the potential for carbon mitigation of forest management, *Forest Ecology and Management*, **154**, 85–104.

Duckham, M., Mason, K., Stell, J. and Worboys, M., 2001, A formal approach to imperfection in geographic information, *Computers, Environment and Urban Systems*, **25**, 89–103.

Dunn, S. M. and Lilly, A., 2001, Investigating the relationship between a soils classification and the spatial parameters of a conceptual catchment-scale hydrological model, *Journal of Hydrology*, **252**, 157–73.

Edwards, G. and Lowell, K. E., 1996, Modelling uncertainty in photointerpreted boundaries, *Photogrammetric Engineering and Remote Sensing*, **62**, 337–91.

Eid, T., 2001, Models for the prediction of basal area mean diameter and number of trees for forest stands in south-eastern Norway, *Scandinavian Journal of Forest Research*, **16**, 467–79.

Eid, T. and Tuhus, E., 2001, Models for individual tree mortality in Norway, *Forest Ecology and Management*, **154**, 69–84.

Ellis, E. C., Li, R. G., Yang, L. Z. and Cheng, X., 2000, Long-term change in village-scale ecosystems in China using landscape and statistical methods, *Ecological Applications*, **10**, 1057–73.

ESRC Global Environmental Change Programme 2000, *Risky Choices, Soft Disasters: Environmental Decision Making Under Uncertainty*, University of Sussex, Brighton, 25pp.

Felguiras, C. A., Fuks, S. D. and Monteiro, A. M. V., 2001, Raster representations of spatial attributes with uncertainty assessment using nonlinear stochastic simulation, *Proceedings of the 6th International Conference on GeoComputation*, University of Queensland, Australia, CD-ROM.

Fisher, P., 2000, Sorites paradox and vague geographies, *Fuzzy Sets and Systems*, **113**, 7–18.

Foody, G. M., 1996, Approaches for the production and evaluation of fuzzy land cover classifications from remotely sensed data, *International Journal of Remote Sensing*, **17**, 1317–40.

Foody, G. M., 2002, Status of land cover classification accuracy assessment, *Remote Sensing of Environment*, **80**, 185–201.

Gluck, M., 2001, Multimedia exploratory data analysis for geospatial data mining: the case for augmented seriation, *Journal of the American Society for Information Science and Technology*, **52**, 686–96.

Goodchild, M. F., 2000, Introduction: special issue on 'uncertainty in geographic information systems', *Fuzzy Sets and Systems*, **113**, 3–5.

Goovaerts, P., 2001, Geostatistical modelling of uncertainty in soil science, *Geoderma*, **103**, 3–26.

Greenland, S., 2001, Sensitivity analysis, Monte Carlo risk analysis and Bayesian uncertainty assessment, *Risk Analysis*, **21**, 579–83.

Guikema, S. and Milke, M., 1999, Quantitative decision tools for conservation programme planning: practice, theory and potential, *Environmental Conservation*, **26**, 179–89.

Gupta, S., 2001, Avoiding ambiguity, *Nature*, **412**, 589.

Gurian, P. L., Small, M. J., Lockwood, J. R. and Schervish, M. J., 2001, Addressing uncertainty and conflicting cost estimates in revising the arsenic MCL, *Environmental Science and Technology*, **35**, 4414–20.

Hallgren, W. S. and Pitman, A. J., 2000, The uncertainty in simulations by a Global Biome Model (BIOMES) to alternative parameter values, *Global Change Biology*, **6**, 483–95.

Halls, P. J., 2001, Geographic information science: innovation driven by application, *Computers, Environment and Urban Systems*, **25**, 1–4.

Hoffman-Riem, H. and Wynne, B., 2002, In risk assessment, one has to admit ignorance, *Nature*, **416**, 123.

Houghton, R. A. and Hackler, J. L., 1999, Emissions of carbon from forestry and land-use change in tropical Asia, *Global Change Biology*, **5**, 481–92.

Houghton, R. A., Lawrence, K. T., Hackler, J. L. and Brown, S., 2001, The spatial distribution of forest biomass in the Brazilian Amazon: a comparison of estimates, *Global Change Biology*, **7**, 731–46.

Hurtt, G. C., Moorcroft, P. R., Pacala, S. W. and Levin, S. A., 1998, Terrestrial models and global change: challenges for the future, *Global Change Biology*, **4**, 581–90.

Jager, H. I., Hargrove, W. W., Brandt, C. C., King, A. W., Olson, R. J., Scurlock, J. M. O. and Rose, K. W., 2000, Constructive contrast between modelled and measured climate responses over a regional scale, *Ecosystems*, **3**, 396–411.

Johnson, P. A. and Brown, E. R., 2001, Incorporating uncertainty in the design of stream channel modifications, *Journal of the American Water Resources Association*, **37**, 1225–36.

Kiiveri, H. T., Caccetta, P. and Evans, F., 2001, Use of conditional probability networks for environmental monitoring, *International Journal of Remote Sensing*, **22**, 1173–90.

Kleman, J. and Hattestrand, C., 1999, Frozen-bed Fennoscandian and Laurentide ice sheets during the last glacial maximum, *Nature*, **402**, 63–6.

Klir, G. J. and Folger, T. A., 1988, *Fuzzy Sets, Uncertainty and Information* (London: Prentice-Hall International).

Lindroth, A., Gelle, A. and Moren, A. S., 1998, Long-term measurements of boreal forest carbon balance reveal large temperature sensitivity, *Global Change Biology*, **4**, 443–50.

Lister, N. M. E., 1998, A systems approach to biodiversity conservation planning, *Environmental Monitoring and Assessment*, **49**, 123–55.

Longley, P. A., Goodchild, M. F., Maguire, D. J. and Rhind, D. W., 2001, *Geographic Information Systems and Science* (Chichester: Wiley).

Loreau, M., Naeem, S., Inchausti, P., Bengtsson, J., Grime, J. P., Hector, A., Hooper, D. U., Huston, M. A., Raffaelli, D., Schmid, B., Tilman, D. and Wardle, D. A., 2001, Ecology – Biodiversity and ecosystem functioning: current knowledge and future challenges, *Science*, **294**, 804–8.

Lowry, W. P., 1998, Urban effects on precipitation, *Progress in Physical Geography*, **22**, 477–520.

Lunetta, R. S., Lyon, J. G., Guidon, B. and Elvidge, C. D., 1998, North American landscape characterization dataset development and data fusion issues, *Photogrammetric Engineering and Remote Sensing*, **64**, 821–9.

Lutz, W., Sanderson, W. and Scherbov, S., 1997, Doubling of world population unlikely, *Nature*, **387**, 803–5.

Lutz, W., Sanderson, W. and Scherbov, S., 2001, The end of world population growth, *Nature*, **412**, 543–5.

Martin, D., 2001, Developing the automated zoning procedure to reconcile incompatible zoning systems, *Proceedings of the 6th International Conference on GeoComputation*, University of Queensland, Australia, CD-ROM.

Martin, D. and Atkinson, P., 2000, Innovation in GIS application? *Computers, Environment and Urban Systems* **24**, 61–4.

Martin-Clouaire, R., Cazemier, D. R. and Lagacherie, P. 2000, Representing and processing uncertain soil information for mapping soil hydrological properties, *Computers and Electronics in Agriculture*, **29**, 41–57.

Matthews, O. P., Scuderi, L., Brookshire, D., Gregory, K., Snell, S., Krause, K., Chermak, J., Cullen, B. and Campana, M., 2001, Marketing western water: can a process based geographic information system improve reallocation decisions? *Natural Resources Journal*, **41**, 329–71.

May, R., 2001, Risk and uncertainty, *Nature*, **411**, 891.

Mitchel, T. D. and Hulme, M., 1999, Predicting regional climate change: living with uncertainty, *Progress in Physical Geography*, **23**, 57–78.

Miura, T., Huete, A. R. and Yoshioka, H., 2000, Evaluation of sensor calibration uncertainties on vegetation indices for MODIS, *IEEE Transactions on Geoscience and Remote Sensing*, **38**, 1399–409.

O'Brien, K. L., 1998, Tropical deforestation and climate change: What does the record reveal?, *Professional Geographer*, **50**(1), 140–153.

O'Riordan, T., 2001, Environmental science on the move, in T. O'Riordan (ed.) *Environmental Science for Environmental Management*, second edition (Harlow: Prentice Hall), pp. 1–27.

Pan, Y., McGuire, A. D., Kicklighter, D. W. and Melillo, J. M., 1996, The importance of climate and soils for estimates of net primary production: a sensitivity analysis with the terrestrial ecosystem model, *Global Change Biology*, **2**, 5–23.

Pittock, A. B. and Jones, R. N., 2000, Adaption to what and why? *Environmental Monitoring and Assessment*, **61**, 9–35.

Phillis, Y. A. and Andriantiatsaholiniaina, L. A., 2001, Sustainability: an ill-defined concept and its assessment using fuzzy logic, *Ecological Economics*, **37**, 435–56.

Remmel, T. K. and Perera, A. H., 2002, Accuracy of discontinuous binary surfaces: a case study using boreal forest fires, *International Journal of Geographical Information Science*, **16**, 287–98.

Richter, D. D., Markewitz, D., Trumore, S. E. and Wells, C. G., 1999, Rapid accumulation and turnover of soil carbon in a re-establishing forest, *Nature*, **400**, 56–8.

Rotmans, J. and Asselt, M. B. A., 2001, Uncertainty management in integrated assessment modelling: towards a pluralistic approach, *Environmental Monitoring and Assessment*, **69**, 101–30.

Roy, D. P., 2000, The impact of misregistration upon composited wide field of view satellite data and implications for change detection, *IEEE Transactions on Geoscience and Remote Sensing*, **38**, 2017–32.

Royal Society, 2001, *The Role of Land Carbon Sinks in Mitigating Global Climate Change*, Policy document 10/01 (London: The Royal Society).

Schott, J. R., 1997, *Remote Sensing: The Image Chain Approach* (New York: Oxford University Press).

Seki, M. and Christ, R., 1995, Selected international efforts to address climate change, *Environmental Monitoring and Assessment*, **38**, 141–53.

Smith, K. A., Dobbie, K. E., Ball, B. C., Bakken, L. R., Sitaula, B. K., Hansen, S., Brumme, R., Borken, W., Christensen, S., Prieme, A., Fowler, D., Macdonald, J. A., Skiba, U., Klemedtsson, L., Kasimir-Klemedtsson, A., Degorska, A. and Orlanski, P., 2000, Oxidation of atmospheric methane in Northern European soils: comparison with other ecosystems, and uncertainties in the global terrestrial sink, *Global Change Biology*, **6**, 791–803.

Stoms, D. M., Davis, F. W. and Cogan, C. B., 1992, Sensitivity of wildlife habitat models to uncertainties in GIS data, *Photogrammetric Engineering and Remote Sensing*, **58**, 843–50.

Sui, D. Z. and Goodchild, M. F., 2001, GIS as media? *International Journal of Geographical Information Science*, **15**, 387–90.

Thome, K., Markham, B., Barker, J., Slater, P. and Biggar, S., 1997, Radiometric calibration of Landsat, *Photogrammetric Engineering and Remote Sensing*, **63**, 853–8.

Torn, M. S., Trumbove, S. E., Chadwick, O. A., Vitousek, P. M. and Hendricks, D. M., 1997, Mineral control of soil organic carbon storage and turnover, *Nature*, **389**, 170–3.

Trudgill, S., 2000, *The Terrestrial Biosphere* (Harlow: Prentice Hall).

Tudhope, A. W., Chilcott, C. P., McCulloch, M. T., Cook, E. R., Chappell, J., Ellam, R. M., Lea, D. W., Lough, J. M. and Shimmield, G. B., 2001, Variability in the El Nino-Southern oscillation through a glacial–interglacial cycle, *Science*, **291**, 1511–17.

Varis, O., Klove, B. and Kettunen, J., 1993, Evaluation of a real-time monitoring-system for river quality – a trade-off between risk attitudes, cost and uncertainty, *Environmental Monitoring and Assessment*, **28**, 201–13.

Veregin, H., 1994, Integration of simulation modelling and error propagation for the buffer operation in GIS, *Photogrammetric Engineering and Remote Sensing*, **60**, 427–35.

Wang, F., 2000, A fuzzy grammar and possibility theory-based natural language user interface for spatial queries, *Fuzzy Sets and Systems*, **113**, 147–59.

Winiwarter, W. and Rypdal, K., 2001, Assessing the uncertainty associated with national greenhouse gas emission inventories: a case study for Austria, *Atmospheric Environment*, **35**, 5425–40.

Worboys, M. F., 2001, Nearness relations in environmental space, *International Journal of Geographical Information Science*, **15**, 633–51.

Zadeh, L. A., 1973, Outline of a new approach to the analysis of complex systems and decision processes, *IEEE Transactions on Systems, Man, and Cybernetics*, **3**, 28–44.

Zhang, J. and Goodchild, M. F., 2002, *Uncertainty in Geographical Information* (London: Taylor and Francis).

Index

Uncertainty in Remote Sensing and GIS. Edited by G.M. Foody and P.M. Atkinson.

ISBN: 0-470-84408-6